Air Pollution: Measurement, Modeling and Mitigation

Air Pollution: Measurement, Modeling and Mitigation

Editor: Chuck Lancaster

STATES
ACADEMIC PRESS
www.statesacademicpress.com

States Academic Press,
109 South 5th Street,
Brooklyn, NY 11249, USA

Visit us on the World Wide Web at:
www.statesacademicpress.com

ISBN: 978-1-63989-038-5 (Hardback)

Cataloging-in-Publication Data

Air pollution : measurement, modeling and mitigation / edited by Chuck Lancaster.
p. cm.
Includes bibliographical references and index.
ISBN 978-1-63989-038-5
1. Air--Pollution. 2. Air--Pollution--Measurement. 3. Air quality management.
4. Atmospheric deposition. 5. Pollution. I. Lancaster, Chuck.

TD890 .A37 2022
628.53--dc23

Table of Contents

Preface

This book has been a concerted effort by a group of academicians, researchers and scientists, who have contributed their research works for the realization of the book. This book has materialized in the wake of emerging advancements and innovations in this field. Therefore, the need of the hour was to compile all the required researches and disseminate the knowledge to a broad spectrum of people comprising of students, researchers and specialists of the field.

Air pollution is caused by the release of harmful gases, particulates, aerosols and biological molecules into the air. These pollutants endanger the life of humans and pose serious health risks. They are responsible for causing allergies, asthma, cardiovascular diseases and lung cancer. This increases the importance of air quality monitoring and management to measure the quality and degradation of air. Continuous monitoring methods, gravimetric particulate methods and passive monitoring methods are some techniques employed to check air quality. The data obtained is thus used to evaluate the air quality index. Higher air quality index signifies a more polluted air. From theories to research, case studies related to all contemporary topics of relevance to this field have been included in this book. Most of the topics introduced herein cover new techniques and the applications of air pollution monitoring and management. As this field is emerging at a rapid pace, the contents of this book will help the readers understand the modern concepts and applications of the subject.

At the end of the preface, I would like to thank the authors for their brilliant chapters and the publisher for guiding us all-through the making of the book till its final stage. Also, I would like to thank my family for providing the support and encouragement throughout my academic career and research projects.

Editor

Preface

Editor

1

Assessment of Areal Average Air Quality Level Over Irregular Areas: A Case Study of PM_{10} Exposure Estimation in Taipei (Taiwan)

Hwa-Lung Yu[1], Shang-Chen Ku[1], Chiang-Hsing Yang[2], Tsun-Jen Cheng[3] and Likwang Chen[4]

[1]*Dept of Bioenvironmental Systems Engineering, National Taiwan University, Taipei,*
[2]*Dept of Health Care Management, National Taipei University of Nursing and Health Sciences*
[3]*Institute of Occupational Medicine and Industrial Hygiene, National Taiwan University*
[4]*Center for Health Policy Research and Development, National Health Research Institutes, Miaoli, Taiwan*

1. Introduction

Assessment of public health risk from the exposure of ambient pollutants has been heavily based upon the analyses of the associations between the pollutant exposure and its potential consequence, e.g. disease occurrences (Abbey et al., 1995; AckermannLiebrich et al., 1997; Beeson et al., 1998; Jerrett et al., 2005; Pope et al., 2002). In many air pollution epidemiologic investigations, individual health datasets at nationwide or regional scales are available to assess the subtle risks of pollution exposure. Among them, the understanding of the spatial or spatiotemporal distribution of ambient pollutants is essential due to their prevalent heterogeneity across space and/or time. In these cases, governmental agencies has established ambient air-quality monitoring networks, such as the Air Quality System (AQS) operated by the U.S. Environmental Protection Agency (EPA), regularly recording important and useful environmental data sources concerning the acute and chronic effects of ambient pollutants (TWEPA, 2006; USEPA, 1992). Based upon these databases, an ideal exposure assessment is performed by applying techniques of spatial or/and spatiotemporal analysis for the estimation of the pollutant level at the locations of individuals in health dataset. However, due to the raising concerns of privacy and confidentiality of personal information, the sensitive personal information, such as residential addresses, of health dataset is usually not allowed to be accessed. As a result, individual information of health dataset obtained from institutes or governmental agencies is often removed or degraded. Among them, the spatial locations of individuals of health dataset are usually aggregated into a larger spatial unit, e.g. higher administrative level, in which no personal identities can be identified. These aggregations raise challenges for environmental epidemiologists to assess the exposure levels and to investigate the potential health risks.

Spatial interpolation techniques have been increasingly used by environmental epidemiologists to estimate the spatiotemporal distribution of air quality levels. Among them, the deterministic approaches such as nearest-neighbor method (NN) and inverse distance weighted method (IDW) are the most widely used methods (Hendryx et al., 2010; Hoek et al., 2001; Michelozzi et al., 2002; Sohel et al., 2010), in which the estimation results are derived from certain assumed functional relationship of geographical distances between the observation and estimation locations without considering the stochastic associations among the air quality measurements. The stochastic techniques, in particular kriging method, have been applied with increasing frequency in exposure assessment studies to address the spatial heterogeneity among the air quality data as well as the estimation uncertainty of predictions (Brauer et al., 2003; Brimicombe, 2000; Buzzelli and Jerrett, 2004; Hoek et al., 2001). Most spatial interpolators are originally developed for the estimations with geographical support of estimation and observation locations at point scale. As a result, most studies of exposure assessment apply spatial techniques for point estimation at the centroid of the geographical unit of interest to characterize its average air pollution level (Chen and Schwartz, 2008; Chen and Schwartz, 2009; Lertxundi-Manterola and Saez, 2009; Maheswaran and Elliott, 2003; Miller et al., 2007). The inconsistency of geographical support between the exposure estimations and aggregated health dataset can potentially distort the results of their associated epidemiological studies (Young et al., 2009; Young et al., 2008).

This study investigates and compares the estimation results of areal-averaged air quality level by several popular spatial mapping techniques, i.e. NN, IDW, ordinary kriging (OK) and block kriging method (BK). Among them, BK is a kriging-based upscaling method which can perform spatiotemporal estimation with the consideration of the stochastic dependence among the locations with irregular sizes and shapes. This comparison is performed on the spatiotemporal PM10 estimation of the townships over Taipei area (Taiwan) during 2004-2006 on the basis of data during 1997-2007.

2. Materials and methods

2.1 Study area

Taipei is the largest metropolitan area in Taiwan including Taipei city and Taipei county with the vehicle density as high as over 6000 vehicles per km^2. Except for traffic emissions, the three incineration plants are the major stationary emission sources in the area (Chang and Lee, 2007). The Taipei area is bounded by mountains, i.e. Yangming mountains to the north, Linkou mesa to the west, and ridge of Snow mountains to the southeast which forms the second largest basin of the island (see Figure 1). Because of the significant variation of the basin topography, the air convection and circulation are generally degraded in this area. In addition, the alluvial plains at basin floor materialized the highly urbanization area of Taipei. As a result, the ambient pollutant concentration across the basin floor is generally higher than that over its surrounding mountain areas. Taiwan Environmental Protection Agency (TWEPA) has established the air quality monitoring network which regularly records the ambient pollutants, including PM10 and other criteria pollutants (TWEPA, 2006), and meteorological covariates throughout the island since September, 1993. Figure 2 shows the eighteen TWEPA monitoring stations in Taipei area. As it can be noticed, the distribution of monitoring locations is spatially imbalanced that most of the stations in Taipei are located at the high urbanized area. In this study, all the PM10 observations from these stations are aggregated into monthly data following the procedure suggested by USEPA (USEPA, 2004). The areal concentration estimations are performed at each of the townships shown in Figure 2 by various spatial interpolation techniques.

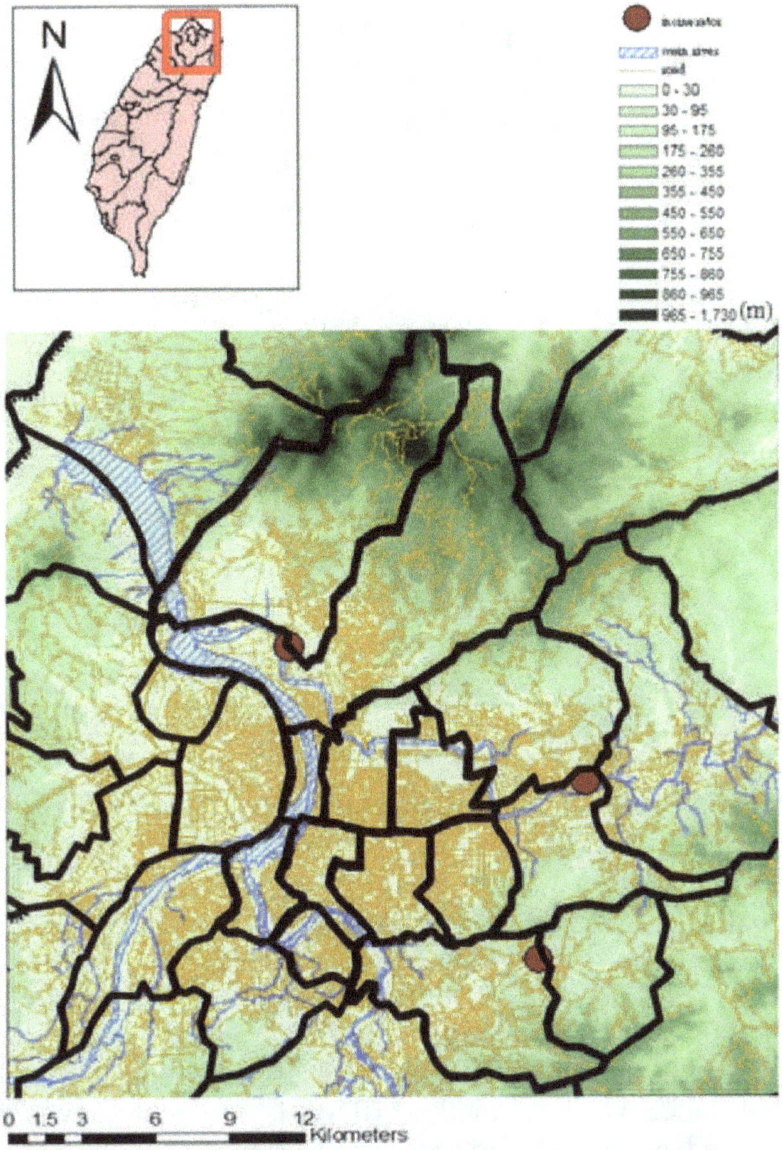

Fig. 1. The highways, rivers and topography in Taipei metropolitan area

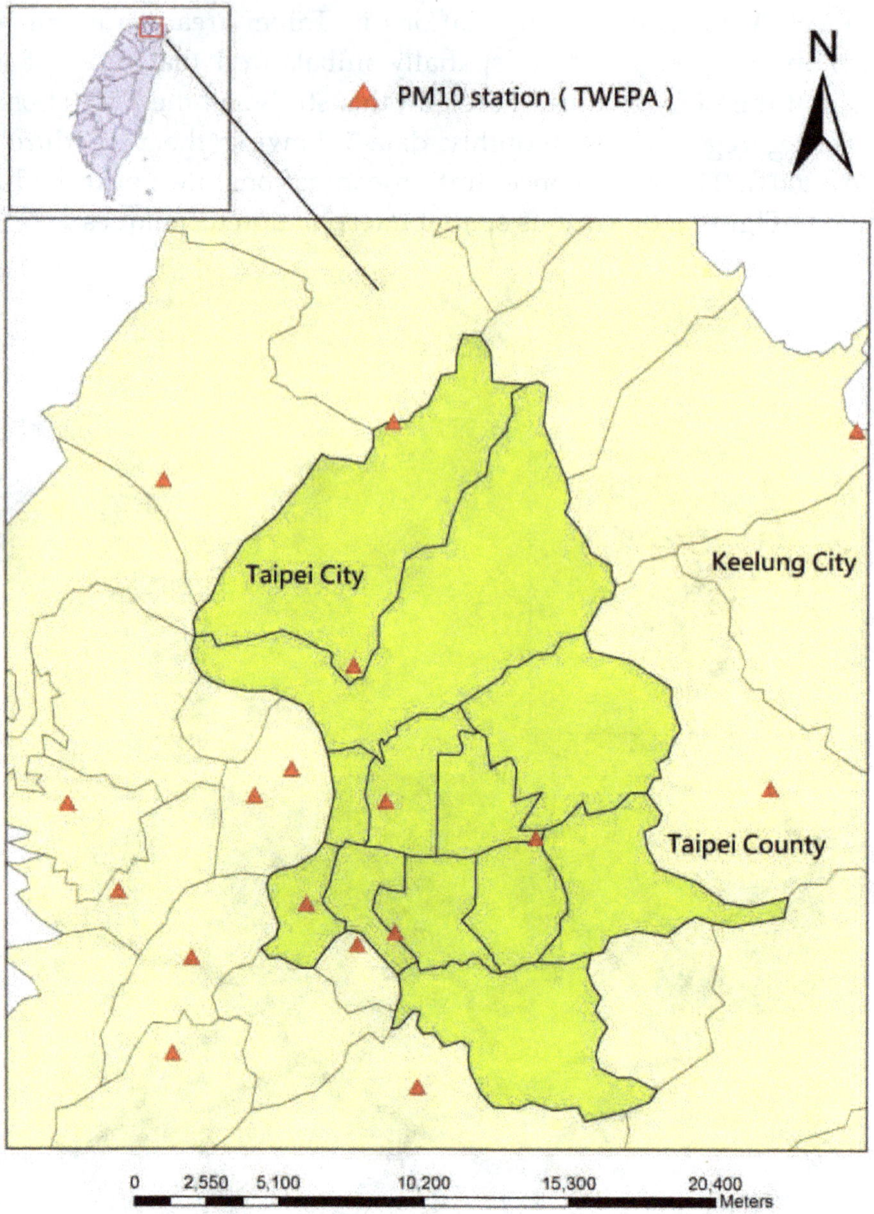

Fig. 2. Spatial distribution of PM10 monitoring stations operated by TWEPA

2.2 Methods

This study investigates several common used spatial interpolation methods for ambient pollutant exposure estimation. Among them, the nearest neighbor method (NN), also called polygon method and Theissen method, is the simplest method for spatial estimation. It assumes the air quality level at an ungauged location is completely determined by its closest observations so that spatial distribution of pollutants is composed of a set of polygon across space. The inverse distance weighted method (IDW) is a weighted average of neighboring observation values. The weights given to each observation is a function of distances between observation and estimation locations (Waller and Gotway, 2004) as below

$$\hat{z}(s_k,t_k)=\sum_{i=1}^{N}z(s_i,t_i)d_{ki}^{-p}\Big/\sum_{i=1}^{N}d_{ki}^{-p} \quad (1)$$

where d_{ki} is the space-time distance between the estimation location (s_k, t_k) to the observation locations $z(s_i, t_i)$. p is the modeler-specified degree. The increase of p can decreases the weights of distant observations. The space-time distance is assumed to be the function of the geographical distance and time interval (Christakos, 2000; Christakos et al., 2000). It can be expressed into the form of $d_{ki} = |s_k - s_i| + m_{st}|t_k - t_i|$, where m_{st} is the space-time metric to account for the relationship between spatial and temporal metric (Christakos et al., 2002). Both NN and IDW methods are deterministic methods which provide the estimations without the information of estimation uncertainty.

Stochastic techniques, on the other hand, can account for the uncertainty of the space-time data under the framework of random field theory which forms a multivariate joint distribution among the space-time attribute of interest. Among them, kriging method is the most popular technique in space-time mapping which is so-called best linear unbiased estimator (BLUE) in the sense of providing the minimum estimation uncertainty (Olea, 1999). Various types of kriging method have been developed in terms of different assumptions and analytical goals. Among them, ordinary kriging (OK) is the most widely used one which assumes the unknown mean among the space-time dependent data. The basic equations of ordinary kriging are shown below (Goovaerts, 1997)

$$\begin{aligned} &\sum_{i=1}^{n(p_k)} \lambda_{i,p_k} C(p_i, p_j) + \mu_{OK,p_k} = C(p_k, p_j) \\ &\sum_{i=1}^{n(p_k)} \lambda_{i,p_k} = 1 \quad i, j = 1, 2\ldots, n(p_k) \end{aligned} \tag{2}$$

where λ_{i,p_k} are the OK weights for the observation at $p_i = (s_i, t_i)$; $C(p_i, p_j)$ denotes the covariance of the attribute between p_i and p_j, $n(p_k)$ is the number of observations used for the kriging estimation at p_k, and μ_{OK,p_k} is the Lagrange multiplier. The OK estimation at p_k can be the linear combination of observations expressed as form of $\hat{z}(p_k) = \sum_{i=1}^{n(p_k)} \lambda_{i,p_k} z(p_i)$. Block kriging (BK) inherits the framework of kriging method and considers the geographical support of observation and estimation locations, i.e. their size and shape. The average value of attribute of concern over a block V can be represented as $\hat{z}_V(p_k) = \int_{V(p_k)} du' z(u') \big/ |V|$, where $|V|$ is the areal size of block V and u' is the point locations within the block V. The block kriging system is very similar to Eq. (2) with the covariance $C(p_k, p_j)$ replaced by the covariance between point observations and estimation block $\overline{C}(V(p_k), p_j)$ (Goovaerts, 1997), where $\overline{C}(p_k, p_j) = \int_{V(p_k)} du' C(u', p_j) \big/ |V|$.

This study compares the township-based estimations of PM10 level by the methods above, i.e. NN, IDW, OK and BK. Among them, three different spatial resolutions are used in BK method, i.e. dividing the irregular areas into 5x5, 10x10 and 40x40 grids. The estimations are performed at all townships in Taipei during 2006 with the support of space-time PM10 observations during 1997-2007. Among them, the township-level PM10 estimations by NN, IDW and OK methods follow the conventional approach that uses the estimations at geographical centroid to characterize the areal average level of pollutant concentration (Chen and Schwartz, 2009; Lertxundi-Manterola and Saez, 2009; Maheswaran and Elliott, 2003). In addition, to assure the comparison is performed at the same basis, the nonstationarity of PM10 process in space and time is removed by locally weighted smoothing regression method (LOESS) (Cleveland, 1979; Cleveland and Devlin, 1988) in advance to the applications of the spatial-time interpolation techniques.

3. Results

The observed PM10 levels across the stations in Taipei vary significantly as shown in Figure 3. The average and variance of monthly PM10 observations at the stations located in highly urbanized areas are generally higher than those in city's surrounding areas. Temporal variation of averaged PM10 across stations is shown in Figure 4 in which the increased PM10 variability happens at the time of higher average PM10 value, i.e. proportional effect. In general, the average of observed PM10 is higher during the seasons of spring and winter. In this study, the spatiotemporal processes of monthly PM10 is decomposed into an nonstationary trend in space and time to account for the general pattern of PM10 observations, e.g. PM10 variation resulting from the changes of seasons and distribution of emissions, and stationary residuals which account for the spatiotemporal dependence among the PM10 transport process. The spatiotemporal trend is estimated by using LOESS method which applies low-order polynomials to obtain the general pattern of PM10 variation at each space-time locality (Cleveland, 1979; Cleveland and Devlin, 1988). The estimation of spatiotemporal trend is performed by R software package. To characterize the spatiotemporal dependence, a space-time seperable function is used as shown below (see Figure 5)

$$c(h,\tau) = c_0 \exp(-\frac{3h}{a_r})\exp(-\frac{3\tau}{a_t}) \tag{3}$$

where $c_0 = 150$ and $[a_r, a_t] = [2000\text{m}, 3\text{ month}]$. Eq. (3) is used in kriging methods, i.e. OK and BK, to account for the space-time covariance at point scale in space and time. In addition, the covariance function identifies the influential ranges of PM10 data in space and time, and therefore characterizes the space-time metric which is used to estimate the space-time distance for IDW and kriging methods as discussed above.

The estimations of the areal PM10 level during the period of 2004-2006 at all townships of irregular sizes and shapes are compared in this study and their differences are shown in Figure 6 in which BK1, BK2, and BK denote the results of block kriging with spatial resolution of 5 by 5, 10 by 10, and 40 by 40 elements of each township. Results show that the spatiotemporal dependence is important to the estimations, i.e. significant differences between the results from deterministic methods, i.e. NN and IDW, and stochastic methods, OK and BK. The spatial distribution of the mean squared difference of estimations between the results from BK and other methods are shown in Figure 7. Figures 7 (a) and (b) shows the comparison between BK and the deterministic methods, i.e., NN and IDW, in which the mean squared differences (MSD) are generally high and increase as the estimation location further apart from the cluster of monitoring stations located at city central. The comparisons among spatial patterns of kriging results in Figures 7 (c) and (d), for OK-BK and BK1-BK

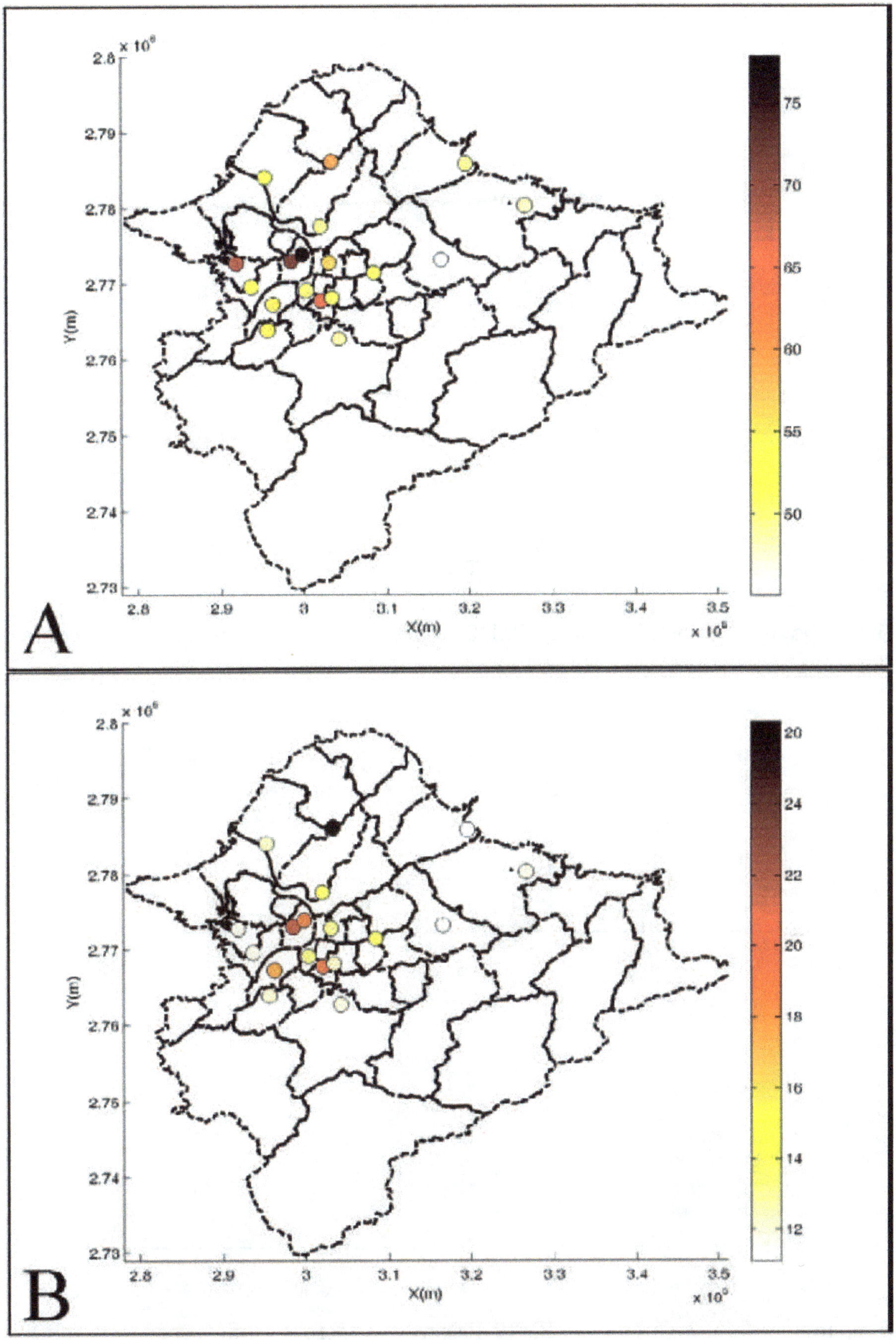

Fig. 3. Spatial distribution of (a) mean and (b) standard deviation of PM10 observations

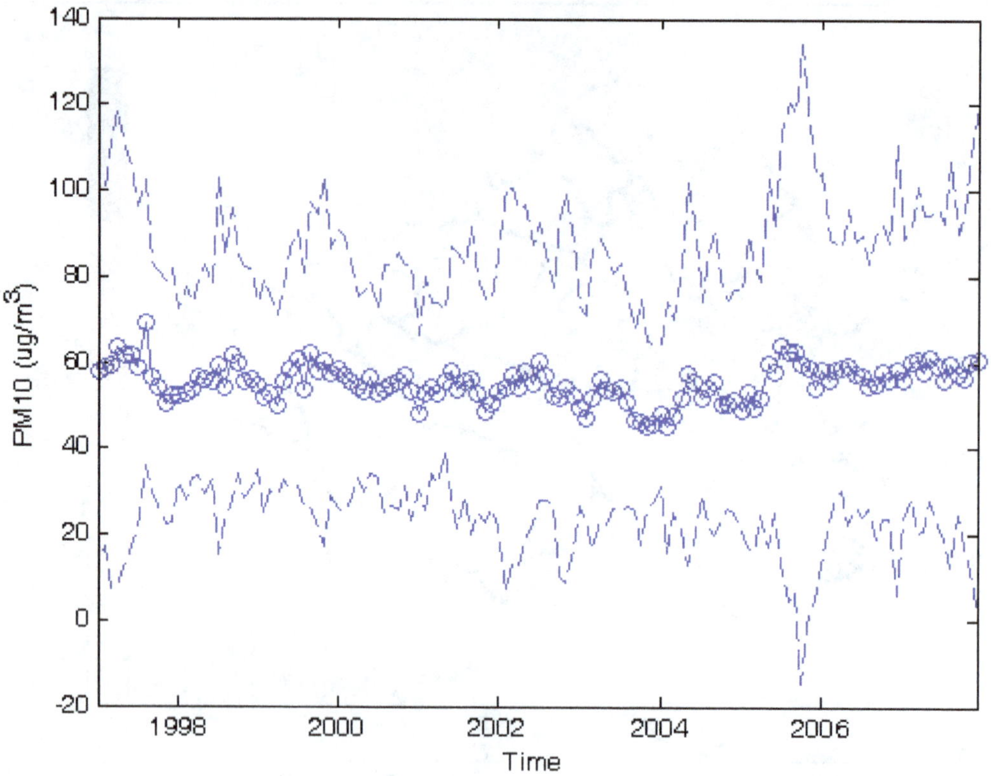

Fig. 4. Temporal variation of averaged PM10 observations and its associated 95% confidence interval.

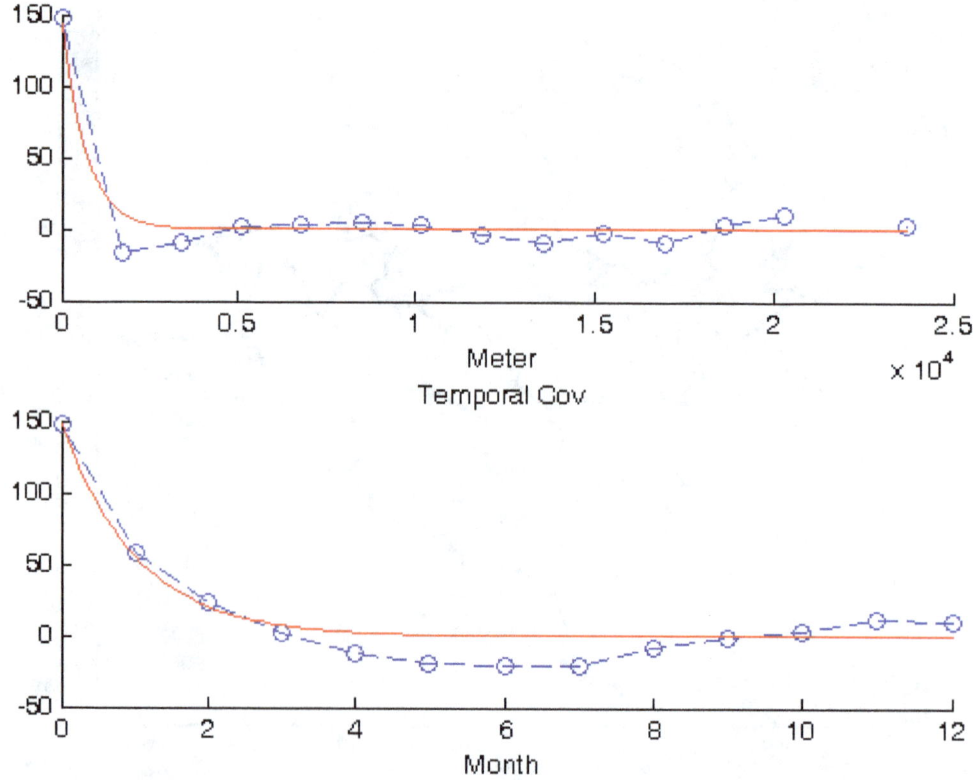

Fig. 5. Spatiotemporal covariance of PM10 observations across (upper) space and (bottom) time.

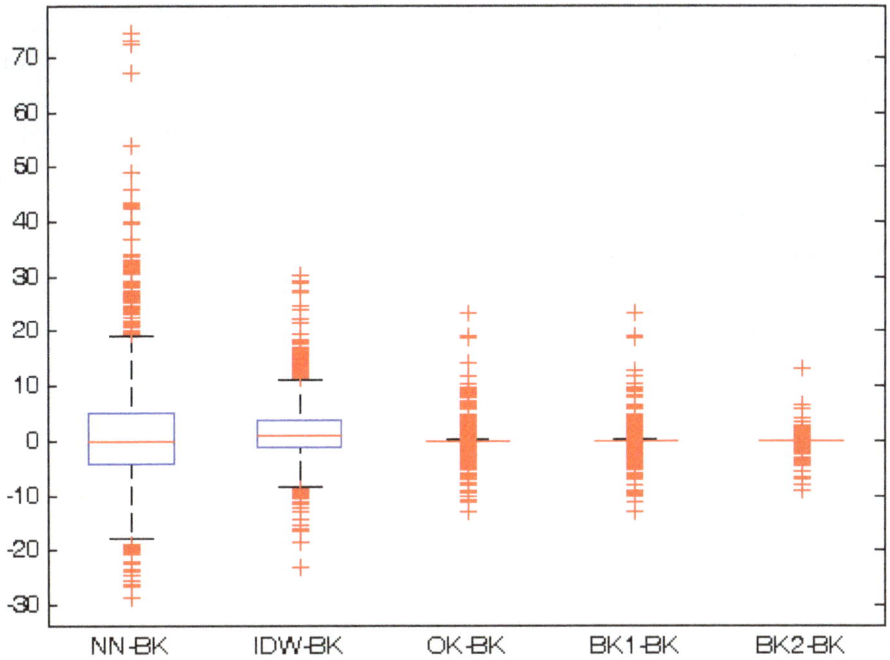

Fig. 6. Boxplot of the estimation differences between the results of block kriging with highest resolution and other methods.

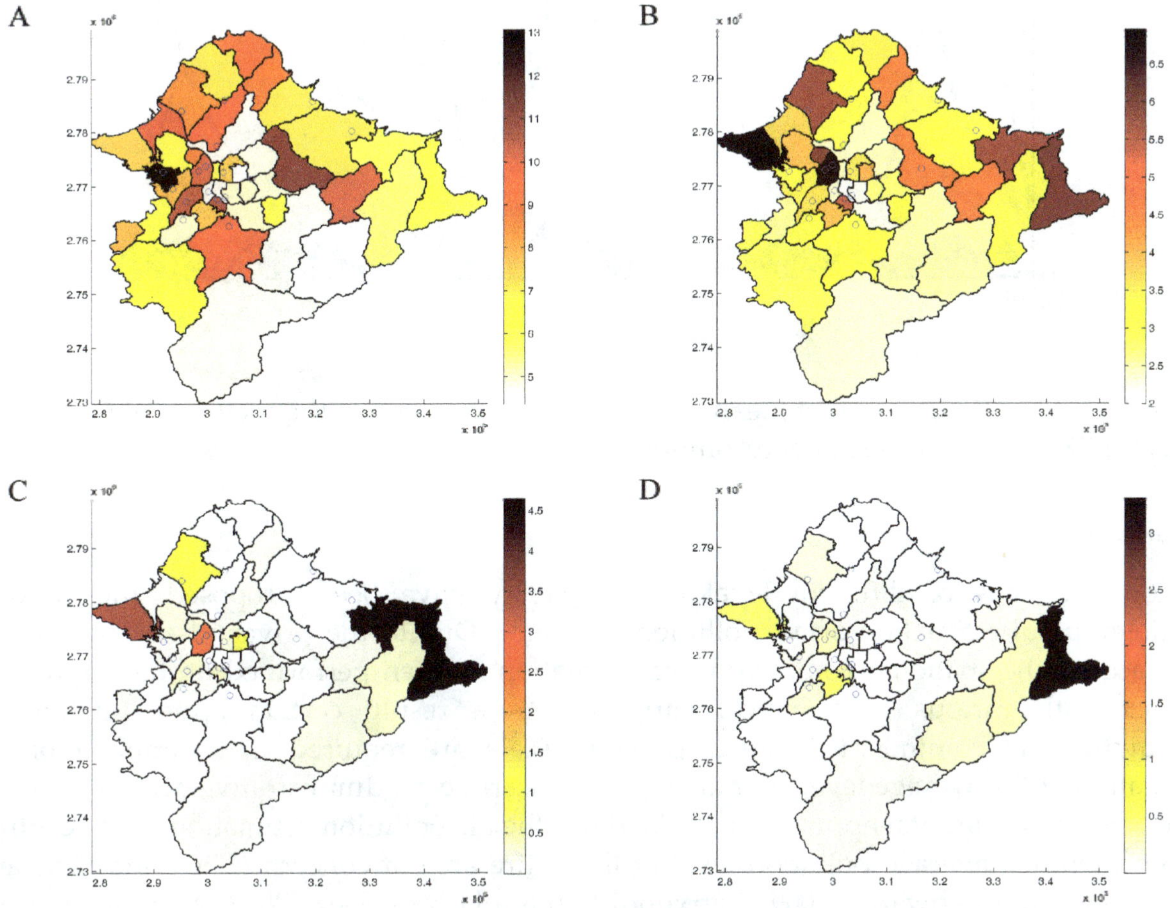

Fig. 7. Spatial distribution of the mean squared differences between the methods of (a) NN-BK, (b) IDW-BK, (c) OK-BK, and (d) BK1-BK

respectively, show the relatively higher MSDs are located at the boundary of study area and townships surrounding by multiple stations with relatively drastic variations of their observations. It should be noted that estimation differences between deterministic methods and BK are generally higher than those in the comparisons between kriging methods. Similar results are shown in the comparisons of over time in Figure 8 in which the seasonality of MSDs is shown in all the comparisons that elevated estimation variability is elevated in spring and winter.

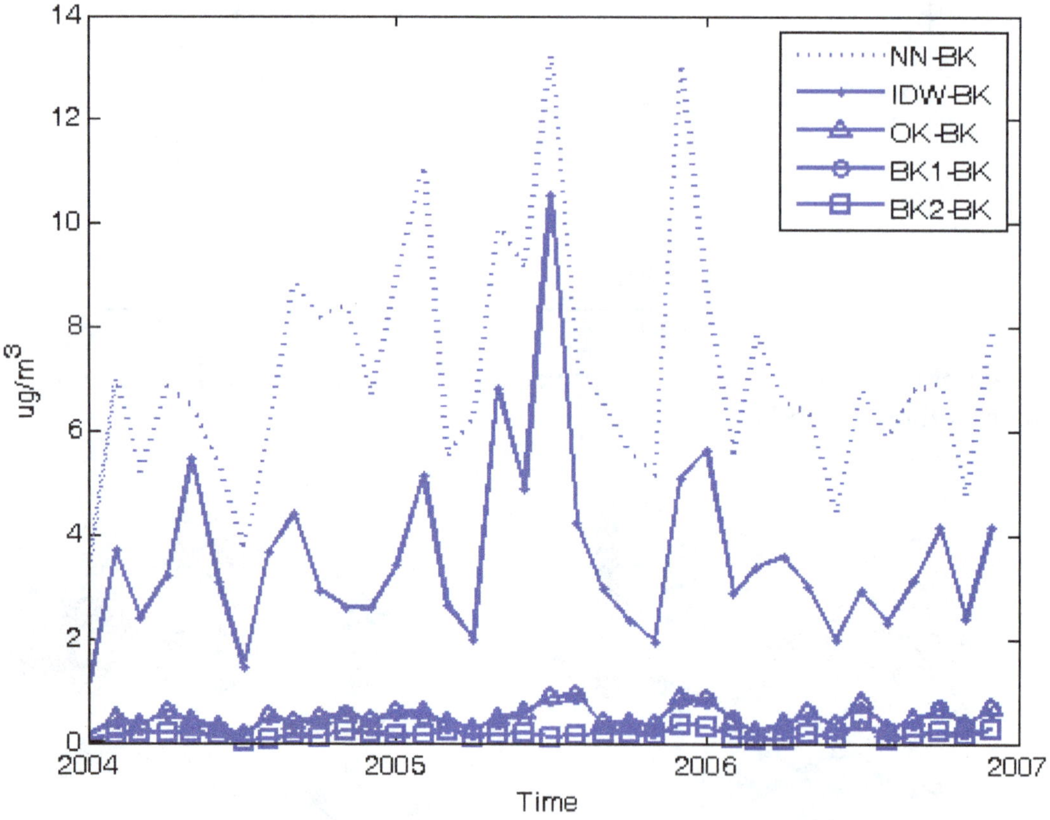

Fig. 8. Temporal distribution of mean squared differences among the methods of NN, IDW, OK, and BK with various spatial resolutions.

4. Discussions

Various methods of different level of complexity have been proposed and used in spatiotemporal estimation of air pollution exposure. Due to the privacy issue of personal data access, the spatiotemporal exposure estimation is often performed on the aggregated dataset with various geographical supports. As a result, certain approximations or assumptions of common techniques of point scale are required to be applied on the estimation of the average level over an area of concern, e.g. administrative division. Among them, the most common approximation is to use the air pollution estimation at the centroid to represent the average pollution level of the entire area of concern (Chen and Schwartz, 2008; Chen and Schwartz, 2009; Lertxundi-Manterola and Saez, 2009; Maheswaran and Elliott, 2003; Miller et al., 2007). The impact of the size and shape of geographical units to the estimation results is seldom considered (Goovaerts, 2008). The assumption of the similar spatial resolution among observation and Estimation areas can be dubious. In addition, the estimation ignores the variability of pollution levels within each of the studied units. Such approximation can be inadequate, especially, in the exposure estimation in metropolitan area where, in addition to its complex topography, various emission sources spreading across space elevates the spatiotemporal variability of air pollution. Though it is theoretical defective, exposure estimation at centroids usually provides a quick and convenient way to assess the air quality level in the corresponding areas of interest. The impact of this assumption of spatial prediction is worthwhile to be examined because the potential biased estimation results of air quality can distort their associated environmental epidemiology studies (Son et al., 2010).

As shown in Figure 6, comparing with the results of BK, the inconsistency is clearly shown between the estimations obtained from the two groups of methods, i.e. deterministic and stochastic, that the variability of the between-group differences are significantly larger than the within group differences. Among them, the results from NN method, i.e. a very common used method for ambient exposure estimation in environmental health studies (Basu et al., 2004; Miller et al., 2007), are significantly variable compared to those from other methods, i.e. its standard deviation is about 20% of common level of PM10. Figure 7 (a) shows that the discrepancy levels of NN results can vary across space and its generally high variability can be reduced at the areas of the higher density of monitoring stations. The periodic feature of the variation in Figure 8 shows that the temporal characteristics of NN results are still distinct from those of the results by other methods. It may resort to the fact that NN method is the only method disregarding the space-time dependence among the PM10 dataset in this study. Despite of its deterministic assumption, IDW accounts for the spatiotemporal dependence by considering the space-time metric among the observation and estimation locations. As a result, IDW reduces the discrepancy in both space and time across Taipei area between its results and those by kriging methods which considering the spatiotemporal dependence by not only space-time metric but the similarity among the observations. The standard deviation of the estimation variability compared to BK is about 10% of common PM10 level in Taipei. In addition, the results of deterministic approach are overestimated, especially at the townships with scarce monitors around, due to the preferential sampling at the highly polluted areas of Taipei central area (see Figure 3), i.e. the observations are mostly sampled at high concentration areas.

Figures 6-8 show the results from kriging methods, i.e. OK and BK with two spatial resolutions, are relatively similar. Among them, the standard deviation of estimation variability is lower than 5% of PM10 level. The distributions of MSDs among kriging methods in both space and time are relatively close to each other shown in Figure 7 (c) and (d). Among them, the areal estimation variability is further reduced by using BK instead of OK that their 95% confidence ranges are 2.8 and 6.8 $\mu g / m^3$, respectively. Because of the short spatial influential range of local PM10 transport, i.e. 2000m (see Eq. (3)), it implies the high spatial variability of PM10 distribution within the townships, i.e. point estimation results can vary significantly from location to location within townships. As a result, the consideration of geographical characteristics can effectively improve the understanding of the areal average level of PM10 concentration. This effect can be especially obvious at the townships containing and surrounding by multiple stations, e.g. San-chung township (the township with two stations), because the nature of kriging methods completely appreciates the observations and therefore result in the higher variability of estimation results among these townships. It should be noted that all the estimations by the present spatial interpolation techniques only depend upon the PM10 observations. The general characteristics of these methods can only provide reasonable estimations within the convex of data locations (Olea, 1999).

5. Conclusions

This study applies several popular spatial techniques to assess the average areal PM10 level of townships in Taipei area. The comparison shows that the importance of the inclusion of space-time metric among the observation and estimation locations as well as the consideration of spatiotemporal dependence among the observations for the estimations. This study shows the consideration of the shape and size of the townships is important to the performance of the estimations of areal average concentration. The MSDs between the estimations by kriging methods with and without considering the geographical characteristics of townships are up to 6% and 2% of the common PM10 level across space and time, respectively. This study provides insights for the impact of geographical support to the exposure estimation in Taipei for environmental epidemiological studies.

6. Acknowledgement

This research was supported by the National Science Council of Taiwan (NSC 99-2625-M-002-007-).

7. References

Abbey DE, Lebowitz MD, Mills PK, Petersen FF, Beeson WL, Burchette RJ. Long-Term Ambient Concentrations of Particulates and Oxidants and Development of Chronic Disease in a Cohort of Nonsmoking California Residents. Inhalation Toxicology 1995; 7: 19-34.

AckermannLiebrich U, Leuenberger P, Schwartz J, Schindler C, Monn C, Bolognini C, et al. Lung function and long term exposure to air pollutants in Switzerland. American Journal of Respiratory and Critical Care Medicine 1997; 155: 122-129.

Basu R, Woodruff TJ, Parker JD, Saulnier L, Schoendorf KC. Comparing exposure metrics in the relationship between PM2.5 and birth weight in California. Journal of Exposure Analysis and Environmental Epidemiology 2004; 14: 391-396.

Beeson WL, Abbey DE, Knutsen SF. Long-term concentrations of ambient air pollutants and incident lung cancer in California adults: Results from the AHSMOG study. Environmental Health Perspectives 1998; 106: 813-822.

Brauer M, Hoek G, van Vliet P, Meliefste K, Fischer P, Gehring U, et al. Estimating long-term average particulate air pollution concentrations: Application of traffic indicators and geographic information systems. Epidemiology 2003; 14: 228-239.

Brimicombe A. Constructing and evaluating contextual indices using GIS: a case of primary school performance tables. Environment and Planning A 2000; 32: 1909-1933.

Buzzelli M, Jerrett M. Racial gradients of ambient air pollution exposure in Hamilton, Canada. Environment and Planning A 2004; 36: 1855-1876.

Chang SC, Lee CT. Evaluation of the trend of air quality in Taipei, Taiwan from 1994 to 2003. Environmental Monitoring and Assessment 2007; 127: 87-96.

Chen JC, Schwartz J. Metabolic syndrome and inflammatory responses to long-term particulate air pollutants. Environmental Health Perspectives 2008; 116: 612-617.

Chen JC, Schwartz J. Neurobehavioral effects of ambient air pollution on cognitive performance in US adults. Neurotoxicology 2009; 30: 231-239.

Christakos G. Modern Spatiotemporal Geostatistics. New York, NY: Oxford Univ. Press, 2000.

Christakos G, Bogaert P, Serre ML. Temporal GIS: Advanced Functions for Field-Based Applications. New York, NY: Springer-Verlag, 2002.

Christakos G, Hristopulos DT, Bogaert P. On the physical geometry concept at the basis of space/time geostatistical hydrology. Advances in Water Resources 2000; 23: 799-810.

Cleveland WS. Robust Locally Weighted Regression and Smoothing Scatterplots. Journal of the American Statistical Association 1979; 74: 829-836.

Cleveland WS, Devlin SJ. Locally Weighted Regression - an Approach to Regression-Analysis by Local Fitting. Journal of the American Statistical Association 1988; 83: 596-610.

Goovaerts P. Geostatistics for natural resources evaluation. New York: Oxford University Press, 1997.

Goovaerts P. Kriging and semivariogram deconvolution in the presence of irregular geographical units. Mathematical Geosciences 2008; 40: 101-128.

Hendryx M, Fedorko E, Anesetti-Rothermel A. A geographical information system-based analysis of cancer mortality and population exposure to coal mining activities in West Virginia, United States of America. Geospatial Health 2010; 4: 243-256.

Hoek G, Fischer P, Van den Brandt P, Goldbohm S, Brunekreef B. Estimation of long-term average exposure to outdoor air pollution for a cohort study on mortality. Journal of Exposure Analysis and Environmental Epidemiology 2001; 11: 459-469.

Jerrett M, Burnett RT, Ma RJ, Pope CA, Krewski D, Newbold KB, et al. Spatial analysis of air pollution and mortality in Los Angeles. Epidemiology 2005; 16: 727-736.

Lertxundi-Manterola A, Saez M. Modelling of nitrogen dioxide (NO2) and fine particulate matter (PM10) air pollution in the metropolitan areas of Barcelona and Bilbao, Spain. Environmetrics 2009; 20: 477-493.

Maheswaran R, Elliott P. Stroke mortality associated with living near main roads in England and Wales - A geographical study. Stroke 2003; 34: 2776-2780.

Michelozzi P, Capon A, Kirchmayer U, Forastiere F, Biggeri A, Barca A, et al. Adult and childhood leukemia near a high-power radio station in Rome, Italy. American Journal of Epidemiology 2002; 155: 1096-1103.

Miller KA, Siscovick DS, Sheppard L, Shepherd K, Sullivan JH, Anderson GL, et al. Long-term exposure to air pollution and incidence of cardiovascular events in women. New England Journal of Medicine 2007; 356: 447-458.

Olea RA. Geostatistics for engineers and earth scientists. Boston: Kluwer Academic Publishers, 1999.

Pope CA, Burnett RT, Thun MJ, Calle EE, Krewski D, Ito K, et al. Lung cancer, cardiopulmonary mortality, and long-term exposure to fine particulate air pollution. Jama-Journal of the American Medical Association 2002; 287: 1132-1141.

Sohel N, Kanaroglou PS, Persson LA, Haq MZ, Rahman M, Vahter M. Spatial modelling of individual arsenic exposure via well water: evaluation of arsenic in urine, main water source and influence of neighbourhood water sources in rural Bangladesh. Journal of Environmental Monitoring 2010; 12: 1341-1348.

Son JY, Bell ML, Lee JT. Individual exposure to air pollution and lung function in Korea Spatial analysis using multiple exposure approaches. Environmental Research 2010; 110: 739-749.

TWEPA. Air quality in Taiwan annual report. Taiwan Environmental Protection Agency, Taipei, 2006.

USEPA. Technology Transfer Network:Air Quality System(AQS). U.S. Environmental Protection Agency, 1992.

USEPA. *AQS Raw Data Summary Formulas Draft*. In: Standards OoAQPa, editor. U.S. Environmental Protection Agency, Washington, DC, 2004.

Waller LA, Gotway CA. Applied spatial statistics for public health data. Hoboken, N.J.: John Wiley & Sons, 2004.

Young LJ, Gotway CA, Kearney G, Duclos C. Assessing Uncertainty in Support-Adjusted Spatial Misalignment Problems. Communications in Statistics-Theory and Methods 2009; 38: 3249-3264.

Young LJ, Gotway CA, Yang J, Kearney G, DuClos C. Assessing the association between environmental impacts and health outcomes: A case study from Florida. Statistics in Medicine 2008; 27: 3998-4015.

2

Air Quality and Bioclimatic Conditions within the Greater Athens Area, Greece-Development and Applications of Artificial Neural Networks

Panagiotis Nastos[1], Konstantinos Moustris[2], Ioanna Larissi[3] and Athanasios Paliatsos[4]
[1]Laboratory of Climatology and Atmospheric Environment, Faculty of Geology and Geoenvironment, University of Athens,
[2]Department of Mechanical Engineering, Technological Educational Institute of Piraeus,
[3]Department of Electronic-Computer Systems Engineering, Technological Educational Institute of Piraeus,
[4]General Department of Mathematics, Technological Educational Institute of Piraeus, Greece

1. Introduction

Over the past few decades the phenomenon of urbanization resulted in severe problems. The quality of human life has been deteriorated in the megacities around the world. This chapter deal with the Artificial Networks (ANNs) forecasting ability in predicting the air quality as well as the bioclimatic conditions in an urban environment.

For this purpose, different ANNs are demonstrated in this chapter. These ANNs have been developed in order to predict the air quality as well as the bioclimatic conditions within the Greater Athens Area (GAA), Greece. The prognosis for both air quality and bioclimatic conditions within GAA concerns the next three days (24 to 72 hours prediction).For the proper ANNs training for both air quality and bioclimatic conditions, hourly values of specific meteorological parameters such as air temperature, relative humidity, wind speed, wind direction, air pressure, sunshine and solar radiation, as well as hourly values of air pollutants concentrations have been used. These hourly data have been recorded in many different sites within GAA from the network of the Greek Ministry of Environment Energy and Climatic Change (GMEECC) during the period 2001-2005. Hourly values of barometric pressure and total solar irradiance for the same time period were acquired from the National Observatory of Athens (NOA).

This chapter is divided into nine sections. The first section is brief introduction concerning ANNs. The second section presents air quality indices that have been used in this work in order to describe the air quality within GAA. The third section presents bioclimatic indices, which describe the human thermal comfort-discomfort due to meteorological conditions. The fourth section presents statistical performance indices that have been used in order to investigate the predictive ability and reliability of the developed ANNs models. The fifth section demonstrates the examined sites within the GAA and the data/methodology used in this study.

The sixth section presents the ANNs that were developed in order to predict the maximum daily value of the air pollution indices as well as the persistence of the phenomenon, namely the number of consecutive hours within the day with high/strong air pollution. The seventh section presents the ANNs that were developed in order to predict the daily values of the bioclimatic indices as well as the number of consecutive hours within the day with dangerous bioclimatic conditions for humans' health. The eighth section includes the spatial variation for both air quality levels and human comfort/discomfort levels within GAA. The ninth is the last section summarizing briefly the extracted results by the performed analysis and how these results can contribute positively to the economy, energy, environment and quality of human life in general.

Finally the results of this work have shown that the ANNs could give an adequate forecast for both air quality and bioclimatic conditions within the urban environment of the GAA for the next three days at a statistical significant level of $p<0.01$.

2. Artificial Neural Networks

Artificial Neural Networks (ANNs) are a branch of artificial intelligence developed in the 1950s aiming at imitating the biological brain architecture. They are an approach to the description of functioning of human nervous system through mathematical functions. Typical ANNs use very simple models of neurons. These artificial neurons models retain only very rough characteristics of biological neurons of the human brain (McCulloh & Pitts, 1943). ANNs are parallel-distributed systems made of many interconnected non-linear processing elements (PEs), called neurons (Hecht-Nielsen, 1990). A renewal of scientific interest has grown exponentially since the last decade, mainly due to the availability of appropriate hardware that has made them convenient for fast data analysis and information processing (Viotti et al., 2002).

Figure 1. presents the structure of a biological neuron (upper graph) as well as the structure of an artificial neural (lower graph).

ANNs have been applied in time series prediction (Lapedes & Farber, 1987; Werbos, 1988). Although their behaviour has been related to non-linear statistical regression (Bishop, 1995), the big difference is that ANNs seem naturally suited for problems that show a large dimensionality of data, such as the task of identification for systems with great number of state variables. Over the last years, black box approaches have been recognized to constitute a viable alternative to conceptual models for input-output simulation and forecasting and also to allow shortening the time required for the model development. In particular, ANNs concentrated a general consensus in predicting different pollutants time series, as shown by the review of Gardner & Dorling (1998a, 1998b).

Many ANNs were developed for very different environmental purposes. Heymans & Baird (2000) have used network analysis to evaluate the carbon flow model built for the northern Benguela upwelling ecosystem in Namibia. Antonic et al. (2001) have estimated the forest survival after building the hydroelectric power plant on the Drava River, Croatia by means of GIS constructed database and a neural network. Karul et al. (2000) used a three-layer Levenberg-Marquardt feedforward neural network to model the eutrophication process in three water bodies in Turkey. Besides, Moustris et al. (2011) used ANNs for long term precipitation forecast, using long-term monthly precipitation time series of four meteorological stations in Greece.

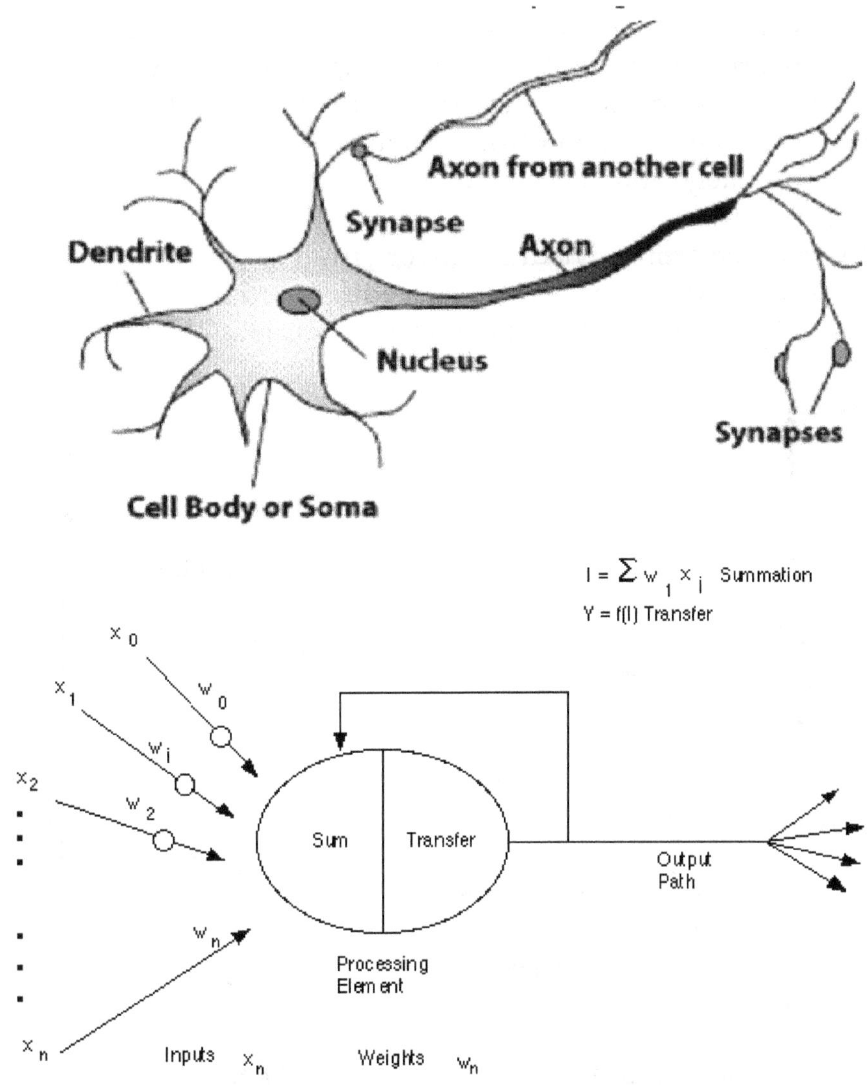

Fig. 1. Biological (upper graph) and artificial (lower graph) neuron structure.

Viotti et al. (2002) used ANNs to forecast short and middle long-term concentration levels for some of the well-known pollutants at the urban area of Perugia, Italy. The ANNs approach proved to be viable also for O_3, PM_{10}, NO_2, NO_x forecasting, outperforming alternative techniques in different case studies (Nunnari et al., 1998; Prybutok et al., 2000; Kolehmainen et al., 2001; Balaguer Ballester et al., 2002; Schlink et al., 2003; Corani, 2005; Slini et al., 2006; Dutot et al., 2007; Papanastasiou et al., 2007).

2.1 Multi-Layer Perceptron and feed-forward ANNs

The Multi-Layer Perceptron (MLP) is the most commonly used type of ANNs. Its structure consists of Processing Elements (PEs) and connections (Hecht-Nielsen, 1991). PEs, which are called neurons, are arranged in layers. The first layer is the input layer, one or more hidden layers follow and the final layer is the output layer. An input layer serves as buffer that distributes input signals to the next layer, which is a hidden layer. Each neuron of the hidden layer communicates with all the neurons of the next hidden layer, if any, having in each connection a typical weight factor. So, each unit-artificial neuron in the hidden layer sums its input, processes it with a transfer function and distributes the result to the output

layer. It is also possible that there are several hidden layers connected in the same fashion. The units-artificial neurons in the output layer compute their output in a similar manner. Finally, the signal reaches the output layer, where the output value from the ANN is compared to the target value and an error is estimated. Thus, the values of weight factors are amended appropriately and the training cycle repeats until the error is acceptable, depending on the application.
Since data flow within the artificial neural network from a layer to the next one without any return path, such kind of ANNs are defined as feed-forward ANNs. The structure of a feed-forward Multi-Layer Perceptron artificial neural network can be represented as in Figure 2.

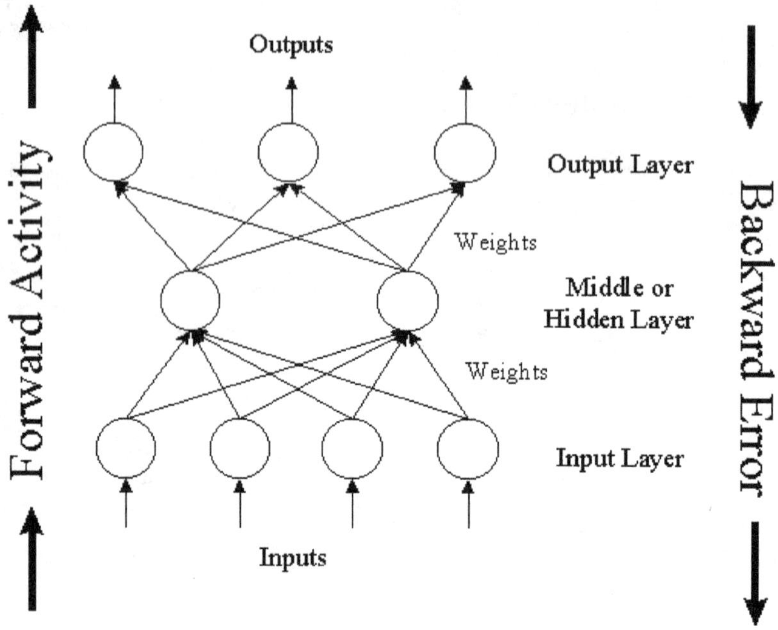

Fig. 2. Typical MLP feed-forward Artificial Neural Network Structure (Caudill & Butler, 1992).

2.2 Feed-forward ANNs training and the Back-propagation training algorithm

The training-learning process of ANNs can be far from the ensemble optimum in some cases, and the problem can be solved only with a very good database, a best choice of the input configuration for training, or using most powerful learning algorithms (Viotti et al., 2002).
The back-propagation learning algorithm consists of two steps of computation: a forward pass and a backward pass. In the forward pass, an input pattern vector is applied to the sensory nodes of the network, i.e. to the units in the input layer. The signals from the input layer are propagated to the units in the first layer and each unit produces an output. The outputs of these units are propagated to the units in the subsequent layers and this process continues until, finally, the signals reach the output layer, where the actual response of the network to the input vector is obtained.
During the forward pass, the synaptic weights of the network are fixed. During the backward pass, on the other hand, the synaptic weights are all adjusted in accordance with an error signal, which is propagated backward through the network against the direction of synaptic connections.

The mathematical analysis of the algorithm is as follows (Viotti et al., 2002). In the forward pass, given an input pattern vector $y^{(p)}$, each hidden node-neuron *j* receives a net input:

$$x_j^{(p)} = \sum_k w_{jk} y_k^{(p)}$$

where w_{jk} represents the weight between the hidden neuron j and the input neuron k. Thus, the hidden neuron j produces an output:

$$y_j^{(p)} = f\left(x_j^{(p)}\right) = f\left[\sum_k w_{jk} y_k^{(p)}\right]$$

where f(x) is the activation faction of the hidden layer. Different kinds of activation functions are referenced in the literature, such as linear, sigmoid, hyperbolic tangent, logistic, etc. (Norgaard et al., 2000). In the following, we consider a hyperbolic tangent activation function for the neurons in the hidden layer, hence, the value returned by the activation function of neuron *j* of the hidden layer is:

$$f(x_j) = \frac{e^{2x_j} - 1}{e^{2x_j} + 1}$$

Each output neuron receives the input from the preceding hidden layer by the forecasted value, so that the entry to the output neuron can be written as:

$$x^{(p)} = \sum_j w_j y_j^{(p)} = \sum_j w_j f\left[\sum_k w_{jk} y_k^{(p)}\right]$$

where w_j represents the weight between the output neuron and the hidden neuron *j*. It therefore produces the final output:

$$y^{(p)} = f\left(x^{(p)}\right) = f\left(\sum_j w_j y_j^{(p)}\right) = f\left[\sum_j w_j f\left(\sum_k w_{jk} y_k^{(p)}\right)\right]$$

The presentation of all the patterns is usually called *epoch.* Many epochs are generally needed before the error becomes acceptably small. In the batch neuron the error signal is calculated for each input pattern but the weights are modified only when the input patterns have been presented. The error function is calculated referring to the Mean Square Error (MSE) and the weights are modified accordingly:

$$E = \frac{1}{2}(d - y)^2 = \frac{1}{2}\left\{d - f\left[\sum_j w_j f\left(\sum_k w_{jk} y_k\right)\right]\right\}^2$$

where *d* is the desired or real output (monitored variable value) and *y* is the ANN output or the forecasted value. In the batch mode, *E* is equal to the sum of all MSEs on all the patterns of the training set. *E* is obviously a differentiable function of all weights (and thresholds) and therefore we can apply the gradient descent method. For the hidden to output connections the gradient descent rule gives:

$$\Delta w_j = -\eta \frac{\partial E}{\partial w_j}$$

where η is a number called *learning rate*. The learning rate is a parameter that determines the size of the weights adjustment each time the weights are changed during the training process. Small values for the learning rate cause small weight changes and large values cause large changes (Attoh-Okine, 1999). The best learning rate is not obvious. If the learning rate is 0.0, the network will never learn.

Refenes et al. (1994) reported that one and tow layered network with a learning rate of η=0.2 and a momentum rate of $0.3<\alpha<0.5$ yield the best combination of convergence. The momentum term is a factor used to speed network training. It adds a proportion of the previous weight changes to the current weight changes.

Using the chain rule it can be written as:

$$\frac{\partial E}{\partial w_j} = \frac{\partial E}{\partial y^{(p)}} \frac{\partial y^{(p)}}{\partial x^{(p)}} \frac{\partial x^{(p)}}{\partial w_j} = -\sum_p \left(d^{(p)} - y^{(p)}\right) f\left(x^{(p)}\right) y_j^{(p)}$$

Thus, the hidden to output connections are updated according to the following equation:

$$\Delta w_j = \eta \sum_p \left(d^{(p)} - y^{(p)}\right) f\left(x^{(p)}\right) y_j^{(p)} = \eta \sum_p \delta^{(p)} y_j^{(p)}$$

where $\delta^{(p)} = \left(d^{(p)} - y^{(p)}\right) f\left(x^{(p)}\right)$

For the input to hidden layer connections the gradient descent rule is:

$$\Delta w_{jk} = -\eta \frac{\partial E}{\partial w_{jk}}$$

Then using the chain rule, we obtain:

$$\frac{\partial E}{\partial w_{jk}} = \frac{\partial E}{\partial y_j^{(p)}} \frac{\partial y_j^{(p)}}{\partial x_j^{(p)}} \frac{\partial x_j^{(p)}}{\partial w_{jk}} = \frac{\partial E}{\partial y_j^{(p)}} f\left(x_j^{(p)}\right) y_k^{(p)}$$

Particularly, with reference to $\frac{\partial E}{\partial y_j^{(p)}}$, it can be written as:

$$\frac{\partial E}{\partial y_j^{(p)}} = -\sum_p \left(d^{(p)} - y^{(p)}\right) \frac{\partial\left[f\left(x^{(p)}\right)\right]}{\partial y_j^{(p)}}$$

and after simple passages, we obtain:

$$\frac{\partial E}{\partial y_j^{(p)}} = -\sum_p \left(d^{(p)} - y^{(p)}\right) f\left(x^{(p)}\right) w_j$$

Therefore, with reference to $\frac{\partial E}{\partial w_{jk}}$, it can be written as:

$$\frac{\partial E}{\partial w_{jk}} = -\sum_p \left(d^{(p)} - y^{(p)}\right) f\left(x^{(p)}\right) w_j f\left(x_j^{(p)}\right) y_k^{(p)}$$

from which the input to hidden connections updating is obtained as:

$$\Delta w_{jk} = \eta \sum_p \left(d^{(p)} - y^{(p)}\right) f\left(x^{(p)}\right) w_j f\left(x_j^{(p)}\right) y_k^{(p)}$$

and finally we get:

$$\Delta w_{jk} = \eta \sum_p \delta^{(p)} w_j f\left(x^{(p)}\right) y_k^{(p)} = \eta \sum_p \delta_j^{(p)} y_k^{(p)}$$

with $\delta_j^{(p)} = \delta^{(p)} w_j f\left(x^{(p)}\right)$

It is worthwhile noting that, a network architecture having just one hidden layer, and activation functions arranged as described above, constitutes a universal predictor and it can theoretically approximate any continuous function to any degree of accuracy. In practice, such degree of flexibility is not achievable because parameters must be estimated from sample data, which are both finite and noisy (Barazzetta & Corani, 2004).

The ANNs work on a matrix containing more patterns. Particularly, the patterns represent the rows while the variables are the columns. This data set is a sample. To be more precise, giving the ANN three different subsets of the available sample we can get the forecasting model; the three subsets concern the training, the validation and the test subsets. These subsets are briefly described:

- *Training subset,* the group of data with which we train-educate the network according to the gradient descent for the error function algorithm, in order to reach the best fitting of the nonlinear function representing the phenomenon.
- *Validation subset,* the group of data, given to the network still in the learning phase, by which the error evaluation is verified, in order to update the best thresholds and weights effectively.
- *Test subset,* one or more sets of new and unknown data for the ANN, which are used to evaluate ANN generalization, i.e. to evaluate whether the model has effectively approximated the general function representative of the phenomenon, instead of learning the parameters uniquely.

3. Air quality indices

Urban air pollution is a growing problem in big cities with large urbanization, where adverse health effects have been established. Bad city design combined with specific topographical and meteorological conditions allowing poor circulation, are associated with frequent episodes of critically high atmospheric pollution, enforcing in some cases extreme actions by the authorities, such as restriction of motor vehicles circulation within large area of the city.

For a better and more effective monitoring and analysis of air quality in big cities, air pollution indices are often used. Most of them have resulted after a series of epidemiological studies, which investigated the impact of air pollution on public health. In this work, two air pollution indices are presented and applied in order to forecast the air quality within GAA using ANNs.

3.1 Description of the European Regional Pollution Index (ERPI)

The European Regional Pollution Index (ERPI) has been proposed and developed by Moustris (2009). This air quality index is based on the air pollution index that is known as Regional Pollution Index (RPI). The New South Wales government in Sydney, Australia used RPI since the mid 1990s (NSW-EPA 1998, 2006).

The calculation of ERPI was performed using the thresholds prescribed by the European Community (EC) based on the framework directive 1996/62/EC and the three affiliated directives 1999/30/EC, 2000/69/EC, and 2002/3/EC (Table 1). Due to the way of calculation of ERPI, based on EC air pollution thresholds, the Australian RPI was renamed as European Regional Pollution Index (ERPI).

In this work, ERPI was calculated for five main air pollutants. Concretely, the air pollutants concern nitrogen dioxide (NO_2), sulfur dioxide (SO_2), carbon monoxide (CO), ozone (O_3) and particulate matter with aerodynamic diameter less than or equal to 10 μm (PM_{10}). For any observed concentration C_i, the value of the sub-index I_i is given by:

$$I_i = \frac{C_i}{Limit_i} \times 50$$

Air Pollutant	Limit values
NO_2	**Hourly value: 200 μg/m³**
SO_2	**Hourly value: 350 μg/m³**
CO	**Maximum daily mean value for 8 hours: 10 mg/m³**
O_3	**Maximum daily mean value for 8 hours: 120 μg/m³**
PM_{10}	**Mean daily value: 50 μg/m³**

Table 1. Limit concentration values of ambient air pollutants according to EC directives.

Once a sub-index I_i is obtained for each air pollutant (Table 1), the overall ERPI is simply taken as the maximum of all the I_i values according the formula:

$$ERPI = \max\{I_1, I_2, I_3, I_4, I_5\}$$

where I_1, I_2, I_3, I_4, and I_5 are the sub-indices whose values are defined by the NO_2, SO_2, CO, O_3 and PM_{10}, respectively. If ERPI ≥ 50 this means that at least one of the pollutants is over its limit value (Table 2). Table 3 presents the classification of air quality according to ERPI values (Moustris, 2009; Moustris et al., 2010).

ERPI	ERPI Class	Classification
0 - 2	1	Very Good
2 - 21	2	Good
21 - 40	3	Satisfactory
40 - 60	4	Sufficient
60 - 79	5	Poor
≥79	6	Very Poor

Table 2. Air quality classification according to ERPI values.

3.2 Description of Daily Air Quality Index (DAQx)

A new impact-related air quality index obtained on a daily basis and abbreviated as DAQx (Daily Air Quality Index) has been recently developed and tested by the Meteorological Institute of Freiburg, Germany, and the Research and Advisory Institute for Hazardous Substances, Freiburg, Germany (Mayer et al., 2002a, 2002b; Makra et al., 2003). DAQx considers the air Pollutants SO_2, CO, NO_2, O_3 and PM_{10}. To enable a linear interpolation between index classes, DAQx is calculated for each pollutant by:

$$DAQx = \left[\left(\frac{DAQx_{up} - DAQx_{low}}{C_{up} - C_{low}}\right) \times \left(C_{inst} - C_{low}\right)\right] + DAQx_{low}$$

with C_{inst}: highest daily 1 hour concentration of SO_2, NO_2, and O_3, highest daily running 8 hours concentration of CO, and mean daily concentration of PM_{10}. C_{up} is the upper threshold of specific air pollutant concentration range; C_{low} is the lower threshold of specific air pollutant concentration range; $DAQx_{up}$ is the value of DAQx according to C_{up}; $DAQx_{low}$ is the value of DAQx according to C_{low} (Table 4).
The daily value of DAQx is considered the highest value extracted by the calculated values for each pollutant.

SO_2 (μg/m³)	CO (mg/m³)	NO_2 (μg/m³)	O_3 (μg/m³)	PM_{10} (μg/m³)	DAQx value	DAQx Class	Classification
0-24	0.0-0.9	0-24	0-32	0.0-9.9	0.5-1.4	1	Very Good
25-49	1.0-1.9	25-49	33-64	10.0-19.9	1.5-2.4	2	Good
50-119	2.0-3.9	50-99	65-119	20.0-34.9	2.5-3.4	3	Satisfactory
120-349	4.0-9.9	100-199	120-179	35.0-49.9	3.5-4.4	4	Sufficient
350-999	10.0-29.9	200-499	180-239	50.0-99.9	4.5-5.4	5	Poor
≥1000	≥30.0	≥500	≥240	≥100	≥5.5	6	Very Poor

Table 3. Upper and lower limits for air pollutant concentrations and DAQx values, DAQx classes and classification according to Mayer et al. (2002a, 2002b).

4. Bioclimatic indices

The growth of the city of Athens during the last decades and the phenomenon of urbanization (Philandras et al., 1999) have established the well known Urban Heat Island (UHI) at a great areal extent of the city, resulting in explicit effects on human thermal comfort-discomfort. Thermal comfort is defined as the condition of mind, which expresses satisfaction with the thermal environment, absence of thermal discomfort, or conditions in which 80% or 90% of humans do not express dissatisfaction (Givoni, 1998).
Several indices, which describe the human thermal comfort-discomfort, have been developed worldwide. In this chapter three bioclimatic indices will be presented. The Discomfort Index (DI), the Cooling Power index (CP) and the Physiologically Equivalent Temperature (PET). In the process, these indices are briefly described.

4.1 Discomfort Index (DI)

The Discomfort Index (DI) was originally developed by Thom (Thom, 1959) and was supported by later works (Clarke & Bach, 1971; Giles et al., 1990). This index describes the degree of thermal load under various meteorological conditions, suitable for both outdoor and indoor environments. It is useful to evaluate how current temperature and relative humidity can affect the sultriness or discomfort sensation and cause health danger in the population.

DI (°C)	Classification of human comfort-discomfort sensation
DI<21	No discomfort feeling
21≤DI<24	Less than 50% of the total population feels discomfort
24≤DI<27	More than 50% of the total population feels discomfort
27≤DI<29	Most of the population feels discomfort
29≤DI<32	The discomfort is very strong and dangerous
DI≥32	State of medical emergency

Table 4. Classification of human comfort-discomfort sensation for DI.

Several formulas of the index have been proposed for use along with tables of boundary values that indicate degrees of comfort-discomfort. In the present work we used the following formula of DI, calculated as a combination of air temperature T (°C) and relative humidity RH (Giles et al., 1990):

$$DI = T - (0.55 \times 0.0055 \times RH) \times (T - 14.5), \qquad \textbf{(°C)}$$

The classification of the DI values with the equivalent feeling of thermal comfort–discomfort is given in Table 4.1.1 (Giles et al., 1990).

4.2 Cooling Power index (CP)

The Cooling Power Index (CP) was developed by Siple & Passel (1945) and describes the loss of energy, per unit of time and body surface, which a human organism can tolerate. The CP index, in contrast to the DI index, takes into consideration the wind speed instead of relative humidity. It describes the heat flux per surface unit of the human body towards the environment and the vice versa. For the calculation of the CP index hourly values of air temperature (T, °C) and wind speed (V, m/sec) were used. The calculation of CP is based on the following formula (Tzenkova et al., 2003):

$$CP = 1.163 \times (10.45 + 10 \times V^{0.5} - V) \times (33 - T), \qquad \textbf{(W/m}^2\textbf{)}$$

The classification of the CP index values modified by Besancenot et al. (1978), with the equivalent feeling of thermal comfort–discomfort is given in Table 5.

CP (W/m²)	Classification of human comfort
CP<0	Endothermal - very hot discomfort
0<CP≤174	Atonic – hot discomfort
175≤CP≤349	Hypotonic – hot sub comfort
350≤CP≤699	Neutral - comfort
700≤CP≤1049	Tonic – cold sub comfort
CP≥1050	Cold discomfort

Table 5. Classification of human comfort-discomfort sensation for CP.

4.3 Physiologically Equivalent Temperature (PET)

The thermal index Physiologically Equivalent Temperature (PET) is based on the total energy balance of the human body. PET values were evaluated (Mayer & Höppe, 1987; Höppe, 1999), in order to interpret the grade of the thermophysiological stress (Table 6).
It describes the effect of the thermal environment as a temperature value (°C) and can be quantified easier for non specialists in this topic. For night time situation, air temperature corresponds very close to the PET value. It has been applied in heat waves and climatic variability studies (Nastos & Matzarakis 2008, Matzarakis & Nastos 2010) and weather impacts on health (Nastos & Matzarakis, 2006).
The PET analysis was performed by the use of the radiation and bioclimate model, RayMan, which is well-suited to calculate radiation fluxes and human biometeorological indices (Matzarakis et al., 1999, 2010) and was chosen for all our calculations of mean radiant temperature and PET. The RayMan model, developed according to the Guideline 3787 of the German Engineering Society (VDI, 1998) calculates the radiation flux in easy and complex environments on the basis of various parameters, such as air temperature, air humidity, degree of cloud cover, time of day and year, albedo of the surrounding surfaces and their solid-angle proportions (Matzarakis et al., 2010).

PET (°C)	Thermal sensation	Physiological stress level
< 4	very cold	extreme cold stress
	....	....
8	cold	strong cold stress
	....	....
13	cool	moderate cold stress
	....	....
18	slightly cool	slight cold stress
	....	....
23	comfortable	no thermal stress
	....	....
29	slightly warm	slight heat stress
	....	....
35	warm	moderate heat stress
	....	....
41	hot	strong heat stress
	....	....
> 41	very hot	extreme heat stress

Table 6. Physiologically Equivalent Temperature (PET) for different grades of thermal sensation and physiological stress on human beings (during standard conditions: heat transfer resistance of clothing: 0.9 clo, internal heat production: 80 W) (Matzarakis & Mayer, 1996)

5. Statistical performance indices

The quality and reliability of the developed ANNs, concerning their ability to forecast both air quality and bioclimatic conditions within GAA, were tested using several statistics indices that have already been applied in similar studies (Moustris et al., 2010). The statistical performance indices that used in this work are presented and described briefly:

$$\text{Mean Bias Error: } MBE = \frac{1}{N}\sum_{i=1}^{N}(P_i - O_i)$$

where N is the number of the data points, O_i is the observed data and P_i is the predicted data. The MBE represents the degree of correspondence between the mean forecast (P_i) and the mean observation (O_i). MBE is used to describe how much the model underestimates or overestimates the observed data. Positive/negative values indicate over estimated/under estimated prediction.

$$\text{Root Mean Square Error: } RMSE = \sqrt{\frac{1}{N}\sum_{i=1}^{N}(P_i - O_i)^2}$$

RMSE provides a measure of how well future outcomes are likely to be predicted by the model.

The coefficient of determination (R^2) indicates how much of the observed variability is accounted by the estimated model (Kolehmainen et al., 2001). The coefficient of determination is a number between 0 and +1 and measures the degree of association between two variables. The coefficient of determination is calculated according to the equation (Comrie, 1997):

$$R^2 = \frac{\sum_{i=1}^{N}(P_i - O_{iave})^2}{\sum_{i=1}^{N}(O_i - O_{iave})^2}$$

where O_{iave} is the average of the observed data.

A relative measure of error, called the index of agreement (IA), is also discussed in Willmott et al. (1985). Index of agreement is calculated according to the formula:

$$IA = 1 - \frac{\sum_{i=1}^{N}(P_i - O_i)^2}{\sum_{i=1}^{N}\left(\left|P_i - O_{iave}\right| + \left|O_i - O_{iave}\right|\right)^2}$$

where O_{iave} is the average of the observed data. This is a dimensionless measure that is limited to the range of 0-1. If IA=0, that means no agreement between prediction and observation and if IA=1, that means perfect agreement between prediction and observation.

6. Data and methodology

For the calculation of the bioclimatic indices as well as the air quality indices, appropriate meteorological data in hourly basis were used. More specifically, hourly values of air temperature (°C), relative humidity (%) and wind speed (m/s were used for DI and CP calculation. In addition to the aforementioned meteorological parameters, total cloudiness cover (octas) was taken into consideration for PET calculation, using the RayMan model (Matzarakis et al., 1999, 2010). The appropriate meteorological parameters used as inputs in the RayMan model were acquired from the National Observatory of Athens, for the period 2001-2004. Besides, hourly values of air pollutants concentrations (NO_2, SO_2, CO, O_3 and PM_{10}) were used in order to estimate the two air quality indices ERPI and DAQx. All the above datasets have been recorded by the network of the GMEECC covering the period 2001- 2005 and concern nine (9) different regions within the GAA, namely the regions: Agia Paraskevi, Thrakomakedones, Lykovrissi, Maroussi, Liossia, Galatsi, Patission, Aristotelous, and Geoponiki (Fig. 3). For a better surveillance, the examined regions-stations listed below with the following abbreviations: Agia Paraskevi (APA), Galatsi (GAL), Liossia (LIO), Maroussi (MAR), Patission (PAT), Aristotelous (ARI), Thrakomakedones (THR), Lykovrissi (LYK). The hourly values of air barometric pressure and total solar irradiance for the same time period were obtained from the National Observatory of Athens.

To describe the air quality within the GAA the values of the air quality indices ERPI and DAQx were calculated on an hourly basis in seven different regions-stations (APA, THR, LYK, MAR, LIO, GAL and PAT) with respect to the pollutants NO_2, SO_2, CO, and O_3 and in five different regions-stations (APA, THR, LYK, MAR and ARI) with respect to the particulate matter PM_{10}. The maximum value of the 24 hourly values was considered as the daily value for each one of the two air quality indices. Thus, for each one station-region two daily values for each one of the two examined air quality indices were calculated. The first daily value concerns the air pollutants NO_2, SO_2, CO, and O_3 and the second concerns the particulate matter PM_{10}. This happened because the daily concentrations of particulate matter PM_{10} as well as the daily concentrations of ozone are both high enough. If only one daily value for each of the two air quality indices was calculated, then, we will not be able to know if that value is due to ozone or PM_{10}. Thereafter, an appropriate number of ANN models were developed and trained in order to predict for the next three days the daily value for each one of the two air quality indices as well as the number of consecutive hours during the day where the value of the index is greater than a threshold value.

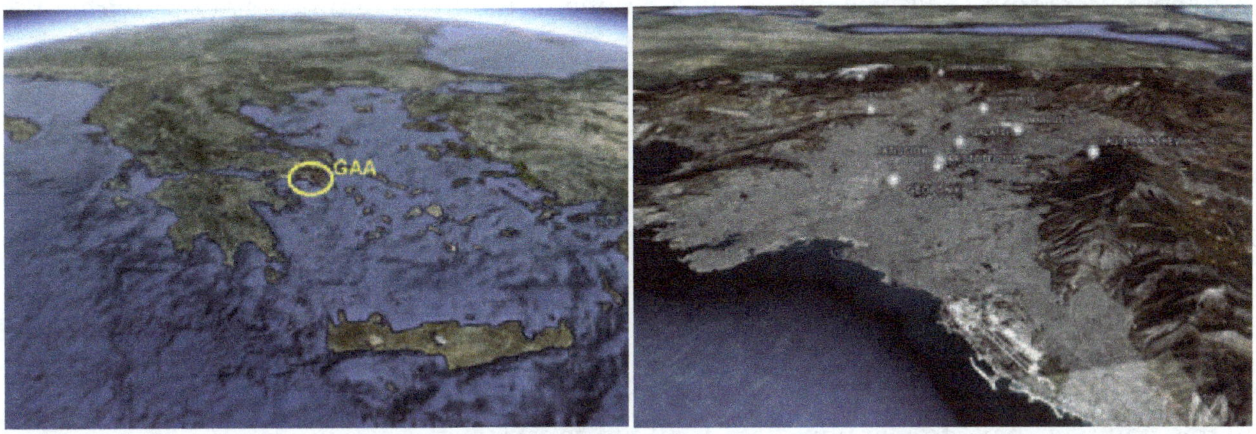

Fig. 3. Map of Greece (left graph) and spatial distribution of the GMEECC network's stations within GAA (right graph).

The bioclimatic conditions within the GAA are interpreted by the use of the bioclimatic indices DI and CP, which were calculated on hourly basis for eight different regions-stations (APA, THR, LYK, MAR, LIO, GAL, GEO and PAT). In the process, the daily value for each index in each region-station was calculated. The calculation was carried out only during the warm period of the year (May-September) in order to describe the human discomfort due to heat stress weather conditions. Then, an appropriate number of ANNs were developed and trained in order to predict for the next three days the daily value for each one of the two bioclimatic indices as well as the number of consecutive hours during the day, where the value of the index is greater than a threshold value (DI ≥ 24 ^{o}C) or less than a threshold value (CP ≤ 174 W/m^2). Furthermore, the mean daily values of PET index were estimated only for the National Observatory of Athens, because of the availability of the total parameters needed as inputs in RayMan model. Thereafter, the developed ANN was evaluated in forecasting PET for the next three days.

7. Air quality forecasting using ANNs

7.1 ANNs description

Six different ANNs were developed in order to forecast the air quality levels within the GAA. The first one (ANN#1) was trained in order to forecast the daily value of ERPI (for the pollutants CO, NO_2, SO_2 and O_3) for the next day at seven different areas of GAA (APA, THR, LYK, MAR, LIO, GAL and PAT). The second one (ANN#2) was trained in order to forecast the daily value of DAQx (for the pollutants CO, NO_2, SO_2 and O_3) for the next day at the above seven different areas within the GAA. The third one (ANN#3) was trained in order to forecast the daily number of the consecutive hours for the next day, with at least one of the pollutants concentrations (CO, NO_2, SO_2 and O_3) above a threshold according to directives of European Community, for each one of the seven examined stations within the GAA. The fourth (ANN#4) was trained in order to forecast the daily value of ERPI (with respect to PM_{10}) for the next day, at five different areas of GAA (APA, THR, LYK, MAR, and ARI). The fifth (ANN#5) was trained in order to forecast the daily value of DAQx (with respect to PM_{10}) for the next day, at the mentioned five different areas within the GAA. Finally the sixth (ANN#6) was trained in order to forecast the daily number of the consecutive hours for the next day with the PM_{10} concentrations above a threshold according to EC directives, for each one of the five examined stations within the GAA.

In each case, the group of data defined as "the training set", used for ANNs training, concerns the time period 2001-2004. The group of data defined as "the validation set", given to the network still in the learning phase, accounts 20% of 'the training set" for each one of the developed ANN models. Finally "the test set" refers to the year 2005. The year 2005 is absolutely unknown to the models, in order to reveal the models forecasting ability. Table 6.1 presents the input and output data for the six developed ANN models.

The combination of selected data for the appropriate ANN models training was done after a series of several tests (trial and error method). At the end, the combination that gave the best forecasting result in each case was selected (Table 7).

In this point, we have to mention that for all the constructed ANN models we have used as input data, in addition to other parameters, the maximum and minimum air temperature, the maximum and minimum wind speed for the next day as well as the mean daily air barometric pressure and the mode daily wind direction for the next day. This may produce a limitation in the forecasting attempt, but it is easy to have access to these forecasted values through the network of the Hellenic National Meteorological Service (HNMS).

INPUT DATA (input layer)	ANN#1	ANN#2	ANN#3	ANN#4	ANN#5	ANN#6
Stations' number (1,2,3,4,5,6,7)	√	√	√	√	√	√
Month (1,2,3,…,12)	√	√	√	√	√	√
Mean daily air pressure (mbar) for the six previous days	√	√	√	√	√	√
Daily sum of the global solar irradiance for the six previous days (W/m^2)	√	√	√	√	√	√
Maximum (T_{max}) and minimum (T_{min}) daily temperature (^{0}C) for the six previous days	√	√	√	√	√	√
Maximum (WS_{max}) and minimum (WS_{min}) daily wind speed (m/sec) for the six previous days	√	√	√	√	√	√
Maximum ($RH\%_{max}$) and minimum ($RH\%_{min}$) daily relative humidity for the six previous days		√			√	
Cosine and sine of the mode daily wind direction for the six previous days	√	√	√	√	√	√
ERPI daily value for the six previous days	√		√	√		√
DAQx daily value for the six previous days		√			√	
The number of consecutive hours during the day with ERPI≥50 for the six previous days			√			√
Mean daily air pressure (hPa) one day ahead	√	√	√	√	√	√
Maximum (T_{max}) and minimum (T_{min}) daily temperature (^{0}C) one day ahead	√	√	√	√	√	√
Maximum ($RH\%_{max}$) and minimum ($RH\%_{min}$) daily relative humidity one day ahead		√			√	
Maximum (WS_{max}) and minimum (WS_{min}) daily wind speed (m/sec) one day ahead	√	√	√	√	√	√
Cosine and sine of the mode daily wind direction one day ahead	√	√	√	√	√	√
OUTPUT DATA (output layer)						
ERPI daily value for the next day	√			√		
DAQx daily value for the next day		√			√	
The number of consecutive hours with ERPI≥50 for the next day			√			√

Table 7. Input and output data for the appropriate training of the six developed ANN models.

7.2 Forecasting of daily ERPI and DAQx values for the next day

The global fit agreement statistical indices as well as the excess statistical indices for the observed and predicted ERPI and DAQx values were calculated and demonstrated for the eight examined stations respectively. More specifically, O_{ave}, P_{ave}, MBE, RMSE, IA and R^2 values for ERPI index are presented in Table 8.

	ANN#1						ANN#2						ANN#3					
	O_{ave}	P_{ave}	MBE	RMSE	IA	R^2	O_{ave}	P_{ave}	MBE	RMSE	IA	R^2	O_{ave}	P_{ave}	MBE (hours)	RMSE (hours)	IA	R^2
APA	44.0	42.5	-1.460	6.629	0.925	0.752	3.4	3.4	0.013	0.396	0.853	0.557	2.8	3.2	0.374	3.015	0.877	0.605
GAL	35.6	35.7	0.118	0.644	0.903	0.697	3.3	3.3	0.060	0.471	0.751	0.378	0.5	0.7	0.199	1.415	0.657	0.231
LIO	37.7	37.8	0.114	6.081	0.919	0.726	3.3	3.3	0.032	0.463	0.746	0.358	0.7	0.9	0.168	1.761	0.671	0.238
MAR	38.3	37.1	-1.272	5.852	0.920	0.738	3.3	3.3	0.000	0.428	0.795	0.446	0.8	0.8	-0.008	1.825	0.708	0.292
PAT	33.1	32.2	-0.944	9.233	0.717	0.381	3.7	3.7	-0.024	0.343	0.791	0.442	0.3	0.3	0.042	1.365	0.299	0.017
THR	42.4	40.6	-1.810	7.290	0.922	0.760	3.3	3.3	-0.002	0.450	0.876	0.637	3.9	3.7	-0.208	4.924	0.829	0.516
LYK	38.9	38.6	-0.317	7.317	0.937	0.826	3.3	3.3	0.021	0.431	0.889	0.686	2.0	1.4	-0.590	3.092	0.742	0.401
	ANN#4						ANN#5						ANN#6					
	O_{ave}	P_{ave}	MBE	RMSE	IA	R^2	O_{ave}	P_{ave}	MBE	RMSE	IA	R^2	O_{ave}	P_{ave}	MBE (hours)	RMSE (hours)	IA	R^2
APA	40.0	40.0	0.356	18.718	0.674	0.266	3.7	3.6	-0.006	0.639	0.911	0.673	6.0	6.0	-0.310	4.259	0.764	0.377
MAR	46.0	46.0	-0.225	19.441	0.699	0.306	3.9	3.9	0.000	0.646	0.930	0.629	8.0	7.0	-0.141	4.908	0.791	0.425
THR	29.0	31.0	2.026	13.285	0.792	0.487	3.0	3.1	0.076	0.636	0.895	0.689	2.0	3.0	0.485	4.002	0.690	0.361
LYK	53.0	53.0	-0.195	19.927	0.729	0.381	4.2	4.2	-0.066	0.634	0.965	0.611	10.0	10.0	-0.303	5.594	0.816	0.492
ARI	53.0	53.0	0.292	18.352	0.721	0.295	4.3	4.3	0.023	0.528	0.779	0.578	10.0	9.0	-1.084	5.845	0.785	0.430

Table 8. Global fit agreement indices for ERPI predicted values for the next day.

Concerning the pollutants CO, NO_2, SO_2 and O_3, the R^2 values show a very satisfactory prediction for ERPI-ANN#1 ($0.381 \leq R^2 \leq 0.826$) as well as for the DAQx-ANN#2 ($0.378 \leq R^2 \leq 0.686$) during the test year 2005. Besides, IA values show also a very good prediction for ERPI-ANN#1 ($0.717 \leq IA \leq 0.937$) and the DAQx-ANN#2 ($0.746 \leq IA \leq 0.889$). In all cases, it seems that the prediction for the pollutants CO, NO_2, SO_2 and O_3 is much more successful using the ERPI, which is according to the European Community directives, instead of the DAQx. But using both predictions we can have a better and safe "picture" about air quality one day ahead within the GAA. As far as the air pollution persistence (for the pollutants CO, NO_2, SO_2 and O_3) is concerned, it seems that ANN#3 gives an adequate prediction. The R^2 values range between 0.017 and 0.605 while IA range between 0.299 and 0.877.

Finally, the worst prediction with respect to the air quality index ERPI appears for the region-station PAT (city centre) against the region-station LIO (urban area) concerning the air quality index DAQx. Generally, it seems that the prediction for the stations, which are closer to the GAA's downtown, is not so good compared to the prediction of the peripheral regions-stations. This is likely due to the traffic load and the bad air circulation within the city's centre, meaning that, more relevant data, associated with the above mentioned factors, are needed for a better ANNs training.

Figure 3 presents the best prediction (LYK) and the worst prediction (PAT) for ERPI concerning the pollutants CO, NO_2, SO_2 and O_3, while the best prediction (LYK) and the worst prediction (LIO) for DAQx concerning the same pollutants are depicted in Figure 4 Accordingly, Figure 5 presents the best prediction (THR) and the worst prediction (APA) for ERPI concerning the pollutant PM_{10}, and Figure 6 shows the best prediction (LYK) and the worst prediction (ARI) for DAQx with respect to the pollutant PM_{10}. During the warm period of the year (May-September) the values of ERPI (Figure 4) are greater than 50, meaning that at least one pollutant's concentration is above its threshold according to the EC directives. In most cases (more than 90%) the corresponding pollutant for these high values of ERPI is the ozone. The same results revealed from Figure 5 regarding DAQx, where during the warm period of the year the daily values of DAQx are greater than 3.5, meaning that a bad air quality exist in most cases. As far as the PM_{10} concentrations are concerned (Figures 5 and 6), it is shown that, for almost half of the days throughout the year are above the threshold concentration value, indicating bad air quality in the most of the examined stations-regions.

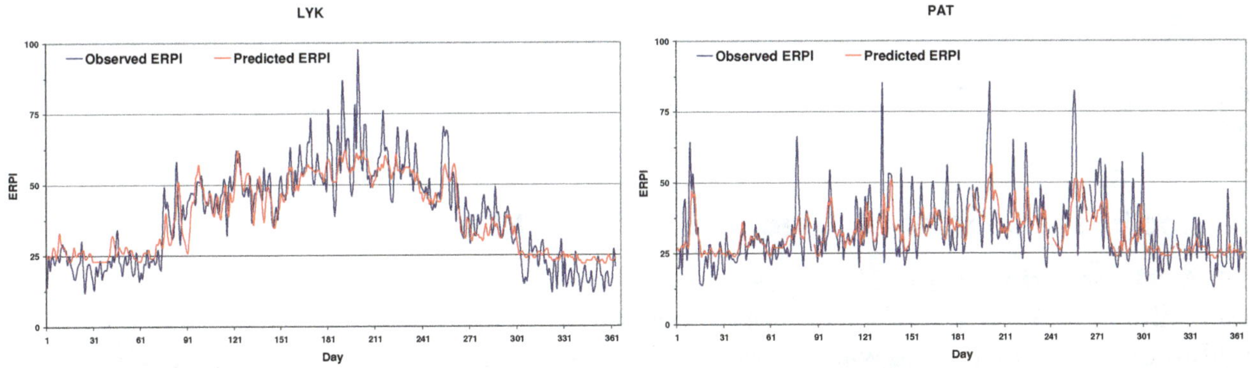

Fig. 3. Predicted vs. observed ERPI values for the CO, NO_2, SO_2 and O_3, pollutants during the test year 2005.

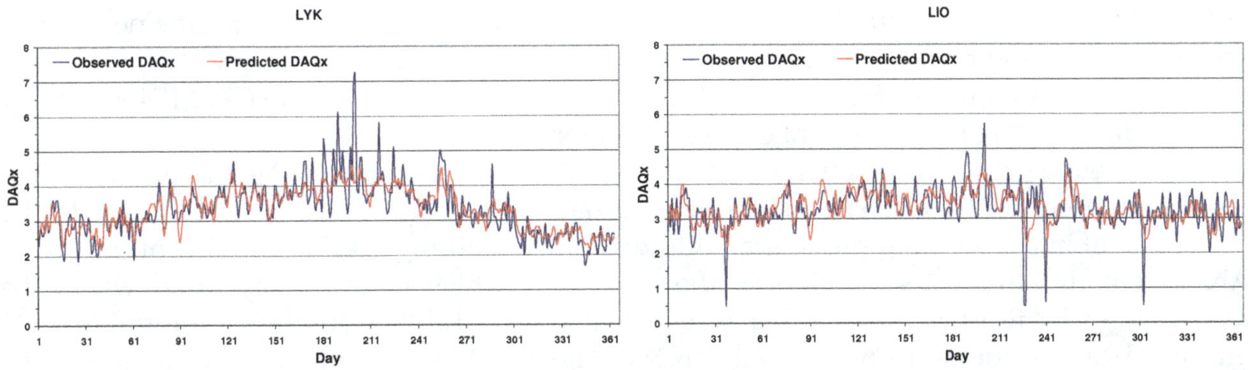

Fig. 4. Predicted vs. observed DAQx values for the CO, NO_2, SO_2 and O_3, pollutants during the test year 2005.

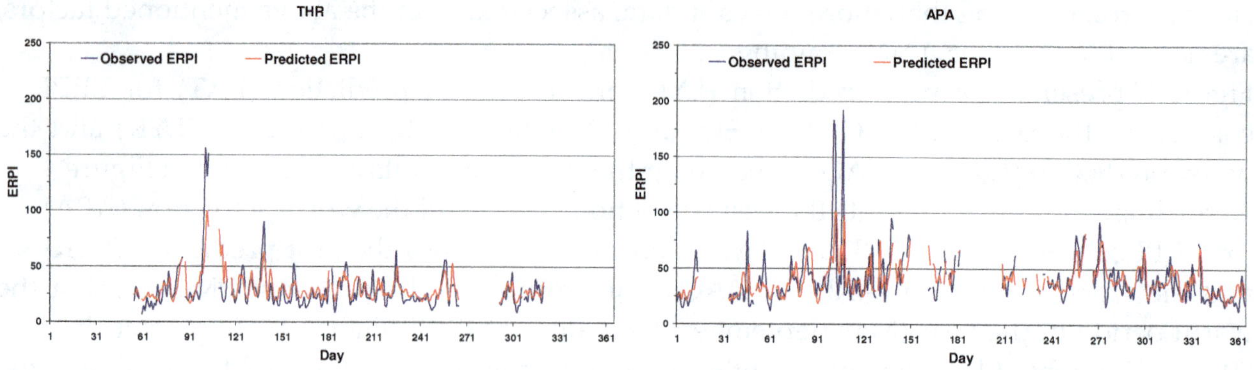

Fig. 5. Predicted vs. observed ERPI values for the PM_{10}, pollutant during the test year 2005.

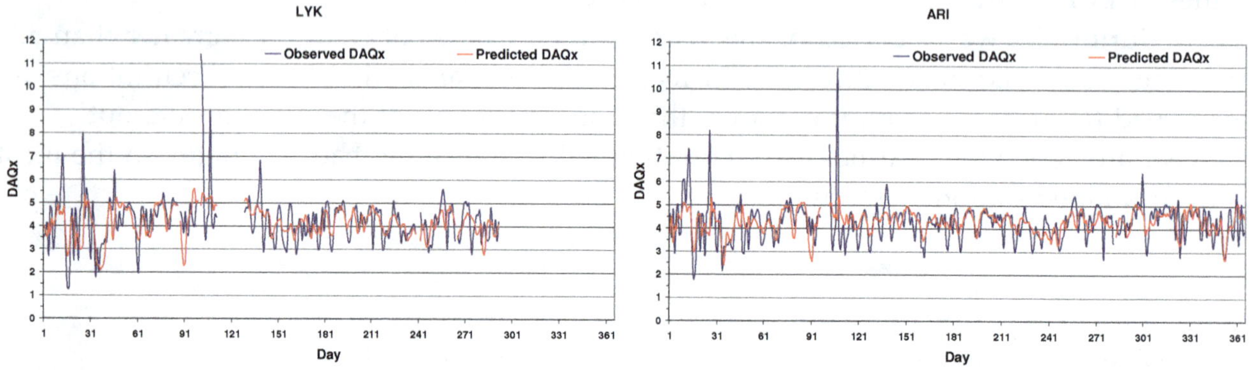

Fig. 6. Predicted vs. observed DAQx values for the PM_{10}, pollutant during the test year 2005

8. Bioclimatic conditions forecasting using ANNs

8.1 ANNs description for DI and CP forecasting

Four different ANN models were developed in order to forecast the bioclimatic conditions within the GAA during the warm period of the year (May-September). The first one (ANN#7) was trained in order to forecast the daily value of Thom's DI index for the next day at eight different areas of GAA (APA, THR, LYK, MAR, LIO, GAL, GEO and PAT). The second one (ANN#8) was trained in order to forecast the daily value of CP index for the next day at the above mentioned eight different areas within the GAA. The third one (ANN#9) was trained in order to forecast the daily number of the consecutive hours with DI ≥ 24 °C for the next day at each one of the eight examined stations within the GAA. Finally, the fourth (ANN#10) was trained in order to forecast the daily number of the consecutive hours with CP ≤ 174 W/m^2 for the next day at each one of the eight examined stations within the GAA.

In each case the group of data named as "the training set" used for ANNs training concerns the time period 2001-2004. The group of data named as "the validation set" given to the network still in the learning phase accounts 20% of the training set for each one of the above ANNs. Finally "the test set" refers to the year 2005, which is absolutely unknown to the models in order to reveal the models forecasting ability. Table 9 presents the input and output data for the four developed ANNs. The combination of selected data for the appropriate ANN models training was done after a series of several tests (trial and error method). At the end, the combination that gave the best forecasting result in each case was selected (Table 8.1.1).

INPUT DATA (input layer)	ANN#7	ANN#8	ANN#9	ANN#10
Stations' number (1,2,3,4,5,6,7)	√	√	√	√
Month (5,6,7,8,9)	√	√	√	√
The maximum (T_{max}) daily temperature for the six previous days.	√	√	√	
The maximum (RH_{max}) daily relative humidity for the six previous days.	√		√	
The maximum (DI_{max}) daily value of DI for the six previous days.	√		√	
The daily number of consecutive hours with DI≥24 ⁰C for the six previous days.	√		√	
The maximum (V_{max}) daily wind speed for the six previous days.		√		
The minimum (CP_{min}) daily value of CP for the six previous days.		√		
The daily number of consecutive hours with CP≤174 W/m^2 for the six previous days.		√		√
The maximum (T_{max}) and minimum (T_{min}) daily temperature for the six previous days.				√
The maximum (V_{max}) and minimum (V_{min}) daily wind speed for the six previous days.				√
The maximum (CP_{max}) and minimum (CP_{min}) daily value of CP for the six previous days.				√
OUTPUT DATA (output layer)				
The maximum (DI_{max}) daily value of DI for the next day.	√			
The minimum (CP_{min}) daily value of CP for the next day.		√		
The daily number of consecutive hours with DI≥24 ⁰C for the next day.			√	
The daily number of consecutive hours with CP≤174 W/m^2 for the next day.				√

Table 9. Input and output data for the appropriate training of the four developed ANNs.

8.2 DI and CP daily value forecasting for the next day

The global fit agreement statistical indices as well as the excess statistical indices for the observed and predicted values were calculated and demonstrated for the eight examined stations respectively. More specifically, O_{ave}, P_{ave}, MBE, RMSE, IA and R^2 values for DI are presented in Table 10.

The R^2 values show a very satisfactory prediction for DI-ANN#7 ($0.676 \leq R^2 \leq 0.841$) during the test year 2005 as well as for the CP-ANN#8 ($0.591 \leq R^2 \leq 0.814$). Concerning the IA values, a very satisfactory prediction for DI-ANN#7 ($0.849 \leq IA \leq 0.956$) as well as for the CP-ANN#8 ($0.813 \leq IA \leq 0.948$) appears. Taking into consideration the persistence of the phenomenon with respect to the daily number of consecutive hours with high discomfort conditions, due to strong heat stress, it seems that ANN#9 and ANN#10 give an adequate prediction. Additionally, the R^2 values show a very satisfactory prediction for ANN#9 (0.140

	ANN#7						ANN#8					
	O_{ave}	P_{ave}	MBE (°C)	RMSE (°C)	IA	R^2	O_{ave}	P_{ave}	MBE (W/m²)	RMSE (W/m²)	IA	R^2
APA	23.2	23.3	0.082	0.972	0.942	0.794	141	135	-6.196	49.754	0.920	0.731
GAL	24.1	24.0	-0.057	0.889	0.952	0.824	98	97	-1.137	39.863	0.948	0.814
GEO	23.2	23.3	0.111	0.831	0.956	0.841	143	130	-12.739	44.964	0.942	0.806
LIO	23.5	23.4	-0.069	0.929	0.944	0.801	126	126	0.145	48.369	0.930	0.764
MAR	24.0	23.9	-0.145	1.046	0.934	0.773	115	113	-2.154	48.893	0.926	0.743
PAT	25.0	25.1	0.090	0.889	0.953	0.832	82	78	-4.338	46.999	0.936	0.792
THR	19.9	20.7	0.779	1.584	0.849	0.676	292	246	-45.137	80.978	0.813	0.591
LYK	22.5	22.7	0.247	1.051	0.929	0.771	167	160	-6.791	55.554	0.878	0.664
	ANN#9						ANN#10					
	O_{ave}	P_{ave}	MBE (hours)	RMSE (hours)	IA	R^2	O_{ave}	P_{ave}	MBE (hours)	RMSE (hours)	IA	R^2
APA	3.6	4.2	0.654	2.610	0.930	0.760	7.0	8.0	1.333	4.372	0.907	0.702
GAL	6.6	6.4	-0.170	3.135	0.951	0.832	12.0	12.0	0.255	4.158	0.946	0.812
GEO	4.0	4.3	0.301	2.526	0.946	0.810	8.0	9.0	0.667	4.292	0.931	0.762
LIO	4.1	4.3	0.204	2.475	0.943	0.795	8.0	9.0	0.849	3.959	0.932	0.764
MAR	6.1	6.0	-0.112	3.254	0.943	0.797	10.0	10.0	0.430	4.436	0.934	0.766
PAT	10.9	11.0	0.103	3.601	0.901	0.861	13.0	14.0	0.524	4.923	0.928	0.751
THR	0.1	0.6	0.541	1.126	0.368	0.104	1.0	2.0	1.281	2.716	0.750	0.443
LYK	2.0	2.6	0.669	2.083	0.897	0.680	5.0	6.0	1.118	3.836	0.903	0.689

Table 10. Global fit agreement indices for DI predicted values for the next day.

$\leq R^2 \leq 0.832$) as well as for the CP-ANN#10 ($0.443 \leq R^2 \leq 0.812$) during the test year 2005. Besides, the IA values, show a very satisfactory prediction regarding ANN#9 ($0.368 \leq IA \leq 0.951$) and ANN#10 ($0.750 \leq IA \leq 0.946$). The worst prediction for the daily number of consecutive hours with high discomfort conditions, due to strong heat stress, refers to the region-station of THR (suburban region-station). This may be attributed to the fact that, in this suburban region (Thrakomakedones) the bioclimatic conditions are better than all the other examined regions within the GAA due to lower temperature values. Both discomfort indices, DI and CP, present daily values over their thresholds for a short period of time during the examined period. Thus, there is not a "memory-experience" of the persistence in THR, so the developed ANN models cannot have the appropriate training in order to forecast the number of consecutive hours with strong discomfort.

Figure 7. reveals that within the city's centre (PAT), the strong discomfort conditions (DI $\geq$ 24 °C) appear from the end of June to the first half of September. At the suburban station (THR) there is not a significant discomfort, according to DI values. Just a few days during the warm period of the year appear to be over the threshold of DI $\geq$ 24 °C; namely at least 50% of the population feels discomfort due to heat stress.

Figure 8 illustrates that close to the city's center (urban area of Galatsi), the hot sub comfort conditions according to CP values (CP $\leq$ 174 W/m²) appear from the middle of June until the first half of September. At the suburban station (THR), the discomfort due to heat stress conditions starts at the beginning of July until the middle of August. In all the above cases it seems that the prediction of bioclimatic conditions one day ahead with the use of ANN models is very satisfactory and realizable.

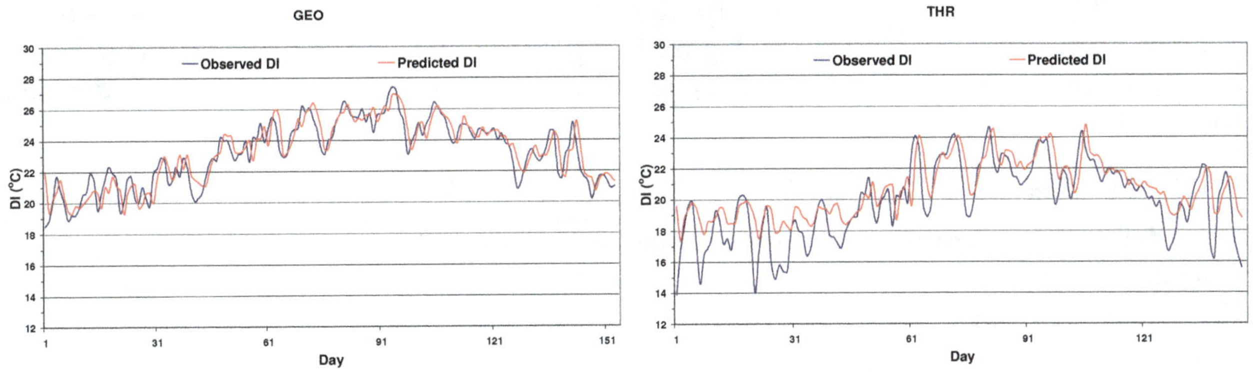

Fig. 7. Predicted vs. observed values of DI for the next day, concerning the best (GEO) and the worst (THR) prediction for DI daily maximum value, during the warm period of the test year 2005.

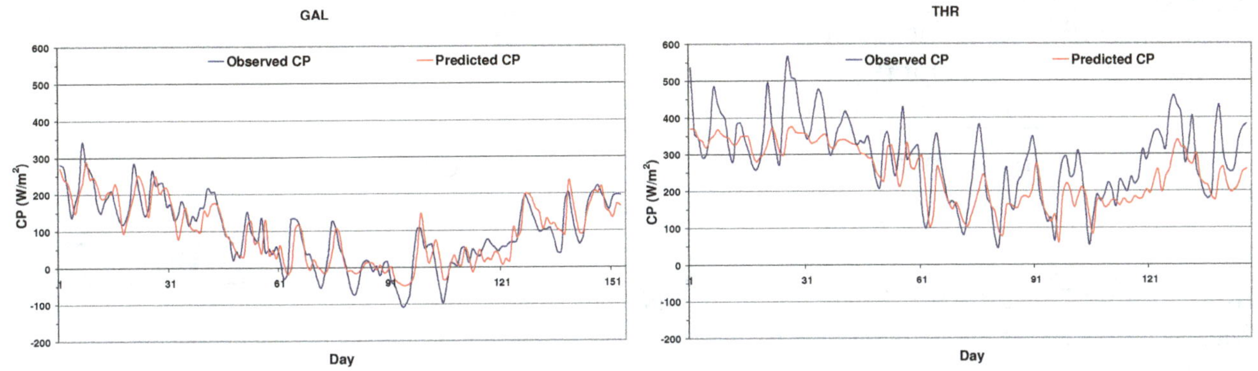

Fig. 8. Predicted vs. Observed values of CP for the next day, concerning the best (GAL) and the worst (THR) prediction for CP daily minimum value, during the warm period of the test year 2005.

8.3 ANNs description for PET forecasting

Three developed ANNs were trained using back-propagation algorithm to forecast the mean daily PET value for the next day (ANN#11), two next days (ANN#12) and three next days (ANN#13). The training dataset concern the period 2001-2003, while the validation dataset concern the year 2004, which was absolutely unknown to the constructed model, in order to test the predictive ability of the model. Superposed epoch analysis on the training datasets indicated that three days before the incidence of strong heat/cold stress are adequate to forecast PET value for the next days. Thus, the input data (Table 11) which were taken for ANNs training concern the mean daily air temperature, relative humidity, wind speed and sunshine for the previous three days from the National Observatory of Athens.

Table 12 presents the fit agreement indices between the observed and the predicted PET values, for the validation year 2004. It is remarkable the high values of IA and R^2, which indicate that the constructed ANNs have an excellent forecasting ability of PET for the next three days. This gives evidence that the developed ANNs, taking into account simple meteorological parameters recorded in the previous three days, are capable to predict a bioclimatic index, which is not easily calculated (PET was estimated using the RayMan model), while the most remarkable finding is that of pronounced agreement between observed and predicted PET values. Figure 9 depicts the predicted and observed mean daily PET time series for the next day (a), the next two days (b) and the next three days (c), along with the respective scatter plots.

INPUT DATA (input layer)	ANN#11	ANN#12	ANN#13
Mean daily air temperature (^{0}C) for the three previous days	√	√	√
Mean daily wind speed (m/s) for the three previous days	√	√	√
Mean daily relative humidity (RH%) for the three previous days	√		
The sunshine duration (hours) for the three previous days	√	√	√
OUTPUT DATA (output layer)			
Mean daily PET value for the next day	√		
4Mean daily PET value for the next two days		√	
Mean daily PET value for the next three days			√

Table 11. Input and output data for the appropriate training of the developed ANN#11.

	MBE	RMSE	IA	R^2
Mean daily PET value for the next day (ANN#11)	+0.5	2.8	0.982	0.933
Mean daily PET value for the next two days (ANN#12)	+0.5	3.8	0.966	0.874
Mean daily PET value for the next three days (ANN#13)	+0.4	4.3	0.956	0.839

Table 12. Global fit agreement indices for PET predicted values for the next one, two and three days.

9. Spatial distribution of air quality and bioclimatic conditions in the GAA

9.1 Spatial variation of air quality within GAA

The mean annual value for both air quality indices ERPI and DQAx was calculated at all the examined regions within GAA, during the time period 2001-2005. Figure 10 shows the spatial variation of air quality levels within GAA. As far as the air quality index ERPI is concerned, only the station THR appears a satisfactory air quality level in annual basis (ERPI < 40). The stations MAR, APA and GAL appear a tolerable air quality level (ERPI < 50). Moreover, the air quality levels at LYK, LIO and PAT stations are very close to the limit value of ERPI ≥ 50. Finally, the air quality level appears to be poor in the city centre station ARI. This may be attributed to the high PM_{10} concentration levels almost during the whole year. In this point, we have to mention that the station PAT is also in the centre of the city and very close to the ARI station, but unfortunately for this station we don't have any PM_{10} observations. Similar conclusions are extracted with respect to the air quality index DAQx. The only exception is the LIO station in which the air quality levels seems to be much closer to the stations GAL, MAR and APA.

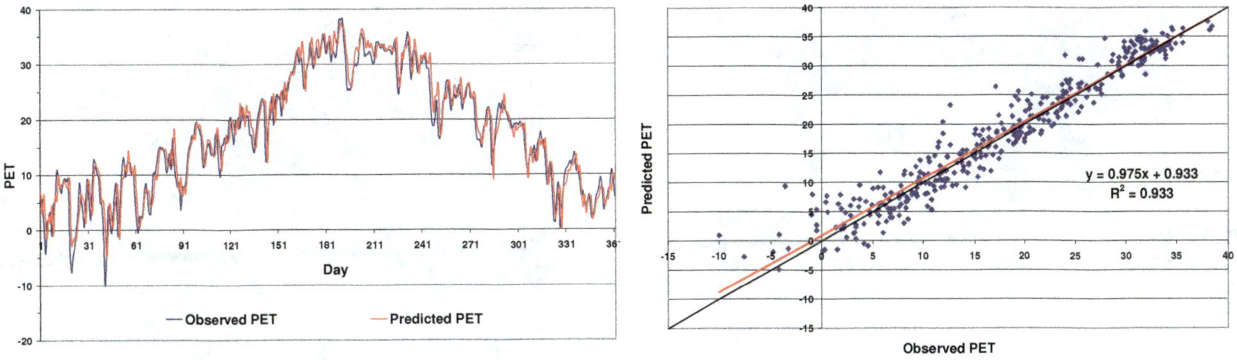

(a)

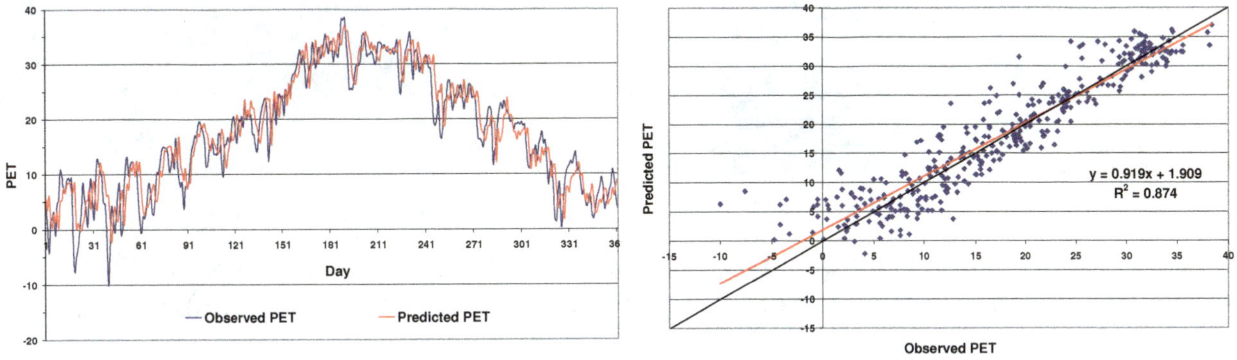

(b)

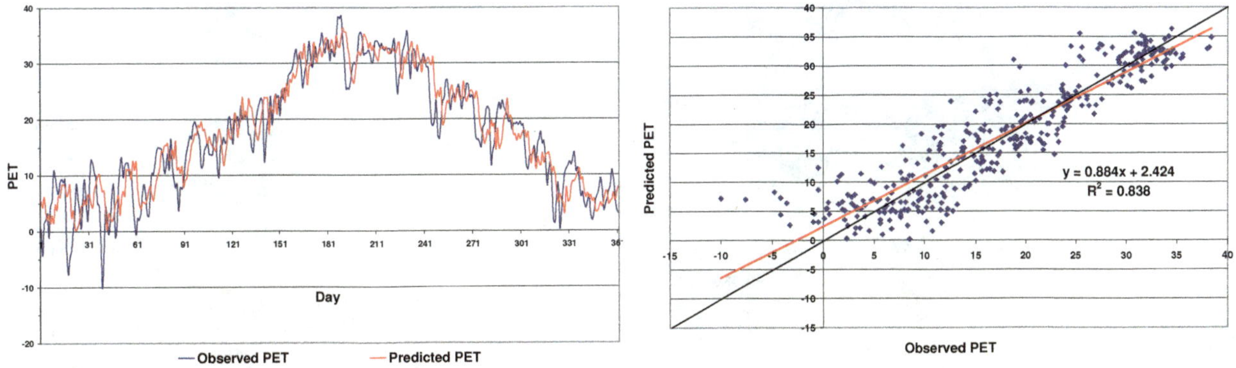

(c)

Fig. 9. Mean daily PET predicted values vs. observed values, for the next day (a), next two days (b) and next three days (c) three days, for the test year 2004.

9.2 Spatial variation of bioclimatic conditions within GAA

During the period 2001-2005, the mean annual value for both bioclimatic indices DI and CP was calculated at all the examined regions within the GAA. Figure 11 depicts the spatial variation of the bioclimatic conditions within the GAA during the warm period of the year (May-September), where three different bioclimatic zones appear. The first zone is the north suburban zone (THR), which can be characterized as a comfortable zone. The second zone extends peripherally the city's center (LIO, LYK, MAR and APA) and can be marginally characterized as a comfortable zone or warm zone. Finally, the third zone concerns the city's center (GAL, PAT and GEO), which can be characterized as an uncomfortable zone or a strong heat stress zone.

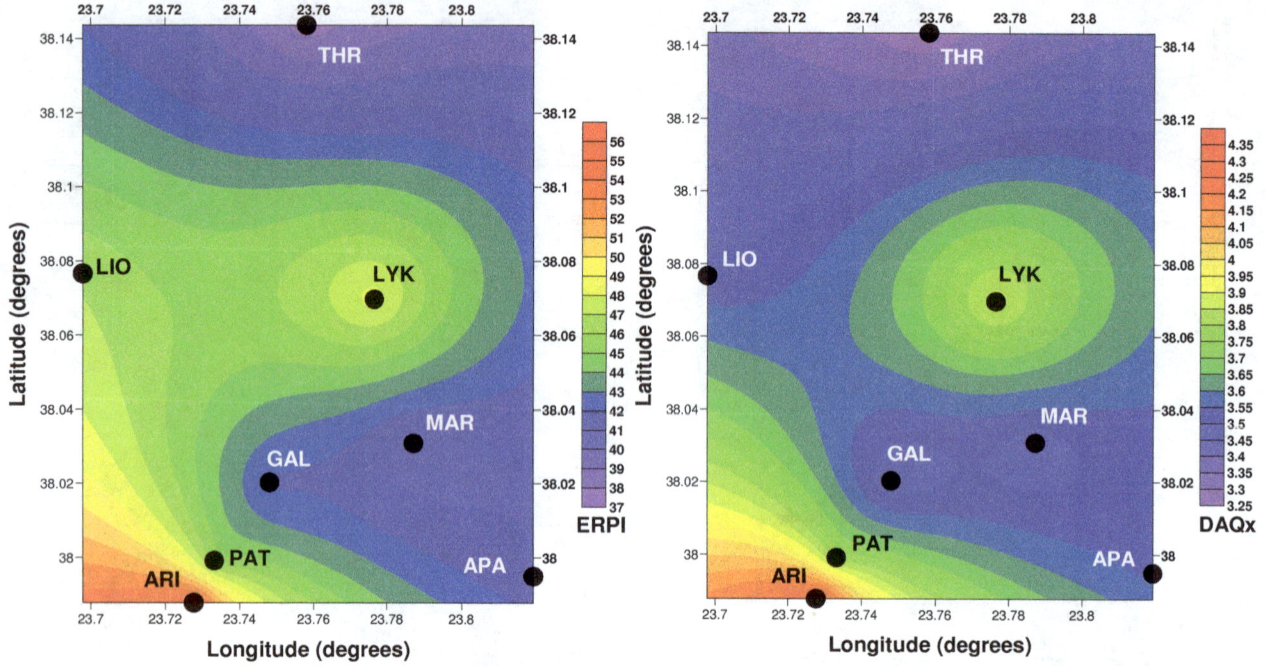

Fig. 10. Spatial variation of air quality levels within GAA for the pollutants NO_2, SO_2, CO, O_3 and PM_{10}. Mean annual values of ERPI (left graph) and DAQx (right graph), during 2001-2005.

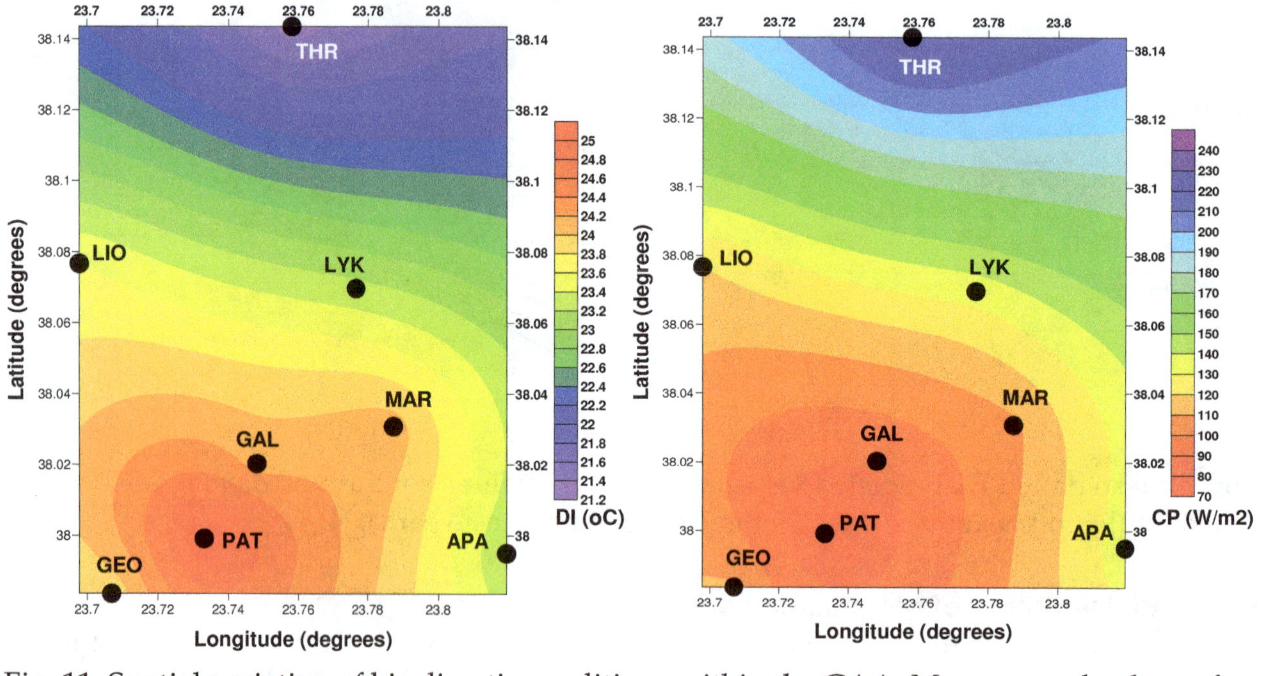

Fig. 11. Spatial variation of bioclimatic conditions within the GAA. Mean annual values of DI (left graph) and CP (right panel), for the warm period of the year during 2001-2005.

As far as the persistence of discomfort during the examined period 2001-2005 is concerned, the mean seasonal number of consecutive hours during the day with high levels of human discomfort appears in the station PAT; 11.3 and 13.6 consecutive hours with respect to DI and CP, respectively, against 1.0 and 2.7 consecutive hours at the station THR, respectively. All the other examined regions-stations within the GAA present a bioclimatic behavior between PAT and THR. This means that for a given building within the city's center region (PAT), we need 5 to 11 times more energy for cooling during the warm period of the year than the energy for cooling at the north suburban area (THR).

10. Conclusions

In this study an application, which concerns the development and the use of ANN models on environmental issues and generally in environmental management, is presented. A number of ANN models have been developed and trained in order to forecast the air quality levels, as well as the bioclimatic conditions in different regions within the GAA. The findings of this work appoint the ANN models forecasting capacity.

The Results showed that the use of ANN models as forecasting tool is realizable and satisfactory at a statistically significant level of $p<0.01$. In particular for the air quality forecasting for the next day, the R^2 values ranged between 0.381 and 0.826 (ERPI) and between 0.378 and 0.686 (DAQx). Besides, the IA index between the predicted and observed values ranged between 0.717and 0.937 for ERPI forecasting, while it ranged between 0.746 and 0.889 for DAQx forecasting. It seems that in all cases, the air quality forecasting is more sufficient using the ERPI air quality index than the DAQx. In this point we have to mention that the ERPI is according to the European Community directives for the air quality levels. The same results are extracted regarding the forecasting of the persistence of air pollution episodes and especially the number of consecutive hours during the day with poor air quality.

Concerning the forecasting of bioclimatic conditions for the next day, the R^2 values ranged between 0.676 and 0.841 for DI and between 0.591 and 0.814 for CP. The IA values ranged between 0.849 and 0.956 for DI and between 0.813 and 0.948 for CP. Taking into account the persistence of the phenomenon (the number of consecutive hours during the day with high discomfort conditions due to strong heat stress), it seems that ANN#9 (consecutive discomfort hours according to DI values) and ANN#10 (consecutive discomfort hours according to CP values) give an adequate prediction.

A remarkable finding of this research is that the high values of IA (0.956 - 0.982) and R^2 (0.839 - 0.933) with respect to PET forecasting for the next three days indicate that the constructed ANNs have an excellent forecasting ability of PET, a more complex bioclimatic index based on the human energy balance. This gives evidence that the developed ANNs, taking into account simple meteorological parameters recorded in the previous three days, are capable to predict a bioclimatic index, which is not easily calculated (PET was estimated using the RayMan model).

11. References

Antonic, O., Hatic, D., Krian, J. & Bukocev, D. (2001). Modeling groundwater regime acceptable for the forest survival after the building of the hydro-electric power plant. *Ecol. Model.*, Vol. 138, pp. 277-288.

Attoh-Okine, N.O. (1999). Analysis of learning rate and momentum term in backpropagation neural network algorithm trained to predict pavement performance. *Adv. Eng. Softw.*, Vol. 30, No. 4, pp. 291-302.

Balaguer Ballester, E., Valls, G., Carrasco-Rodriguez, J., Soria Oliva, E., Valle-Tascon, S. (2002). Effective 1-day ahead prediction of hourly surface ozone concentrations in eastern Spain using linear models and neural networks. *Ecol. Model.*, Vol. 156, pp. 27-41.

Barazzetta, S., Corani, G. (2004). First results in the prediction of PM10 in Milan: the Air Sentinel project. *Proceedings of the 9th International Conference on Harmonisation within Atmospheric Dispersion Modelling for Regulatory Purposes*, Garmisch, May 2004.

Besancenot, J.P. (1978). Le bioclimat humain de Rio. In: *Recherches de Climatologie en Milieu Tropical et Mediterranean*, Suchel, J.B., Altes, E., Besancenot, J.P. & Maheras, P. (Editors), Cahier No. 6 du Centre de Recherches de Climatologie, Universite de Dijon, Dijon, France.

Bishop, C.M. (1995). *Neural Networks for Pattern Recognition*, Oxford University Press, ISBN 019853864, Oxford, U.K.

Caudill, M. & Butler, C. (1992). Understanding Neural Networks; Computer Explorations, MIT Press-Cambridge, ISBN 02625309996, MA, USA.

Clarke, J.F., Bach, W. (1971). Comparison of the comfort conditions in different urban and suburban microenvironments. *Int. J. Biometeorol.*, Vol. 15, pp. 41–54.

Comrie, A.C. (1997). Comparing neural networks and regression models for ozone forecasting. *J. Air Waste Manage.*, Vol.. 47, pp. 653-663.

Corani, G. (2005). Air quality prediction in Milan: feed-forward neural networks, pruned neural networks and lazy learning. *Ecol. Model.*, Vol. 185, pp. 513-529.

Council Directive 96/92/EC (1996). *On ambient air quality assessment and management*. Official Journal of the European Communities, L296, 21.11.1996, pp. 55-63.

Council Directive 1999/30/EC (1999). *Limit values of sulphur dioxide, nitrogen dioxide and oxides of nitrogen, particulate matter and lead in ambient air*. Official Journal of the European Communities, L163, 29.6.1999, pp. 41-60.

Directive 2000/69/EC of the European Parliament and the Council (2000). *Limit values for benzene and carbon monoxide in ambient air*. Official Journal of the European Communities, L313, 13.12.2000, pp. 12-22.

Directive 2002/3/EC of the European Parliament and the Council (2002). *Ozone in ambient air*. Official Journal of the European Communities, L67, 9.3.2002, pp. 14-31.

Dutot, A.L., Rynkiewicz, J., Steiner, F.E. & Rude, J. (2007). A 24-h forecast of ozone peaks and exceedance levels using neural classifiers and weather predictions. *Environ. Modell. & Softw.*, Vol. 22, No. 9, pp. 1261-1269.

Gardner, M.W., Dorling, S.R. (1998a). Artificial Neural Networks, the multilayer perceptron. A review of applications in the atmospheric sciences. *Atmos. Environ.*, Vol. 32, pp. 2627-2636.

Gardner, M.W., Dorling, S.R. (1998b). Neural network modeling and prediction of hourly NOx and NO2 concentrations in urban air in London. *Atmos. Environ.*, Vol. 33, pp. 709-719.

Giles, B.D., Balafoutis, C.H., Maheras, P. (1990). Too hot for comfort: the heatwaves in Greece in 1987 and 1988. *International Journal of Biometeorology*, Vol. 34, pp. 98-104.

Givoni, B. (1998). *Climatic considerations in building and urban design*, John Wiley & Sons, ISBN 0471291773, New York.

Hecht-Nielsen, R. (1990). *Neurocomputing*, Addison-Wesley, ISBN 0201093553, Reading, M.A.

Heymans, J.J., Baird, A. (2000). A carbon flow model and network analysis of the northern Benguela upwelling system, Namibia. *Ecol. Model.*, Vol. 126, pp. 9-32.

Höppe, P.R. (1999). The physiological equivalent temperature–a universal index for the biometeorological assessment of the thermal environment. *Int. J. Biometeorol.*, Vol. 43, pp. 71–75.

Karul, C., Soyupak, S., Cilesiz, A.F., Akbay, N. & Germen, E. (2000). Case studies on the use of neural networks in eutrophication modeling. *Ecol. Model.*, Vol. 164, pp. 145-152.

Kolehmainen, M., Martikainen, H., Ruuskanen, J. (2001). Neural networks and periodic components used in air quality forecasting. *Atmos. Environ.*, Vol. 35, pp. 815-825.

Lapedes, A. & Farber, R. (1987). *Non-linear signal processing using neural networks.* Technical Report no. LA-UR-2662 Los Alamos National Laboratory.

Makra, L., Mayer, H., Beczi, R. & Borsos, E. (2003). Evaluation of the air quality of Szeged with some assessment methods. *Acta Climatologica et Chorologica,* Universitatis Szegediensis, Tom. 36-37, pp. 85-93.

Matzarakis, A., Mayer, H. (1996). *Another kind of environmental stress: Thermal stress.* WHO Collaborating Centre for Air Quality Management and Air Pollution Control, NEWSLETTERS No. 18, pp. 7-10.

Matzarakis, A., Mayer, H., Iziomon, M. (1999). Applications of a universal thermal index: physiological equivalent temperature. *Int. J. Biometeorol.*, Vol. 43, pp. 76-84.

Matzarakis, A., Rutz, F., Mayer, H. (2010). Modelling Radiation fluxes in simple and complex environments - Basics of the RayMan model. *Int. J. Biometeorol.*, Vol. 54, pp. 131-139.

Mayer, H., Höppe, P. (1987). Thermal comfort of man in different urban environments. *Theor. Appl. Climatol.*, Vol. 38, pp. 3-49.

Mayer, H., Kalberlah, F. & Ahrens, D. (2002a). TLQ-Am impact related air quality index obtained on a daily basis. *Proceedings of the fourth Symposium on the Urban Environment*, Norfolk, Virginia (USA), 20-24 May 2002, pp. 80-81.

Mayer, H., Kalberlah, F., Ahrens, D. & Reuter, U. (2002b). Analysis of indices for the assessment of the air (in German). *Gefahrstoffe-Reinhaltung der Luft*, Vol. 62, pp. 177-183.

McCulloh, W.S., Pitts W. (1943). A logical calculus of ideas immanent in Nervous activity. *Bulletin of Mathematical Biophysics*, Vol. 5, (1943), pp.115-133.

Moustris, K.P. (2009). *Air quality forecasting with the use of Artificial Neural Networks in the greater Athens area.* PhD Thesis, School of Chemical Engineering, National Technical University of Athens, Athens, Greece.

Moustris, K.P., Ziomas, I.C., Paliatsos, A.G. (2010). 3-Day-Ahead Forecasting of Regional Pollution Index for the Pollutants NO_2, CO, SO_2, and O_3 Using Artificial Neural Networks in Athens, Greece. *Water Air Soil Poll.*, Vol. 200, pp. 29–43.

Moustris, K.P., Larissi, I.K., Nastos, P.T., Paliatsos, A.G. (2011). Precipitation forecast using artificial neural networks in specific regions of Greece. *Water Resour. Manag.*, DOI 10.1007/s11269-011-9790.

Nastos, P.T., Matzarakis, A. (2006). Weather impacts on respiratory infections in Athens, Greece. *Int. J. Biometeorol.*, Vol. 50, pp. 358-369.

Nastos, P.T., Matzarakis, A. (2008). Variability of tropical days over Greece within the second half of the twentieth century. *Theor. Appl. Climatol.*, Vol. 93, pp. 75-89.

Matzarakis, A., Nastos, P.T. (2010). Human-Biometeorological assessment of heat waves in Athens. *Theor. Appl. Climatol.*, DOI 10.1007/s00704-010-0379-3.

Norgaard, M., Ravn, O., Poulsen, N.K., Hansen, L.K. (2000). *Neural Networks for Modelling and Control of Dynamic Systems*, Springer-Verlag, ISBN 1852332271, London.

Nunnari, G., Nucifora, M., Randieri, C. (1998). The application of neural techniques to the modeling of time-series of atmospheric pollution data. *Ecol. Model.*, Vol. 111, pp. 187-205.

Papanastasiou, D.K., Melas, D., Kioutsioukis, I. (2007). Development and Assessment of Neural Network and Multiple Regression Models in Order to Predict PM10 Levels in a Medium-sized Mediterranean City. *Water Air Soil Poll.*, Vol. 182, pp. 325-334.

Philandras, C.M., Metaxas, D.A., Nastos, P.T. (1999). Climate variability and urbanization in Athens. *Theor. Appl. Climatol.*, Vol. 63, No. 1-2, pp. 65-72.

Prybutok, R., Junsub, Y., Mitchell, D. (2000). Comparison of neural network models with ARIMA and regression models for prediction of Huston's daily maximum ozone concentrations. *Eur. J. Oper. Res.*, Vol. 122, pp. 31-40.

Refenes, A.N., Zapranis, A. & Francis, G. (1994). Stock performance modeling using neural networks: A comparative study with regression models. *Neural Networks*, Vol. 7, No. 2, pp. 375-388.

Schlink, U., Dorling, S., Pelikan, E., Nunnari, G., Cawley, G., Junninen, H., Greig, A., Foxall, R., Eben, K., Chatterton, T., Vondracek, J., Richter, M., Dostal, M., Bertucco, L., Kolehmainen, M., Doyle, M. (2003). A rigorous inter-comparison of ground-level ozone predictions. *Atmos. Environ.*, Vol. 37, pp. 3237-3253.

Siple, P.A., Passel, C.F. (1945). Measurements of dry atmospheric cooling in subfreezing temperatures. *Proceedings of the American Philosophical Society*, Vol. 89, No. 1, pp. 177–199.

Slini, T., Kaprara, A., Karatzas, K. & Moussiopoulos, N. (2006). PM10 Forecasting for Thessaloniki, Greece. *Environ. Model. Softw.*, Vol. 21, No. 4, pp. 559-565.

Thom, E.C. (1959). The discomfort index. *Weatherwise*, Vol. 12, pp. 57–60.

VDI (1998). VDI 3787, Part I: Environmental Meteorology, Methods for the human biometeorological evaluation of climate and air quality for the urban and regional planning at regional level. Part I: Climate, Beuth, Berlin.

Viotti, P., Liuti, G. & Di Genova, P. (2002). Atmospheric urban pollution: applications of an artificial neural network (ANN) to the city of Perugia. *Ecol. Model.*, Vol. 148, No. 1, pp. 27-46.

Werbos, P. (1988). Generalization of Backpropagation with application to a recurrent gas market model. *Neural Networks*, Vol. 1, pp. 339-356.

Willmott, C.J., Ackleson, S.G., Davis, R.E., Feddema, J.J., Klink, K.M., Legates, D.R., O'Donnell, J. & Rowe, C. (1985). Statistics for the evaluation and comparison of models. *J. Geophys. Res.*, Vol. 90, pp. 8995-9005.

3

Particularities of Formation and Transport of Arid Aerosol in Central Asia

Galina Zhamsueva[1], Alexander Zayakhanov[1], Vadim Tsydypov[1], Alexander Ayurzhanaev[1], Ayuna Dementeva[1], Dugerjav Oyunchimeg[2] and Dolgorsuren Azzaya[2]
[1]*Department of Physical Problems of Buryat Science Center, Siberian Branch of Russian Academy of Sciences, Ulan-Ude,*
[2]*Institute of Meteorology and Hydrology of Mongolia, Ulaanbaatar,*
[1]*Russia*
[2]*Mongolia*

1. Introduction

The climate change is a global problem of humanity. During last decades anthropogenic air pollution is a main cause of the content change of the atmosphere. The situation constantly is redoubled by continuing intensive growth of the air pollution in the areas located near of emission sources and also far off territories.
At the present the square of deserts increases in East Asia, and it affects on increasing of dusty days number. The total area of deserts and desertification lands is about 1.653 million km^2 in Northern China [Wang T. & Zhu Z., 2001; Wang S., 2005]. More than 70% of the pastureland area of Mongolia is under desertification [Natsagdorj et al., 2003].
Gobi Desert covers 1/3 of Mongolia territories and is the driest area in the world. The climate of Gobi Desert is dry, cold, and more continental. Soils are poor with humus, containing a lot of gravel.
Dust storm frequently occurs in arid and semi-arid territories of Central Asia. It is known that Gobi desert is one of the major sources of dust storms occurrence in the East Asia. Asian dust storms significantly influence on air quality [Chan et al., 2005; Gillette, 1986; Lee et al, 2006; Wei & Meng, 2006]. At the intense dust storms in Gobi desert the great many of sand and dust are transferred to Eastern China, Korea and Japan. Sometimes many particles of sand and dust are raised up by strong winds and transfer all over the world [Kim & Chung, 2008]. The total dust emission from sources in East Asia is estimated is 10.4×10^6 ton yr^{-1} for PM_{10}, 27.6×10^6 ton yr^{-1} for PM_{30} and 51.3×10^6 ton yr^{-1} for PM_{50}. The total dust emission from Mongolian sources is 2×10^6 ton yr^{-1} for PM_{10}, 2.9×10^6 ton yr^{-1} for PM_{30} and 8.7×10^6 ton yr^{-1} for PM_{50}. It is suggested that Southern Mongolia is an important dust source region [Xuan et al., 2004].
In recent years the intense dust storms are observed not only in Central Asia but also in different parts of the world. For example, strong dust storm was observed in Australia on 23 October 2002. The dust storm was 2400 km long, up to 400 km across and 1.5-2.5 km in height. The plume area was estimated at 840,860 km^2 and was raised 3.35-4.85 million tones of sands and dust with ground [McTainsh et al., 2005].

The researches of dust storms distribution in Mexico City [Jaurequi, 1989], spatial-temporal distribution of dust storm producing by wind erosion in USA [Gillette & Hanson, 1989], sources of dust storms formation and analysis of their far transport in China [Littmann, 1991; Sun et al., 2001; Wang S. et al., 2005; Zhou, 2001], climatology of dust storms in Mongolia [Natsagdorj et al., 2003; Zhadambaa et al., 1967a, 1967b] were carried out. However the detail researches of dust storms in Gobi Desert (Mongolia) didn't carry out to present days.

The purpose of this chapter is a study of analysis of mass and number concentrations of dust aerosol, their chemical compound, meteorological and turbulent characteristics of atmosphere and the mechanism of dust storm formation in Gobi Desert.

2. Sampling and methodology

Complex researches of aerosol and small gas components, meteorological, turbulent and radiative characteristics of atmosphere of arid territories of Mongolia have been begun by Department of physical problems of Buryat science centre of SB Russian Academy of Sciences in 2005 and continue nowadays. Measurements are conducted in the summer expeditionary periods 2005-2010 at Dornogovi aimag (station Sainshand, $44^0 54'$; $110^0 07'$, station Zamyn-Uud, $43^0 44'$; $111^0 54'$) and Sukhbaatar aimag (station Baruun-Urt, $46^0 41'$; $113^0 17'$) [Zhamsueva et al., 2008]. Observations points are situated at the Hydrometeorological centers located on considerable removal from settlements. Round-the-clock measurements of meteorological parameters are carried out by autonomous ultrasonic meteorological complexes «AMK - 03». Autonomous ultrasonic meteorological complexes measures momentary values of three orthogonal component of wind speed, air temperature, atmosphere pressure and air humidity. Momentary values of these parameters are used for calculations of turbulent air characteristics [Azbukin et al., 2009]. Fine aerosol was selected by diffusion spectrometer DSA in size range from 1.6 to 400 nm. Sampling of aerosol particulates with a size less than or equal to 10 µm was carried out on «Whatman-41" filters by high volume sampler PM-10 of General Metal Works Inc.

3. The research of air mass transport in Central Asia

3.1 The particularities of air mass circulation in Mongolia

Synoptical conditions and winds in desert regions are two main factors influenced on transport of sand, dust and atmosphere impurities. It is known that in the arid and semi-arid regions of Mongolia and Northern China located near 40^0 N the west flow of air transports great many of sand and dust particles to different parts of Earth.

For investigations of pathways of the atmospheric impurities transport in arid and semi-arid territories of Mongolia we have constructed forward and backward trajectories of air mass movement using of the reanalysis model NCEP/NCAR HYSPLIT (http://www.arl.noaa.gov/ready/hysplit4.html) [Draxler & Rolph, 2003] and archival meteorological data (archive FNL) National Oceanic and Atmospheric Administration (USA).

For construction of trajectories we set geographical coordinates of two stations of Gobi Desert (Sainshand and Baruun-Urt). Trajectories of air mass movement are calculated for 2008 with duration 120 h and step on time 6 h at heights of 1500, 2500 and 3500 m. The air mass movement of these heights mainly confirm of regional far transfer of impurities of regional

and global scales. In total 283 trajectories are constructed namely 140 forward trajectories and 143 backward trajectories for all seasons of 2008. The most often repeatable types of trajectories of air transport based on their direction and length are detailed as result of average data. It is established, that east, northeast and southeast carrying out of air mass are prevailed in this region. The episode of forward trajectories of air mass movement from stations Sainshand (13 April 2008) is presented presented on Fig. 1a for illustration. Apparently from figure, air mass from Gobi Desert passed over territories of Korea, Japan and Russia.
The results of the calculation of air mass backward trajectories show the prevalence of northwest, western, southwest and southern direction of winds to Gobi Desert. Under the southern and southwestern directions the conditions for air transport from China to the area of study are formed. This is confirmed by the construction of backward trajectories by the model HYSPLIT. On Fig. 1b is shown the most typical trajectories of air mass drift to Baruun-Urt station calculated for 26 May 2008.

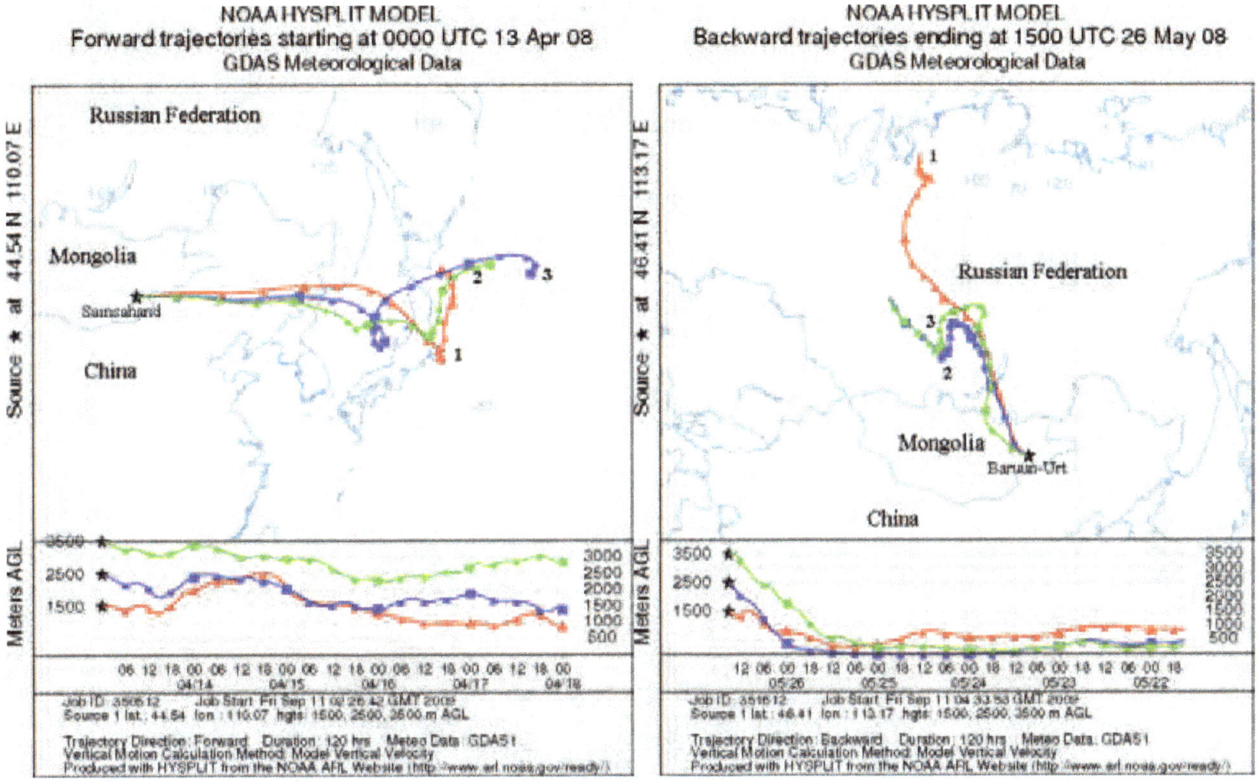

a) b)

Fig. 1. Forward and backward air masses trajectories during observation period at the Sainshand and Baruun-Urt stations by the model HYSPLIT: a) 13 April, b) 26 May

3.2 Study of wind regime in atmosphere of Gobi Desert

The variety of Mongolia relief leads to irregular distribution of air temperature over different locations of terrestrial surface (arid, semi-arid, steppe territories, mountain ecosystems and so on) and it promotes to development of local circulation of wind [Zhadambaa et al., 1967a, 1967b]. In this context great interest presents the study of wind regime in atmosphere of Gobi Desert.
The daily fund meteorological data for 2 years from 2004 to 2005 of meteorological stations (Sainshand and Baruun-Urt) have been analyzed for study of wind regime of Gobi Desert.

Roses of winds constructed on daily fund and experimental meteorological data show basically a northwest direction of wind. However in summer the formation of local atmospheric circulation is observed. Besides in summer the increasing of repeatability of southern (30%), northeast (26.4 %) directions of wind at Sainshand is noted in comparison with other seasons.

It proves to be true also experimental data of meteorological parameters received by means of acoustic meteorological complexes AMK-03 during scientific expeditions in June-July between 2005 and 2009, and also by using of meteorological data of CLIWARE system (http://cliware.meteo.ru/izomap).

4. Dust storm in East Gobi

4.1 Number of dusty days and trend

The analysis of daily fund meteorological data for 18 years between 1991 and 2009 has been carried out for detail research of frequency and duration of dust storms at Sainshand in East Gobi. Fig. 2 shows temporal trend of number of dusty days. It is revealed that last years number of dusty days is increased. For example, frequency of dust storms from 1991 to 2006 is increased in 3 times.

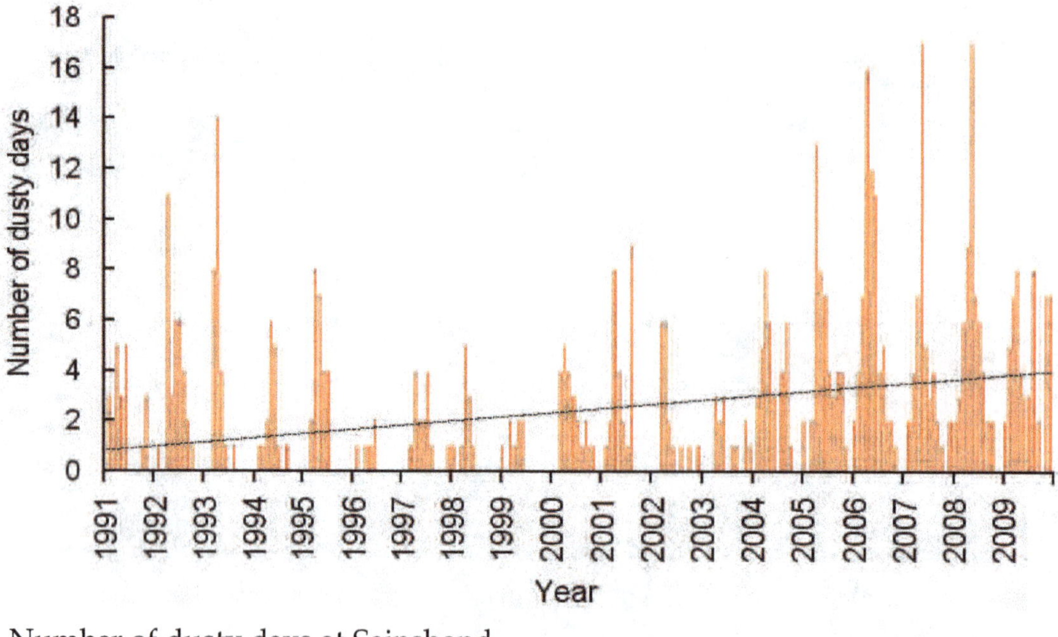

Fig. 2. Number of dusty days at Sainshand

To study the hour duration of dust storms we used daily fund data from 2003 to 2009. It is established that dust storms duration from 2003 to 2007 has increased in 40 times. The most duration of dust storms (573 hours) was observed in 2007. Analysis of number of dusty days has been allowed us to identify seasonal features. It is important to note that since 2004 dust storms began to appear in autumn and winter that it was not observed in previous years (Table 1).

The most repeatability of dust storms in atmosphere of Mongolia arid territories is observed in April-May at day and evening. The repeatability maximum of dust storms is marked in 15-18 hours of local time (Fig. 3).

	January	February	September	October	November	December
1991	0	3	0	1	3	0
1992	0	1	2	1	0	0
1993	0	0	0	0	0	0
1994	0	1	1	0	0	0
1995	0	0	0	0	0	0
1996	0	1	0	0	0	0
1997	0	0	0	0	0	1
1998	1	0	0	0	0	0
1999	1	0	0	0	0	0
2000	0	0	2	1	1	0
2001	0	1	0	0	0	0
2002	0	0	0	1	0	1
2003	0	0	1	0	2	1
2004	0	3	6	1	0	0
2005	2	0	4	4	1	0
2006	2	4	2	2	1	0
2007	0	2	2	1	0	2
2008	2	3	2	2	0	0
2009	2	5	2	0	7	7

Table 1. Number of dusty days in winter and autumn at Sainshand station

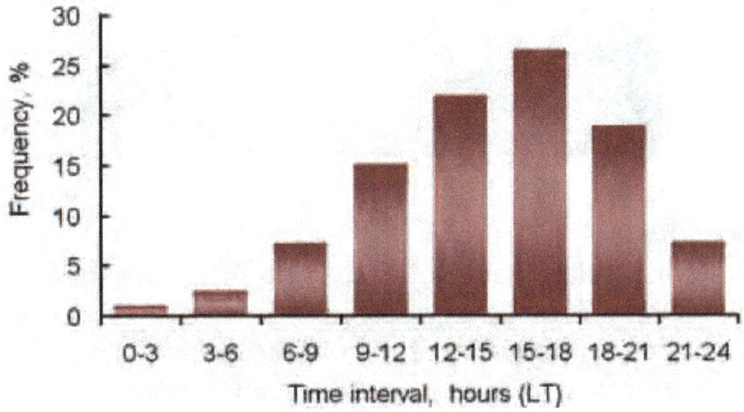

Fig. 3. The daily variation of dust storms for spring in the Gobi

4.2 Analysis of synoptical situation at severe dust storm in May 2008

The most repeatability of dust storms in arid and semi-arid territories of Central Asia is observed in spring, causing by strong winds due to passing of atmospheric fronts. For detail analysis of the episodes of strong dust storm in May 2008 in Mongolia we scrupulously considered the meteorological and weather conditions. In this time strong dust storms are mainly evidenced in East Gobi.

"Black wall" of sand and dust are accompanied by strong winds. Wind force reaches 30-40 m/s in some regions and the tones of sand and dust is transported to regions of China, Korea and Japan. The calculation of forward trajectories air mass movement using of model HYSPLIT confirms this result.

The trajectories of cyclones moving in May 2008 were investigated with help the surface and high altitude (500 hPa) meteorological maps of the Northern hemisphere (http://www.aari.nw.ru/odata/_d0010.php).

We find that the cold Arctic air masses are passed across Lake Baikal mixing with air mass of moderate latitudes in this time. Then it moved to the territories of Mongolia and China creating the significant gradients of the temperature and pressure. These meteorological and synoptical conditions were cause of intensive dust storms in arid and semi-arid territories of Mongolia and China. Therefore Lake Baikal such as huge reservoir of fresh water is a specific climatic zone which is situated on the frontier of arctic and moderate air masses. Baikal influences on pathways of air mass and dust storm occurrence in arid and semi-arid regions of Mongolia and China.

The satellite images of MODIS (http://modis.gsfc.nasa.gov) and the results of calculation of air mass backward trajectories demonstrate this fact and the episodes of severe dust storms in East Gobi on 26-27 May 2008 (Fig. 4).

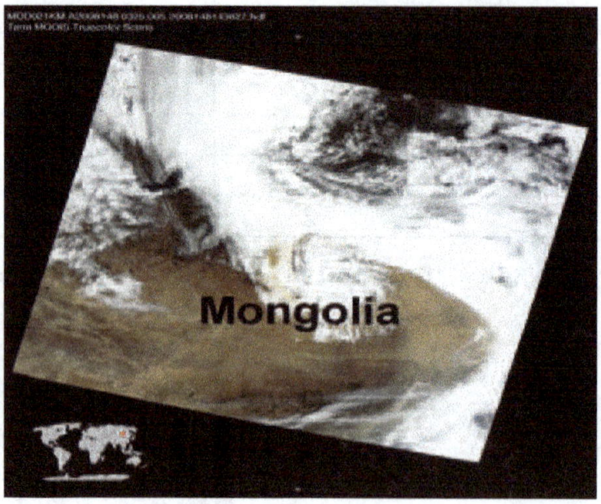

Fig. 4. Satellite picture passage of cold front in Mongolia (26 May 2008)

4.3 Behaviors of meteorological parameters during dust storm episodes

The analysis of behaviors of momentary meteorological parameters during the passage of dust storms presents a special interest for study of dust storm formation.

Usually two types of dust storms are observed such as the short-term connected with passage of atmospheric front and the long-term caused by strong winds due to elevated pressure gradients [Hoffmann et al., 2008].

Two regional dust storms were observed in the evening during complex scientific expedition in Baruun-Urt (semi-arid territory) in July, 2006. Notice, that first dust storm was accompanied by short-term precipitations, second was accompanied by a strong wind, local wind soil erosion, visibility no more than 50 m [Zhamsueva et al., 2008]. High temperature (+27^{0}C) and low relative humidity of air (20%) are registered in the beginning of dust storm. Under dust storm passage the sharp decreasing of air temperature during 15-20 minutes up to 8.8 ^{o}C (July, 10) and up to 7.9 ^{o}C (July, 11) independently of atmospheric turbidity are observed. In this time the increasing of average wind speed up to 21.5 m/s (July, 11), wind flaws are achieved up to 50 m/s and sharp increasing of relative humidity of air was noticed (Fig. 5).

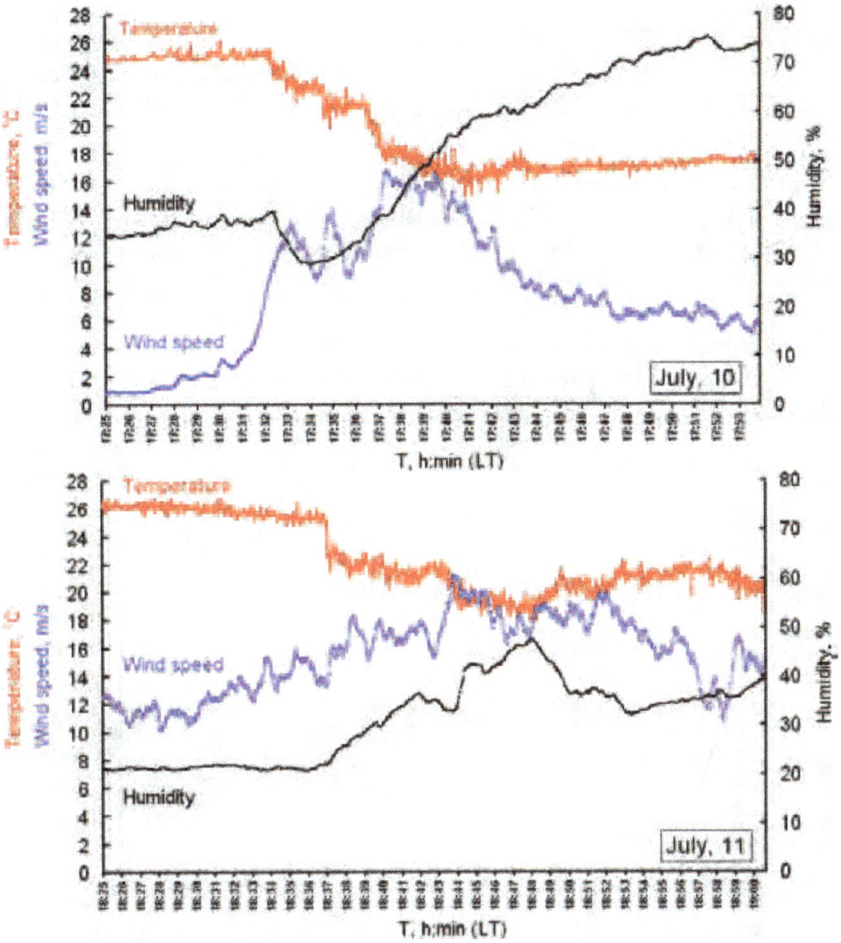

Fig. 5. Temporal variations of wind speed, temperature, humidity in atmosphere of Baruun-Urt during dust storm (10, 11 July 2006)

Under the weakening of dust storm the temperature of air stayed without change. After local dust storm we observed the decreasing of temperature and wind speed and increasing of relative humidity due to cold atmosphere front passage.

4.4 Turbulent characteristics of Gobi Desert atmosphere

The research of turbulent characteristics of atmosphere during dust storm is a great importance.

The turbulence promotes to heat and moisture exchange in the atmosphere of the Earth. Turbulence is caused by topographical heterogeneity and non-uniform heating of the Earth surface. The increasing of atmospheric turbulence conduces to the far transport of dust aerosol [Tverskoi, 1951; Bezuglaya & Berlyand, 1983].

According [Park et al., 2010; Bezuglaya & Berlyand, 1983] the dust storms usually occur under certain critical values of wind speed depending on terrain features and soil properties, which are irregular for different regions.

To get a complete picture of the atmospheric dynamics we analyzed around-the-clock daily measurements of turbulent characteristics obtained using of autonomous ultrasonic meteorological complex AMK-03 in Gobi Desert from 2005 to 2009 in summer. The turbulent characteristics of atmosphere are calculated based on the theory of Monin-Obukhov [Monin & Yaglom, (1965); Obukhov, (1988)]:

$$H = c_p \rho < T^{/} \cdot w^{/} > \quad (1)$$

$$K_h = v^* T^* / (\partial \theta / \partial z) \quad (2)$$

By results of the analysis it is revealed that in days without dust storms average vertical turbulent heat flux (H) in the surface layer in the day is H = 50 W/m². The maximal average daily value of the turbulent exchange coefficient (kh) is kh = 0.8 m²/s. Turbulence data analysis during dust storms showed that the vertical heat flux was directed downward from atmosphere to surface due to a sharp temperature fall. Maximum H value was -281 W/m² and kh = 2.7 m²/s during the passage of dust storms. After passing of dust storm the vertical heat flux varies from -7.3 to 6.0 W/m², and the coefficient of turbulent exchange equals 0.7 m²/s (Fig. 6).

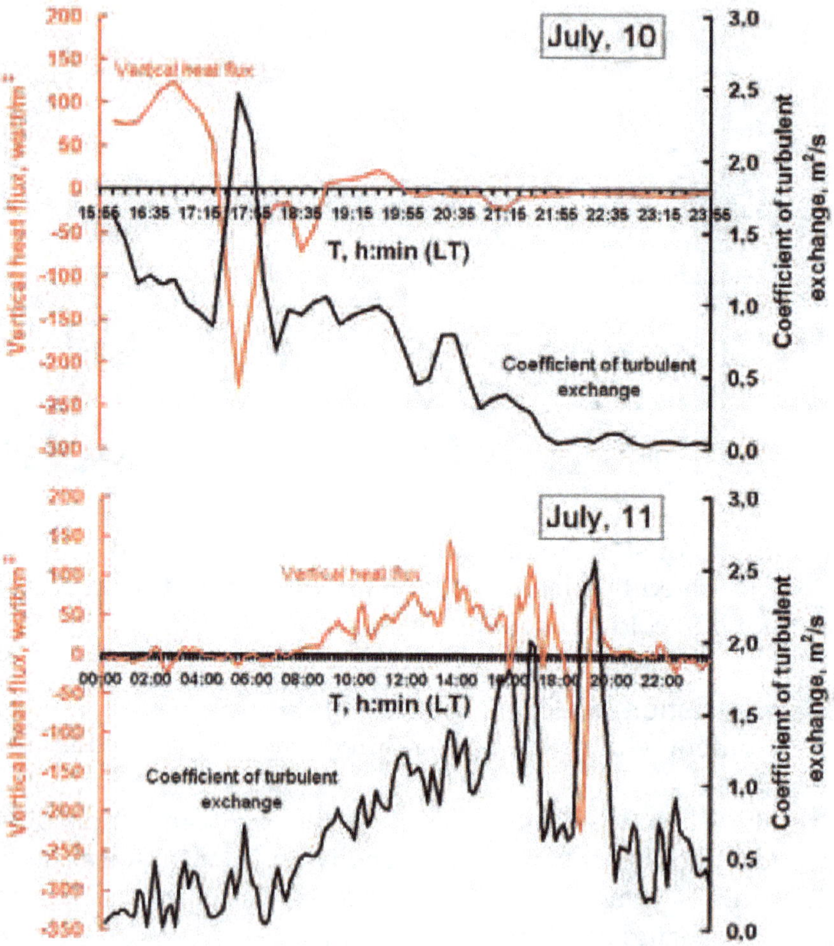

Fig. 6. Temporal variations of vertical heat flux, coefficient of turbulent exchange in atmosphere of Baruun-Urt during dust storm (10, 11 July 2006)

5. Mass concentration of dust aerosol PM10 and PM2.5 in Gobi Desert

In recent decades the problem of global and regional changes of environment and climate are actual due to increasing of air pollution by aerosol or particulate matter (PM). The aerosol changes the radiative balance in the earth-atmosphere system and, consequently, in weather and climate change. Also a harmful effect of particulate matter on human health is noted. The epidemiological studies reveal that under increase $PM_{2.5}$ (particles with an aerodynamic diameter of 2.5 µm or less) and PM_{10} (particles with an aerodynamic diameter of 10 µm or less) concentrations the morbidity, hospitalization and premature mortality of peoples are increased due to particle penetrate into the lower respiratory tract [WHO, 2006, 2007]. Particularly the most attention is devoted to study of $PM_{2.5}$ fine aerosols due to their relatively long lifetime in atmosphere and therefore far transfer. Consequently, the air pollution by suspended particles is not only local but also global problem. Monitoring measurement of mass concentration of PM_{10} and $PM_{2.5}$ at Sainshand station are conducted from 2008 within the framework of the program KOSA Monitor on development of network of dust and sand storms in Central Asia. Samples are collected using high volume samplers Partisol FRM Model 2000 Air Sampler (Japan).

We analyzed daily fund data of 2008 to reveal of daily and annual changes the aerosol mass concentration of PM_{10} and $PM_{2.5}$ at Sainshand and Zamyn - Uud. It is revealed the mean annual course of mass concentration of PM_{10} and $PM_{2.5}$ fine aerosols at Sainshand and Zamyn -Uud. Average annual concentrations of aerosol don't exceed 30 µg/m^3 of PM_{10} and 8 µg/m^3 of $PM_{2.5}$. The elevated monthly average PM_{10} and $PM_{2.5}$ concentrations are observed in spring and winter but minimum is autumn. The maximal average concentrations 30 µg/m^3 of PM_{10} and 20 µg/m^3 of $PM_{2.5}$ are marked in May 2008. Concentrations of these fractions are minimum value in autumn and up to 5 and 3 µg/m^3, according. Under the stable weather the daily average concentrations of 5 – 8 µg/m^3 of PM_{10} and 3 – 5 µg/m^3 of $PM_{2.5}$ are observed usually. But at dust storms the highest hourly concentrations exceed 1400 µg /m^3 (PM_{10}) and 380 µg /m^3 ($PM_{2.5}$). Average annual concentrations of aerosol at Zamyn-Uud located on the border of Mongolia and China considerably exceeds values at Sainshand and is 83 µg/m^3 of PM_{10} and 55 µg/m^3 of $PM_{2.5}$. The maximal average concentrations 100 µg/m^3 of PM_{10} and 59 µg/m^3 of $PM_{2.5}$ are marked in May 2008. Concentrations of these fractions are minimal in July and up to 11 µg/m^3 (PM_{10}) and 13 µg/m^3 ($PM_{2.5}$) in September. The daily average concentrations are varied range 18 - 20 µg /m^3 (PM_{10}) and 16 - 18 µg /m^3 ($PM_{2.5}$) under week wind. Maximum values of 1230 µg /m^3 (PM_{10}) and above, the 700 µg/m^3 ($PM_{2.5}$) and above under dust storms are observed. Data of mass concentration of PM_{10} at Sainshand and Zamyn - Uud good agree with data Erdene station (44^0 27′N, 111^0 05′E). The Erdene station is located in the about 90 km southeast from the station Sainshand and 100 km northwest from the station Zamyn - Uud. According to [Park et al., 2010] 10-day averaged daily maximum concentration of PM_{10} is 140 µg/m^3 in early May 2009.

6. Experimental studies of fine and chemical composition of arid aerosol in the atmosphere of the Gobi Desert

6.1 Study of fine arid aerosol

Studies of the daily dynamics of fine arid aerosol were carried out using the diffusion spectrometer DSA range in size from 1.6 to 400 nm at the Sainshand station (Gobi Desert, Mongolia) in August 2009.

The presence of night minima in the diurnal variation of total number concentration of fine aerosol is founded (Fig. 7). In addition, unlike the diurnal cycle of fine aerosol concentrations with peaks during the daytime in other regions, the maximum of the nuclei mode particles (d <0.01 µm) in the daily dynamics is often observed in the morning due to the rise of the particles of fine fraction due to more rapid heating of the earth's surface in arid territories and intensification of the turbulent processes in these hours [Ayurzhanaev et al., 2009].

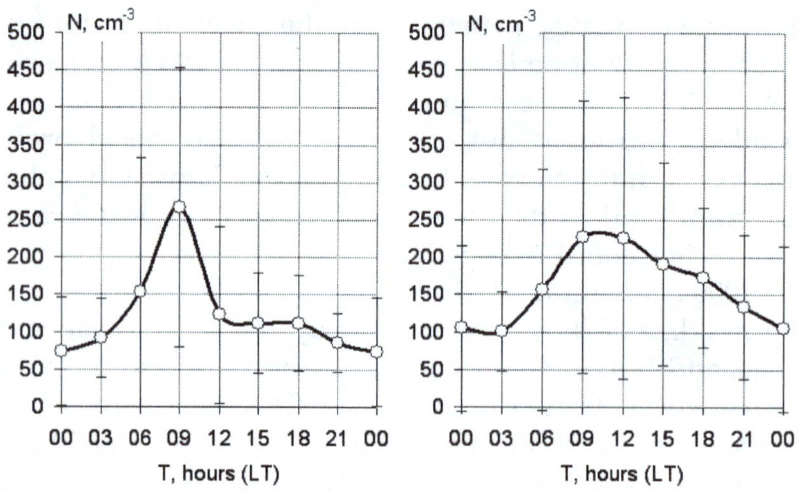

a) b)

Fig. 7. Daily variations of the total number concentration of aerosol (August 2009, Sainshand station, Mongolia): a) nuclei mode; b) the Aitken mode

Reduction of number concentration of nuclei mode particles during the daily hours apparently associate with increased of particle coagulation of this fraction and their transition to the range of Aitken particles fraction (0.01 <d <0.08 μm). The diurnal variations of Aitken particle concentration is confirmed by this conclusion (Fig. 7 b).

Diurnal variation of Aitken particle concentrations is similar to the behavior of the total concentration of fine faction. Number concentration maximum of particles of this fraction is often observed during daily hours, which may be the result of the increased coagulation of nuclei mode particles and the equalization of the generation and destruction rates of this aerosol fraction in this period [Ayurzhanaev et al., 2009; Zhamsueva et al., 2009]. Figure 8 shows a comparison of diurnal variations of soil and air temperatures during the experiments.

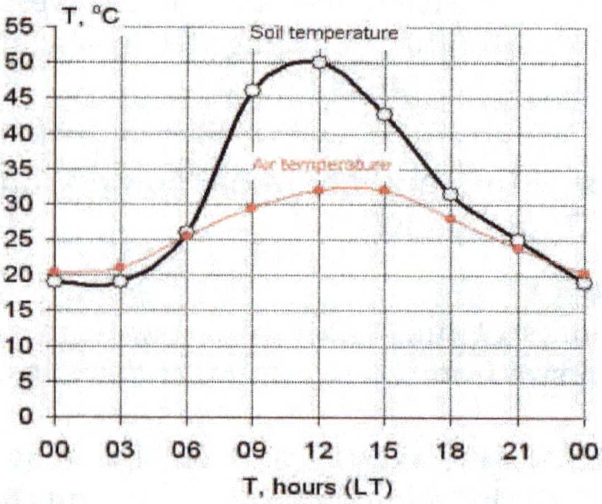

Fig. 8. Daily variations of soil and air temperatures (11 August 2009, Sainshand station, Mongolia).

It is revealed that the number concentration of particles of nuclei mode and Aitken mode differ depend on meteorological and weather conditions of the experiment. Figure 6 shows diurnal variation of the fine particles distribution for days with different weather conditions: under clear weather, with weak winds (10 and 12 August) and in cloudy weather, with gusty winds (13 and 14 August). As can be seen from the figure, the Aitken mode in the distribution of size spectrum particle presents for all days and is a major fraction the fine aerosol.

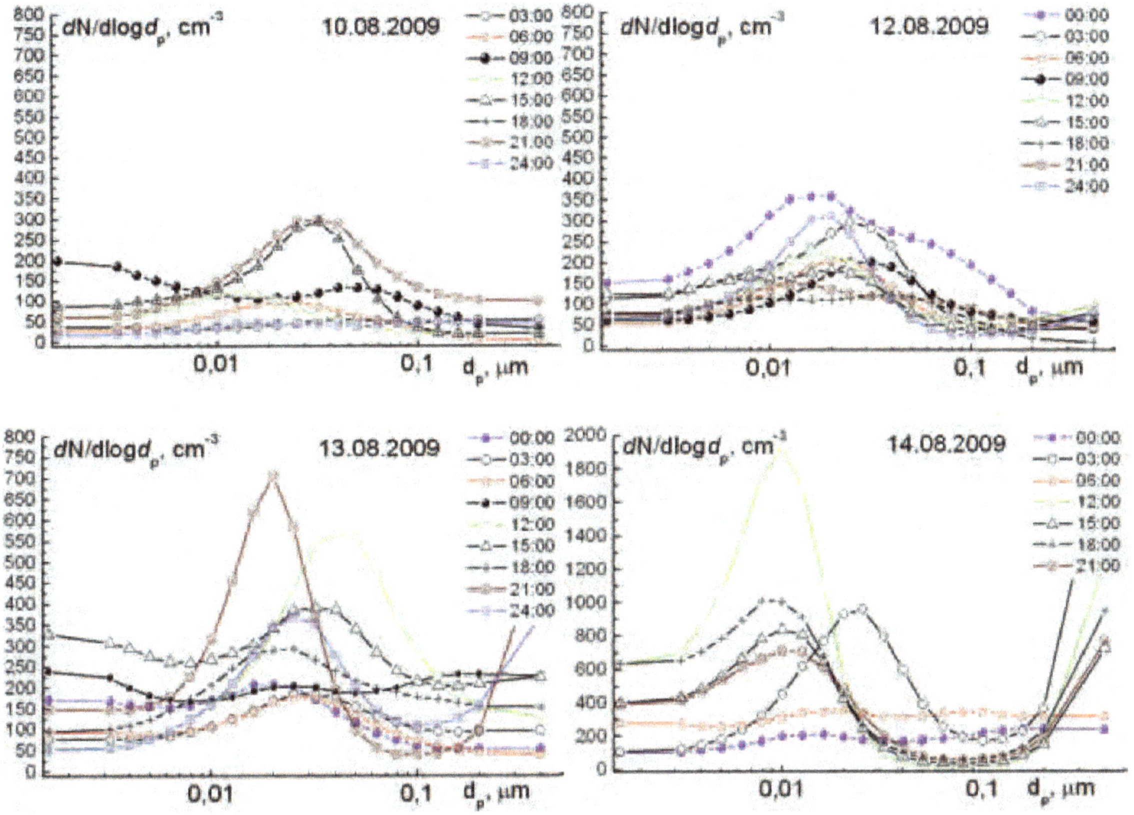

Fig. 9. Diurnal variation of size distribution of the fine aerosol (August 2009, Sainshand station, Mongolia).

6.2 The research of chemical composition of aerosol

The data of the chemical composition of aerosol are important source of information about the processes of transport, distribution and transformation of atmospheric pollutants. As the tracer component composition of atmospheric aerosol is generally conserved during the transport, the chemical composition of aerosol and its microstructure contain the important information about sources and ways of long-range transport of aerosol, including trace gases. Sampling of aerosol particulates with an size less than or equal to 10 μm was carried out on «Whatman-41" filters by high volume sampler PM-10 of General Metal Works Inc. with the volumetric rate of 0.4-1.7 m^3/min and accuracy of ± 0.03 m^3/min in the temperature range 0 ° C - 45 ° C.

The analysis of ionic composition of aerosols was carried out by liquid chromatography Milichrom A-02 on anions and atomic absorption method AAS-30 on cations in the Limnological Institute of SB RAS. The dates of sampling and their duration are presented in Table 2.

Sample number	Date	Time	
		Sampling start	Sampling end
30	11.08.09 – 12.08.09	09:34	08:20
31	12.08.09 – 13.08.09	09:13	08:52
32	13.08.09 – 14.08.09	09:56	07:00
33	15.08.09 – 16.08.09	20:05	17:45
34	16.08.09 – 17.08.09	19:24	16:28
35	17.08.09 – 18.08.09	17:05	15:40
36	18.08.09 – 19.08.09	16:51	16:13

Table 2. Periods of aerosol sampling

Figure 10 shows the results of the analysis of ionic composition of aerosols in the atmosphere of the arid area of Mongolia (Sainshand, August 2009).

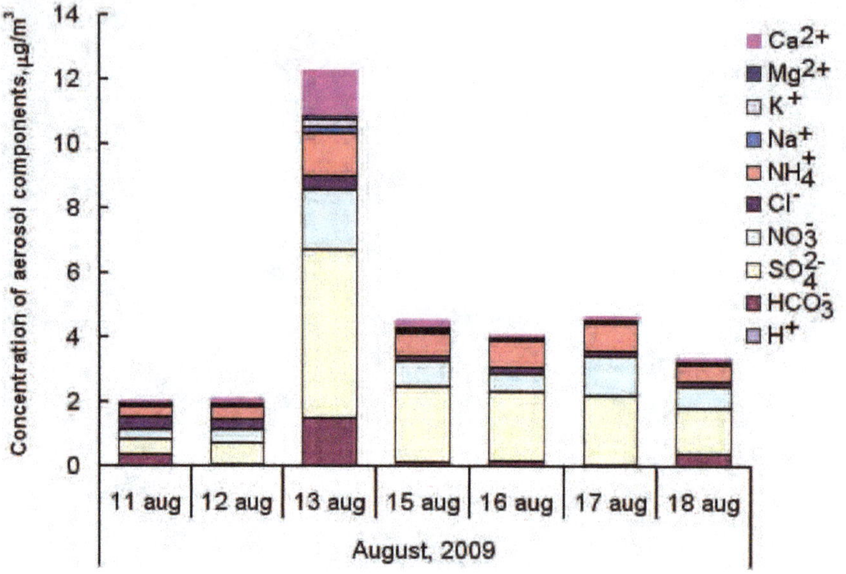

Fig. 10. The chemical composition of aerosols in the atmosphere of Gobi Desert (August 2009)

Apparently from Fig. 7 the anions SO_4^{2-}, NO_3^- and cations NH_4^+ are major components of aerosols in the atmosphere. These anions and cations are the most typical components of industrial emissions. Also the anions HCO_3^+, the cations Ca, Mg, K are contained in significant amount. Samples pH is basically slightly acidic due to the predominance strong anions SO_4^{2-}, NO_3^- in aerosol.

The proportion of these ions is high and is range 15-18%-eq. for nitrate-ions and is range 31-73% eq for sulphate-ions. Also content of chloride is heightened (7-34% eq) (Fig. 11).

Observation site was located in a relatively clean area at a distance of 6-7 km from the Sainshand and 500 km or more from the nearest major industrial centers of China and Mongolia. Influence of local anthropogenic emission sources is minimal in the summer.

Possible mechanism of transport of anthropogenic contaminants to the region of study could be a long-range atmospheric pollutants transfer. The results of calculations of the trajectories of air masses by reanalysis trajectory model HYSPLIT is confirmed this fact (Fig. 12 a, b).

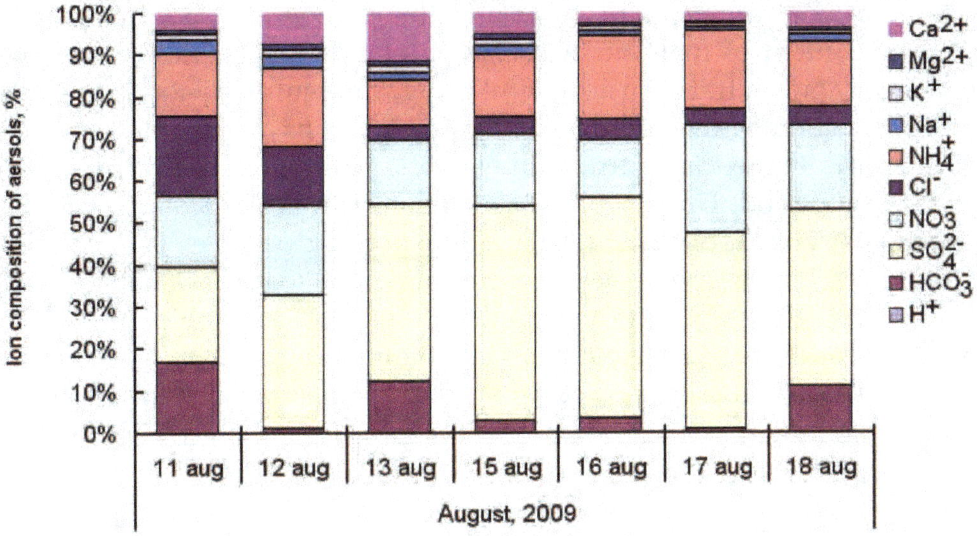

Fig. 11. Contents of the main components in aerosols in the atmosphere of Gobi Desert in August 2009

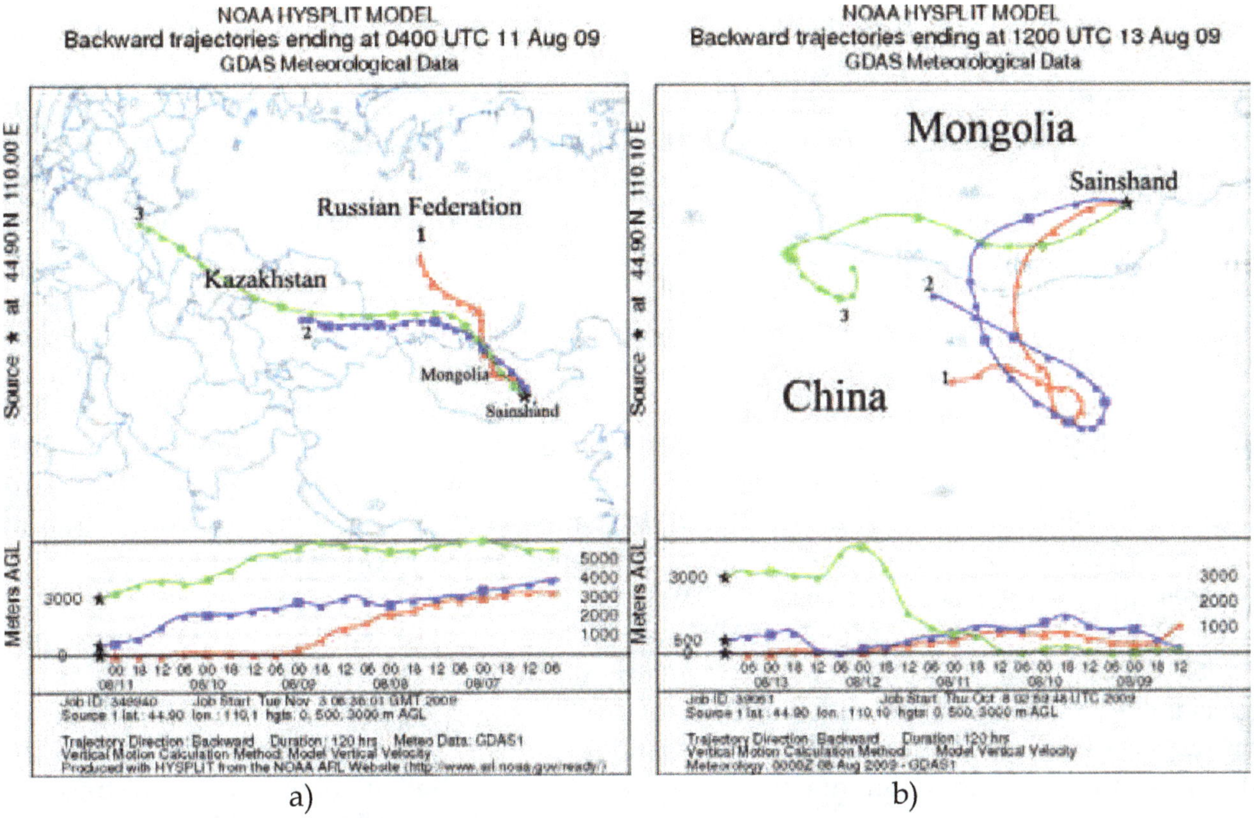

Fig. 12. Backwards air masses trajectories during observation period at the Sainshand in August 2009 on the model HYSPLIT: a) 11August, b) 13 August

The total content of ions at Sainshand are less than 2 μg/m³ under the western and north-western direction of air masses movement from the Eastern Siberia, Kazakhstan and clear stable weather 11-13 August 2009 (Fig. 12). The sulphate-ion (0.46-0.67 μg/m³), nitrate-ion (0.34-0.45 μg/m³), ammonium-ion (0.30-0.40 μg/m³), chloride-ion (0.29-0.39 μg/m³) are predominant ions in aerosols in this time. The total ion content is increased up to 12.3 μg/m³ with the passage of cold atmospheric front and the change of synoptic situation from 13 August 2009. The content of main aerosol components is increased, for example, calcium ions up to 1.4 μg/m^3, nitrate-ion up to 1.8 μg/m^3, sulphate-ion up to 5.2 μg/m^3, hydrocarbonate-ion content up to 1.5 μg/m^3. Calculations of air masses movement by model HYSPLIT indicate the transport of atmospheric pollutants from the industrial regions of China during this period. Under the development of local circulation processes from 15 August to 18 August 2009 the concentration of suspended particles is decreased to 3.4-4.6 μg/m^3. In these days we observed the high proportion of some ions such as sulphate-ions (73%-eq.), nitrate-ions (28%-eq.) and ammonium ions (80%-eq.) (Fig. 11). These ions are typical components of industrial emissions. The obtained data testify to strong influence of anthropogenic sources on the air composition of arid areas of Eastern Gobi due to long-range transport.

7. Conclusion

In this chapter the investigations of pathways and basic directions of the atmospheric impurities transport in arid and semi-arid territories of Mongolia using of the reanalysis model NCEP/NCAR HYSPLIT (http://www.arl.noaa.gov/ready/hysplit4.html) and archival meteorological data (archive FNL) are conducted. It is established that east, northeast and southeast carrying out of air mass prevail in this region.
As a whole the wind regime within a year in Gobi Desert repeats a direction of the general northwest, characteristic for free atmosphere. During the summer the influence of local circulation is significant. It is revealed that the duration and number of dusty days increase in Gobi Desert. The repeatability of dust storms from 1991 to 2006 has increased in 3 times. It is established that dust storms duration from 2003 to 2007 has increased in 40 times. It is noted that dust storms observe in autumn and winter since 2004 that it was not marked in previous years. The most repeatability of dust storms in atmosphere of arid territories of Mongolia is observed in April-May in day and evening with the maximum concentration of fine aerosols exceeding 1400 μg/m^3 (PM_{10}) and 380 μg/m^3 ($PM_{2.5}$) at Sainshand and 1230 μg/m^3 (PM_{10}) and 700 μg/m^3 ($PM_{2.5}$) at Zamyn-Uud station. Under the days with stable, settled weather the mass mean concentration of aerosols PM_{10} and $PM_{2.5}$ changes 5-8 μg/m^3 (PM_{10}) and 3-5 μg/m^3 ($PM_{2.5}$) at Sainshand. In similar conditions daily mean concentration at Zamyn-Uud change within 18-20 μg/m^3 (PM_{10}) and 16-18 μg/m^3 ($PM_{2.5}$).
For research of dust storm formation in the Central Asia in May, 2008 we analyzed the surface and high altitude (500 hPa) meteorological maps (http://www.aari.nw.ru/odata/_d0010.php) of Northern hemisphere for 2008. It is established that Lake Baikal is the special climatic area which lies on frontier of Arctic and moderate air mass and influences on trajectories of air mass moving in arid and semi-arid areas of Mongolia and China. In this time the cold Arctic air masses passed across Lake Baikal mixing with air mass of moderate latitudes. Then they moved to the territories of Mongolia and China creating the significant gradients of the temperature and pressure. These meteorological and synoptical conditions were cause to intensive dust storms in arid and semi-arid territories of Mongolia and China.
The chemical composition of aerosol, number concentration and size distribution of submicron fraction of aerosol are analyzed. It is established that the SO_4^{2+}, NO_3^- and NH_4^+ are major components of aerosol particles in atmosphere of Sainshand. These anions and cations are the most typical components of industrial emissions. Also the anions HCO_3^+, the

cations Ca, Mg, K are contained in significant amount. The revealed high ion concentration SO_4^{2+}, NO_3^- and NH_4^+ in aerosol at station Sainshand located far from industrial centers and results of modeling by HYSPLIT confirms the strong influence of anthropogenic sources of China on the air composition of arid areas of Eastern Gobi due to long-range transport.
The investigation of fine aerosol content revealed that the Aitken mode in the distribution of size spectrum particle is a major fraction and depends on meteorological and weather conditions.

8. Acknowledgment

This work was carried out within the framework of the Complex Integrate Projects № 4.13 and № 75 supported by Russian Academy of Science. The authors gratefully acknowledge the Hydrometeorology and Environment Monitoring Center of Dornogobi and Sukhbaatar aimags for allowing work to be carried out at the study sites. Thanks also to N. Enkhmaa, K. Enkhbayar for help with the fieldwork.

9. References

Arimoto R. (2001). Eolian dust and climate: relationships to sources, tropospheric chemistry, transport and deposition. Earth-Science Reviews, 54, 29-42

Ayurzhanaev A.A., Zhamsueva G.S., Zayakhanov A.S. & Tsydypov V.V. (2009). Microstructure of submicron aerosols in the Gobi Desert. XVI Workshop "Aerosols of Siberia". Tomsk, 24-27 November 2009, 7.

Azbukin A.A., Bogushevich A.Ya., Ilichevskii V.S., Korolkov V.A., Tikhomirov A.A. & Shelevoi V.D. (2006). Automated ultrasonic meteorological complex AMK-03. Meteorology and hydrology, 11, 89-97

Bezuglaya E.Yu., Berlyand M.E., 1983 Climatic characteristics of distribution of impurities conditions of atmosphere. Hydrometeoizdat, pp. 328

Chan Y.C., McTainsh G., Leys J., McGowan H. & Tews K. (2005). Influence of the 23 October 2002 dust storm on the air quality of four Australian cities. Water Air and Soil Pollution, 164, 329-348

Draxler R.R. & Rolph G.D. (2003). HYSPLIT (HYbrid Single-Particle Lagrangian Integrated Trajectory) Model access via NOAA ARL READY Website (http://ready.arl.noaa.gov/HYSPLIT.php). NOAA Air Resources Laboratory, Silver Spring, MD.

Gillette D.A. (1986). Dust production by wind erosion: necessary conditions and estimates of vertical fluxes of dust and visibility reduction by dust. In: El-Baz F., Hassan M.H.A. (Eds.). Physics of Desertification. Martinus Nijhoff Publishers, Dordrecht. 361-371

Kim H. & Chung Y. (2008). Satellite and ground observations for large-scale air pollution transport in the Yellow Sea region. Atmos. Chem. DOI: 10.1007/s10874-008-9111-4.

Lee B.K., Lee H.K. & Jun N.Y. (2006). Analysis of regional and temporal characteristics of PM_{10} during an Asian dust episode in Korea. Chemosphere, 63, 1106-1115

McTainsh G., Chan Y.C., McGowan H., Leys J. & Tews K. (2005). The 23rd October 2002 dust storm in eastern Australia: characteristics and meteorological conditions. Atmospheric Environment, 39, 1227-1236

Monin A.S. & Yaglom A.M. (1965). Statistical hydrodynamics. M.: Publishers "Nauka", 640 p.

Natsagdorj L., Jugder D. & Chung Y.S. (2003). Analysis of dust storms observed in Mongolia during 1937-1999. Atmospheric Environment, 37, 1401-1411.

Obukhov A.M. (1988). Turbulence and atmospheric dynamics. L.: Gidrometeoizdat, 1988, 413 p.

Park S.-U., Park M.-S. & Chun Y. (2010). Asian dust events observed by a 20-m monitoring tower in Mongolia during 2009. Atmospheric Environment, 44, 4964-4972

Sun J., Zhang M., & Liu T. (2001). Spatial and temporal characteristics of dust storms in China and its surrounding regions, 1901–1999: relations to source area and climate. J. Geophys. Res. 106, 10325–10333.

Tverskoy P.N., 1951. Meteorological course. Hydrometeoizdat. - L., pp. 888.

Wang S., Wang J., Zhou Z. & Shang K. (2005). Regional characteristics of three kinds of dust storm events in China. Atmospheric Environment, 39, 509-520

Wang T. & Zhu Z. (2001). Studies on the sandy desertification in China. Chinese Journal of Eco-Agriculture, 9 (2), 7-12

Wei A.L. (2006). Evaluation of micronucleus induction of sand dust storm fine particles (PM2.5) in human blood lymphocytes. Environmental Toxicology and Pharmacology, 22, 292-297

WHO (2006). Health risk of particulate matter from long-range transboundary air pollution- Joint WHO/UNECE Convention Task Force on the Health Aspects of Air Pollution. World Health Organisation, European Centre for Environment and Health, Bonn Office.

WHO (2007). Health relevance of particulate matter from various sources, Report on a WHO Workshop, Bonn, Germany, 26-27 March 2007.

Xuan J., Sokolik I.N., Hao J., Guo F., Mao H. & Yang G. (2004). Identification and characterization of sources of atmospheric mineral dust in East Asia. Atmospheric Environment, 38, 6239-6252

Zhadambaa Sh., Neushkin A.I. & Tuvdendorzh D. (1967a). Circulation factors of Mongolia climate.

Zhadambaa Sh., Neushkin A.I. & Tuvdendorzh D. (1967b). Seasonal particularities of atmospheric circulation above Mongolia. Meteorology and hydrology, 5, 29-32

Zhamsueva G.S. Zayakhanov A.S, Tsydypov V.V, Ayurjanaev A.A Azzaya D. & Oyunchimeg D. (2008). Assessment of small gaseous impurities in atmosphere of arid and semi-arid territories of Mongolia. Atmospheric Environment, 42(3), 582-587.

Zhamsueva G.S., Zayakhanov A.S, Tsydypov V.V., Ayurjanaev A.A., Golobokova L.P., Khodzher T.V., Azzaya D. & Oyunchimeg D. (2009). Temporal variability of small gaseous impurities, chemical and dispersive composition of aerosol in arid territories of Mongolia. XVI Workshop "Aerosols of Siberia". Tomsk, 24-27 November 2009, 70.

4

The Electrical Conductivity as an Index of Air Pollution in the Atmosphere

Nagaraja Kamsali[1], B.S.N. Prasad[2] and Jayati Datta[3]
[1]*Department of Physics, Bangalore University, Bangalore*
[2]*University of Mysore, Mysore*
[3]*Indian Space Research Organization, Bangalore*
India

1. Introduction

Air is never perfectly clean. Many natural sources of air pollution have always existed. Ash from volcanic eruptions, salt particles from breaking waves, pollen and spores released by plants, smoke from forest and brushfires, and windblown dust are all examples of natural air pollution. Human activities, particularly since the industrial revolution, have added to the frequency and intensity of some of these natural pollutants.

Air pollution has considerable effects on many aspects of our environment: visually aesthetic resources, vegetation, animas, soils, water quality, natural and artificial structures, and human health. The effect of air pollution on vegetation include damage to leaf tissue, needles, or fruit, reduction in growth rates or suppression of growth, increased susceptibility to a variety of diseases, pests, and adverse weather. Air pollution can affect human health in several ways. The effects on an individual depend on the dose or concentration of exposure and other factors, including individual susceptibility. Some of the primary effects of air pollutants include toxic poisoning, causing cancer, birth defects, eye irritation and irritation of the respiratory system, viral infections causing pneumonia and bronchitis, heart diseases, chronic diseases, etc. The consequences of air pollution reach beyond health and agriculture, and they influence the weather as well. There is strong evidence that increased atmospheric contamination reduces visibility, modifies electrical conductivity, alters precipitation, and changes the radiation balance.

The very presence of a city affects the local climate, and as the city grows, so does its climate changes. Now-a-days, cities are warmer than the surrounding areas. The temperature increase is a result of the enhanced production of heat energy, may be the heat emitted from the burning of fossil fuels and other industrial, commercial, and residential sources, and the decreased rate of heat loss since the dust in the urban air traps and reflects back into the city as long wave radiation. In addition, particulates in the atmosphere over a city are often at least 10 times more abundant than in rural areas. Although the particulates tend to reduce incoming solar radiation by up to 30% and thus cool the city, this cooling effect of particulates is small in relation to the effect of processes that produce heat in the city.

Particulate matter encompasses the small particles of solid or liquid substances that are released into the atmosphere by many activities and are referred to as aerosols. Modern farming adds considerable amounts of particulate matter to the atmosphere, as do desertification and volcanic eruptions. Nearly all industrial processes, as well as the burning of fossil fuels, release particles into the atmosphere. Much particulate matter is easily visible as smoke, soot, and dust. Emission of aerosols to the atmosphere has increased significantly since the industrial revolution began. An aerosol is a particle with a diameter less than 10μm. Because the effects of collisions with air molecules dominate over gravity, the smaller aerosol particles tend to remain in the atmosphere for a long time. Bigger particles of diameter greater than 10μm drop out of the atmosphere faster because of gravity. Recent research has indicated that aerosols emitted from coal (sulfates) may contribute to global cooling because sulfates act as seeding for clouds, the aerosol particles provide surfaces for water to condense, forming clouds that reflect a significant amount of sunlight directly. The net cooling owing to sulfate aerosols may offset partially the global warming expected from the anthropogenic greenhouse effect.

The concern that man's activity in a world of rapidly growing population may lead to inadvertent modification of the climate on a local basis and even on a global scale, has led to the establishment of a worldwide network of air quality monitoring stations. An important parameter to be monitored is the particulate load of the atmosphere, which may have a direct impact on the climate change. Recently, atmospheric electric scientists have renewed their interest in the possibility of using atmospheric electrical parameters as indices of air pollution trends, despite the complexity of the relationship involved. Comparing fair weather conductivity recorded on the Carnegie expedition during 1967 revealed a decreasing trend of conductivity, which has dropped at least 30% in the North Atlantic. This decreasing trend in conductivity is attributed to a significant rise in the aerosol pollution over the Northern hemisphere.

An attempt has been made (Nagaraja Kamsali et al., 2009) to infer pollution trend from atmospheric electrical conductivity on a regional basis. There seems to be sufficient evidence that atmospheric electrical conductivity, properly deduced and analyzed, with due consideration of the local meteorology and air pollution climatology, may serve as a sensitive and practical tool, capable of documenting aerosol air pollution trends, especially secular changes, in largely populated urban area. Considering the relative simplicity and reliability of the electrical conductivity measurement, which is well suited for continuous and automatic recording, its inclusion in the measurement program of the regional air pollution may be preferred.

1.1 Radioactivity of the Earth's atmosphere

The rocks and soil in the solid Earth contain radioactive elements, ^{238}U and ^{232}Th in the form of minerals Uranite and Monozite; their concentrations vary from region to region. Decay of these elements release α, β and γ radiations along with daughter products, which themselves may or may not be radioactive. Some of the daughter elements are gases such as radon and thoron. These gases get either released into the atmosphere or trapped within the rock soil, depending on the permeability of the surrounding material. The radioactive gases released from the soil get mixed with the atmospheric air and contribute to ionization during their decay. The concentration of these gases varies from place to place, and with altitude at any given place. The distribution of these gases with altitude depends on the atmospheric stability. The decay products of these gases are mostly radioactive elements, and they also contribute to the ionization of air.

The origin of radon and thoron in the Earth's crust stems directly from the radium isotopes and their decay products distributed in minute quantities in the ground within few meters from the Earth's surface. Once the isotopes of the gas are produced they migrate to a significant distance from the site of generation, even during their brief half-lives. A fraction of the radon gas generated in soil near the Earth's surface enters into the atmosphere before undergoing into the radioactive decay. The amount of radon gas that escapes depends on the amount of ^{226}Ra and ^{232}Th in the ground, the type of soil, porosity, dampness and meteorological conditions. The amounts of ^{222}Rn (radon), ^{220}Rn (thoron) and ^{219}Rn (action) in the atmosphere depend primarily on the concentration of ^{238}U, ^{232}Th and ^{235}U. The relative abundances of the isotopes in natural uranium by weight are ^{238}U - 99.28%, ^{235}U - 0.71% and ^{234}U - 0.0054%. On the other hand, ^{232}Th is even more abundant than ^{238}U in the Earth's crust.

At present, twenty isotopes of radon are known. ^{222}Rn, from radium, has a halflife of 3.823 days; ^{220}Rn emanating naturally from thorium and called thoron has a halflife of 55.6s. ^{219}Rn emanating from actinium and called actinon has a halflife of 3.96s. It is estimated that every square mile of soil to a depth of 6 inches contains about 1g of radium, which releases radon in tiny amounts into the atmosphere.

1.2 Radon entry into the atmosphere

For a radon atom to escape from the mineral grain into the pore space, the decay must occur within the recoil distance of the grain surface. The recoil range for ^{222}Rn is 20-70 nm in common minerals, 100 nm in water, and 63 μm in air. Radon atoms entering the pore space are then transported by diffusion and advection through this space until they in turn decay or are gets released into the atmosphere. The escape of radon atoms from soil grains into the pore space is called radon emanation. The transportation of radon atoms from pore space by diffusion and advection through the space and gets released into the atmosphere is called radon exhalation. The amount of radon released per unit surface area per unit time is called radon exhalation rate. The rate of radon exhalation from soil depends on geophysical parameters such as geology of the region, soil porosity, soil texture, humidity and temperature of the soil, and also on meteorological parameters such as temperature, humidity, pressure etc. The influence of environmental parameters on the exhalation rate of ^{222}Rn from soil is not clearly understood. When the soil is frozen or covered with snow, the exhalation rate is reduced because diffusion processes are slowed down. It was observed by George (1980) that, the exhalation rate was linearly correlated with outdoor temperature. However, Arabzedegan et al., (1982) observed that the meteorological parameters had no significant influence on the value of ^{222}Rn exhalation rate. Raghavayya et al., (1982) have found that soil moisture had a strong influence on the ^{222}Rn exhalation rate. ^{222}Rn exhaled from the surface soil reaches the atmosphere and further decays to a series of radionuclides which are heavy metal radionuclides that attach to other atmospheric components to form aerosol particles.

1.3 Ionization in the atmosphere

Cosmic radiation is the primary source of ions over the oceans and above a couple of kilometers over the land (Hoppel et al., 1986). In the lower atmosphere, ions are predominantly produced by radiation emitted from radioactive materials in the soil, in building materials, and in the air (radon and its daughter products). The α and β particles and the γ rays from ^{222}Rn, ^{220}Rn and their decay products give up their energy by producing ions and raising atoms and molecules to excited states. An α particle from the decay of ^{222}Rn

atom (5.49 MeV) will produce about 150,000 ion pairs with the expenditure of about 34 eV per ion-pair along its trajectory through the surrounding air in the atmosphere (Israel, 1970). Cosmic radiation contributes about 10% to the ionization at ground level. However, at higher altitudes, the partitioning shifts drastically both because the radiation from the soil and airborne materials decreases whereas the intensity of the cosmic radiation increases.
The radioactive gases (^{222}Rn and ^{220}Rn) can diffuse from the ground into the atmosphere and contribute to the volume ionization. The concentration of these gases and the rate of ionization in the atmosphere depend upon the amount of uranium and thorium in the ground, and the temperature, dampness and porosity of the soil. Radon emanated from the Earths crust reaches the atmosphere and further decays to a series of short lived radionuclides namely, polonium (^{218}Po), lead (^{214}Pb) and bismuth (^{214}Bi). Radiation from radioactive gases exhaled from the ground and their daughter products causes ionization in the atmosphere. Radon is one of the relatively longer-lived gases. The principal decay modes and half-lives of ^{222}Rn and its short-lived daughters in order are ^{222}Rn – α, 3.82 days; ^{218}Po – α, 3.05 min; ^{214}Pb – β, 26.8 min; ^{214}Bi – β, 19.7 min; and ^{214}Po – α, 2×10^{-4} s. The radiations α , β and γ released during the decay of radon and its progeny cause ionization, and hence these are important in the study of atmospheric electricity. Ionization due to radioactive gases in the air is even more variable and depends not only on the amount exhaled from ground but also on atmospheric dispersion. Direct measurements of ionization due to radioactive gases in the atmosphere are difficult and have not generally been satisfactory. Estimation of ionization rate is therefore based on measurements of radioactive products in the air at that level, while the ionization due to cosmic radiation is almost constant with time. On cool nights with nocturnal temperature inversions the radioactive gases can be trapped in a concentrated layer close to the ground, whereas during unstable convective periods, the gases can be dispersed over an altitude of several kilometers. But during daytime due to convective processes, these gases disperse to higher altitudes. Because of this, the height distribution of radon in the atmosphere has also been used to determine the turbulent diffusion coefficient (Hoppel et al., 1986).

1.4 Influence of meteorological parameters on Ionization

In general, the ionization increases rapidly with height but shows certain changes in the troposphere and stratosphere that appear to be associated with the existence of aerosol layers (Israel, 1971). However, in the lowest region of the atmosphere the rate of ionization greatly depends upon the different meteorological parameters. The ionization rate near the Earth's surface is markedly influenced by variations in meteorological parameters such as wind and precipitation, in addition to the particles and air mass in the atmosphere. The wind speed which is produced under the influence of pressure gradient and Coriolis force is highly variable in time and space. It is an important factor in the transportation of atmospheric nuclei and ions in the atmosphere. It shows diurnal as well as seasonal variations. Its magnitude and direction vary with height and hence the density of ionization in the atmosphere is very much influenced by the wind. However, the precipitation is another important meteorological parameter which has a profound influence on the ion content of the atmosphere. It washes out both positive and negative ions, leaving the air fairly low in ion content. The degree of washout depends on the spectrum of the rain and its duration. Addition to this, the air mass also influences the ionization. The qualities of an air mass depend upon the region over which it stagnates for a sufficiently long time and acquires the characteristics of the environment. Air mass that lies over the densely populated areas and cities will be weak in the ion content because of the large quantity of particulate matter that goes into it from the city environment. Air mass that lies over the rural surroundings is mostly free from such particulate matter and hence it is rich in ion content. Thus the quality of an air mass influences the ion content and consequently columnar resistance of the local generator. Another major parameter is the suspended particles in the atmosphere. These particles always exist in the atmosphere but the number varies with the atmospheric conditions. The visibility of the atmosphere is a good indicator of the particle content in the atmosphere. It enhances the formation of fog and cloud in the atmosphere under certain favourable conditions. They promote attachment of ions to particles and reduce their mobility.

1.5 Particulate matter in the atmosphere

Particulate matter encompasses the small particles of solid or liquid substances that are released into the atmosphere by many activities and are referred to as aerosols. Modern farming adds considerable amounts of particulate matter to the atmosphere, as do desertification and volcanic eruptions. Nearly all industrial processes, as well as the burning of fossil fuels, release particles into the atmosphere. Much particulate matter is easily visible as smoke, soot, and dust. Emission of aerosols to the atmosphere has increased significantly since the industrial revolution began. An aerosol is a particle with a diameter less than 10μm. Because the effects of collisions with air molecules dominate over gravity, the smaller aerosol particles tend to remain in the atmosphere for a long time. Bigger particles of diameter greater than 10μm drop out of the atmosphere faster because of gravity. The aerosols emitted from coal (sulfates) may contribute to global cooling because sulfates act as seeding for clouds, the aerosol particles provide surfaces for water to condense on, forming clouds that reflect a significant amount of sunlight directly. The net cooling owing to sulfate aerosols may offset much of the global warming expected from the anthropogenic greenhouse effect.

1.6 Air pollution

Air is never perfectly clean. Many natural sources of air pollution have always existed. Ash from volcanic eruptions, salt particles from breaking waves, pollen and spores released by plants, smoke from forest and brushfires, and windblown dust are all examples of natural air pollution. Ever since people have been on earth, however, they have added to the frequency and intensity of some of these natural pollutants. As the faster moving fluid medium in the environment, the atmosphere has always been one of the most convenient places to dispose of unwanted materials. Ever since fire was first used by people, the atmosphere has been a sink for waste disposal. As long as a chemical is transported away or degraded relatively rapidly, there is no pollution problem. Chemical pollutants can be thought of as compounds that are in the wrong place or in the wrong concentrations at the wrong time. Pollutants that enter the atmosphere through natural or artificial emissions may be broken down or degraded not only within the atmosphere but also by natural processes in the hydrologic and geochemical cycles; pollutants that leave the atmosphere may become pollutants of water and of geological cycles.

Air pollution has considerable effects on many aspects of our environment: visually aesthetic resources, vegetation, animals, soils, water quality, natural and artificial structures, and human health. The effect of air pollution on vegetation include damage to leaf tissue, needles, or fruit, reduction in growth rates or suppression of growth, increased susceptibility to a variety of diseases, pests, and adverse weather. Air pollution can affect human health in several ways. The effects on an individual depend on the dose or concentration of exposure and other factors, including individual susceptibility. Some of the primary effects of air pollutants include toxic poisoning, causing cancer, birth defects, eye irritation and irritation of the respiratory system, viral infections, causing pneumonia and bronchitis, heart diseases, chronic diseases, etc. Wherever there are many sources emitting air pollutants over a wide area there is a potential for the development of smog. Whether air pollution develops depends on the topography and on weather conditions, because these factors determine the rate at which pollutants are transported away from their sores and converted to harmless compounds in the air. When the rate of production exceeds the rate of degradation and of transport, dangerous conditions may develop. Meteorological conditions can determine the fate of air pollution. Air pollution and meteorology are linked in two ways. One concerns the influence that weather conditions have on the dilution and dispersal of air pollutants. The second connection is the reverse and deals with the effect that air pollution has on weather and climate. Air pollution is a continuing threat to our health and welfare. The cleanliness of air, therefore, should certainly be as important to us as the cleanliness of food and water.

The concern that man's activity in a world of rapidly growing population may lead to inadvertent modification of the climate, on a local basis and even on a global scale, has led to the establishment of a worldwide network of air quality monitoring stations. An important parameter to be monitored is the particulate load of the atmosphere, which may have a direct impact on the climate change. Recently, atmospheric electric scientists have renewed their interest in the possibility of using atmospheric electrical parameters as indices of air pollution trends, despite the complexity of the relationship involved. Comparing fair weather conductivity recorded on the Carnegie expedition during 1967 revealed a decreasing trend of conductivity, which has dropped at least 30% in the North Atlantic. This decreasing trend in conductivity is attributed to a significant rise in the aerosol pollution over the Northern hemisphere. An attempt to infer pollution trend from atmospheric electrical conductivity on a regional basis, due to extensive urbanization has been made. There seems to be sufficient evidence that atmospheric electrical conductivity, properly reduced and analyzed, with due consideration of the local meteorology and air pollution climatology, may serve as a sensitive and practical tool, capable of documenting aerosol air pollution trends, especially secular changes, in largely populated areas.

1.7 Electrical nature of the Earth's atmosphere

The planetary boundary layer is that region of the lower atmosphere in which the influences of the earth's surface are directly felt. The primary influences of the surface are drag, heating (or cooling), and evaporation (or condensation). These processes cause vertical fluxes of momentum, sensible heat, and moisture, which penetrate into the lower atmosphere to a finite height. These fluxes, in turn, generate turbulence, ultimately controlling the mean profiles of wind speed, temperature, and water vapour in the planetary boundary layer.

Atmospheric electrical conductivity, ion mobility and small ion number density, etc. are important parameters for understanding the electrical nature of the atmosphere. The small ions consisting of aggregates of few molecules determine the electrical conductivity over the region (Israel, 1970). The number densities of these ions are controlled by ionizing mechanisms for the production of ions and electrons, and the loss processes for these charged species. The electrical conductivity of the air in an aerosol free atmosphere is mainly due to small ions. However, in a polluted atmosphere these ions soon get attached to the aerosol particles and form the intermediate and large ions. Since the mobility of small ions is at least two orders of magnitude more than that of large ions, the small ions are still considered to be, if not the only, the main contributors to the local electrical conductivity (Dhanorkar and Kamra, 1997). Since the presence of aerosol particles depletes the small ions, the electrical conductivity and aerosol concentration are generally considered to have an inverse relationship in the atmosphere and the electrical conductivity has been often proposed to act as an index of air pollution. The decrease in electrical conductivity with an increase in air pollution has been reported under different meteorological conditions such as in continental air (Misaki, 1964; Mani and Huddar, 1972), near coastlines (Misaki and Takeuti, 1970; Morita et al., 1972; Kamra and Deshpande, 1995), in marine air (Cobb and Wells, 1970), at the level of inversion (Rosen and Hofmann, 1981) and during volcanic eruption (Srinivas et al., 2001).

The electrical conductivity of the air is mainly due to the presence of highly mobile small ions produced in the atmosphere by cosmic rays and local radioactive sources (Retalis et al., 1991). A balance is maintained since the small ions are being removed from the atmosphere at the same rate that they are being produced. They are removed by recombination with oppositely charged small ions and by attachment to larger aerosol particles. These recombination and attachment processes thus determine the life times of small ions in the atmosphere and secondarily of the conductivity, since its value is formed predominantly by the small ion concentration in the air. Ion lifetimes near earth's surface vary from 20 sec in highly polluted air to 300 sec in very clean air (Cobb, 1973).

1.8 The electrical conductivity of the atmosphere

The atmospheric electrical conductivity depends on the existence of positive and negative ions. If a small potential is applied to two electrodes in an ionized gas, a week current flow is induced by the ions which flow in opposite directions to the electrodes where they deliver their charge. The current density (per unit cross section) is defined as the amount of electric charge which flows per unit time through a unit surface area perpendicular to the direction of flow. The current density is composed of two terms corresponding to the respective ion flow, and is given by the relation $i = e\left(n^{+}\mu^{+} + n^{-}\mu^{-}\right)E = \sigma E$, where $\mu^+$and μ^- are the mobilities and n^+ and n^- are the ion concentrations of positive and negative polarity, and e is the electronic charge. The resulting conductivity can be expressed in terms of the number densities and mobilities of the individual species as $\sigma = e\left(n^{+}\mu^{+} + n^{-}\mu^{-}\right)$. For several types of ions of different mobility are present, the above expression become $\sigma = e\sum_{i=1}^{n}\left(n_i^{+}\mu_i^{+} + n_i^{-}\mu_i^{-}\right)$.

Near to the Earth's surface, electrical conductivity is not constant with time. Observations over long periods at various sites have shown that atmospheric conductivity has well defined diurnal and seasonal variations during fair weather periods. The exact nature of variation differs with stations, the mean values and amplitude being different. For a given station, the pattern of variation of conductivity at the surface normally remains more or less a constant during a year. The average behaviour at each station may therefore be characteristic of the 'atmospheric electric climate' of that station. But some similarities are seen in the diurnal behaviour of conductivity at most stations. For instance, conductivity is higher at night than during the day. This is typical for continental stations and makes it probable that, apart from all local and other influences, a more general factor which causes this general tendency in the diurnal variation must exist. This variation is possibly a reflection of the general rhythm of the atmosphere, since nights are usually calmer, with low winds and hardly any convective motion, and therefore conducive for accumulation of radon and other radioactive species near the surface. A sharp fall associated with sunrise is seen at many stations. This is believed to be due to an increase in the aerosol concentration due to onset of circulation and human activity. In marine air, the mean conductivity is slightly higher than that over land but the diurnal amplitude is very low. The average value on land is about $1.8\times10^{-14}\ \Omega^{-1}\ m^{-1}$ while over the ocean it is about $2.8\times10^{-14}\ \Omega^{-1}\ m^{-1}$ (Chalmers, 1967). And unlike over land, conductivity over ocean is minimum in the late afternoon or early evening hours, and remains low during night. It rises to a peak in the forenoon and is followed by a gradual decline. This is almost a mirror image of the pattern seen over continental stations. Further, while the positive conductivity displays a clear variation in the manner described, the variation in the negative conductivity has very low amplitude and is just discernible. The ratio of polar conductivities thus exhibits a clear pattern over ocean. From the life time it is well known that, the mobility of large or intermediate ions are a few orders of magnitude smaller than those of small ions. So, their contribution to the polar conductivity is relatively smaller than that of the small ions. Small ions of opposite polarity recombine and cause a decrease in the small ion concentration and consequently a reduction in the conductivity of air. There exist a close correlation between the concentration of small ions and polar conductivity near the surface of Earth. In the presence of aerosol particles losses in the small ion concentration are caused not only due to the ion–ion recombination process, but also due to attachment of small ions to the aerosol particles. Attachment of small ions to the aerosol particles makes them almost immobile and causes a further decrease in conductivity of the atmosphere. This develops inverse relationship between the aerosol concentration and the electrical conductivity (Cobb and Wells, 1970). Mani and Huddar (1972) observed the decrease in conductivity presumably caused due to increase in dust particle content at Pune, India over a period of 20 years. Hogan et al., (1973) pointed out that the conductivity inversely follows the product of aerosol size and number density better than the number density alone. Similar type of results was obtained by Srinivas et al., (2001) and they concluded that the conductivity of the atmosphere depends on the number density of aerosols rather than their concentration alone. Cobb (1973) suggested that this product is really close to what is meant by pollution. Due to this inverse relationship between the electrical conductivity and the pollution, the electrical conductivity has been proposed to act as a pollution index.

1.9 Need for theoretical estimation of electrical conductivity of the atmosphere

Measurements of atmospheric electrical conductivity are in general difficult to interpret because of large variety of influencing factors. Therefore, a thorough theoretical and experimental analysis is necessary in order to conduct research in atmospheric electricity. In the atmospheric surface layer, particularly the lowest few meters above the ground, large number of factors will dominate, for example, radioactive emanation from the ground; porosity, dampness and temperature of the soil; aerosol concentration; atmospheric electric field and mobility of small ions. The electrical conductivity is very sensitive to the presence of aerosols. Thus, the aerosol loading has a bearing on the conductivity of the atmosphere. The aerosols reduce the conductivity of the atmosphere by (i) converting the highly mobile small ions into less mobile aerosol ions through ion - aerosol attachment and (ii) neutralizing the small ions through the aerosol ion - small ion recombination. Another process that makes the ion-aerosol attachment rate faster is the charged aerosol - aerosol recombination. In the present communication ion - aerosol interactions are taken into account to develop a model for the electrical conductivity of the atmosphere. It is validated by comparing the computed values of conductivity with the experimental values.

Several model studies on the electrical conductivity of the atmosphere of higher altitude are available (Datta et al., 1987; Prasad et al., 1991; Srinivas and Prasad, 1993, 1996). Most of the model studies have considered the loss of small ions as solely due to a) ion - ion recombination b) ion - attachment to aerosols. The other two types of loss of small ions arise from a) the recombination of molecular ions with oppositely charged aerosol and b) charged aerosol - aerosol recombination, and hence further deplete the small ions. Obviously, addition of these two terms results in more realistic values (Srinivas et al., 2001). The equilibrium density of small ions is governed by the equations of continuity for the production and loss of these ions, where the gain and loss due to transport are negligible. The effective recombination of small ions is altered in the presence of aerosols, since these aerosols interact with the ions through various attachment and recombination processes. The attachment of small ions to neutral aerosols produce charged aerosols referred to as 'large ions', which are less mobile than the smaller molecular ions. The subsequent recombination of charged aerosols with ions as well as oppositely charged aerosols would result in the depletion of small ion concentration more rapidly than in the absence of aerosols. Therefore, the formation of less mobile aerosol ions and the reduction of more mobile molecular ions alter the electrical conductivity of the atmosphere. Thus, the pollution due to aerosols/dust in the atmosphere can considerably reduce the atmospheric conductivity. Hence, the atmospheric pollution level can be monitored through electrical conductivity measurements and conductivity is used as index of air pollution.

2. Instrumentation and methodology

2.1 Radon concentration in air

Concentration of radon in air at a height of 1 m above the surface is measured using the Low Level Radon Detection System. The procedure briefly consists of sampling the air in a collection chamber and exposing a circular metallic disc to the radon inside the collection chamber. A delay of at least 10-min is normally allowed for any thoron, which may be present in the chamber to decay completely. The positively charged ^{218}Po (RaA) atoms created in the chamber get collected on the metallic plate maintained at an optimum negative potential that should be sufficient to force all the RaA atoms onto the plate. The collection is carried out for an optimized period and thereafter the charged plate is removed from the chamber and alpha-counted. The concentration of radon, R_n, (in Bq m^{-3}) is calculated (Nagaraja et al., 2003a) with the expression: $Rn = 1000C/(EFVZ)$ where C is the total number of counts, E is the efficiency of alpha counting system, F is the efficiency of collection of RaA-atoms on the metallic disc and is empirically related to the relative humidity (H) by $F = 0.9 * [1 - \exp(0.039 * H - 4.118)]$, V is the volume of chamber, Z is the correction factor for build up and decay of radon daughter atoms on the metallic disc during the exposure and counting period.

2.2 Radon progeny concentration in air

An air flow meter kept at a height of 1 m above the surface is used to measure the radon progeny concentration. Air is drawn through a glass fiber filter paper by means of a suction pump at a known flow rate. The radon progeny in air sample are retained on the filter paper. The filter paper is then alpha-counted at any specific delay time. Total activity on the filter paper is measured at three different counting intervals of 2 - 5, 6 - 20 and 21 - 30 minutes. Activities of polonium (RaA), lead (RaB) and bismuth (RaC) (in Bq m^{-3}) are calculated using the modified equations given by Raghavayya (1998):

$$RaA = \frac{+4.249019 \times C_1 - 2.062417 \times C_2 + 1.949949 \times C_3}{V \times E}$$

$$RaB = \frac{-0.355129 \times C_1 + 0.006232 \times C_2 + 0.240618 \times C_3}{V \times E}$$

$$RaC = \frac{-0.215175 \times C_1 + 0.371319 \times C_2 - 0.502945 \times C_3}{V \times E}$$

$$R_d = \frac{+0.048445 \times C_1 - 0.019335 \times C_2 + 0.037053 \times C_3}{V \times E}$$

where C_1, C_2 and C_3 are the gross counts during the three counting intervals, E is the efficiency of alpha counting system, V is the sampling rate in liters per minute (LPM), R_d is the concentration of radon progeny (in Working Level)

2.3 Ion - pair production rate due to radioactivity

The ion-pair production rate due to radon and its progeny is calculated using the expression: $Q = \varepsilon / I$ and $\varepsilon = 5.49 \times 10^6 Rn + 6.00 \times 10^6 RaA + 0.85 \times 10^6 RaB + 7.69 \times 10^6 RaC$ where Rn is the concentration of radon (in Bq m^{-3}), RaA, RaB and RaC are the concentration of radon progenies i.e., polonium, lead and bismuth (in Bq m^{-3}), respectively, ε is the total energy (in eV) released by the decay of radon and its progeny, I is the energy required to produce one ion-pair = 32 eV, Q is the ion-pair production rate (ion-pairs $cm^{-3}s^{-1}$).

2.4 Aerosol concentration near the earth's surface

The concentrations of aerosols were measured with an Electrical Aerosol Analyzer in the size range from 3 to 750 nm diameter. The instrument works on the principle of diffusion charging-mobility analysis of particles. The ambient air is sampled at a rate of 50 litre per minute. The samples is first exposed to a Kr-85 radioactive neutralizer and then passed through a mobility analyzer that contains concentric cylindrical electrodes and a central collector rod. A pre-determined voltage is applied between the electrodes to produce an electric field in the condenser. The charged particles are deflected towards the collector rod by the electric field and the concentration of aerosols is measured in terms of mobility spectrum (Liu and Pui, 1975). A computer is interfaced with analyzer to measure the size distribution.

2.5 Electrical conductivity of the atmosphere

The Gerdien condenser is basically a cylindrical capacitor that collects atmospheric ions and provides air's electrical conductivity, which is shown in Fig. 1 and plate 1. It consists of two coaxial cylinders between which air is allowed to flow. A voltage is applied to one cylinder, known as the driving electrode, with respect to the other. This driving voltage repels ions of one polarity towards the other electrode where ions get collected. It is measured using a sensitive electrometer. For voltages that are sufficiently low, the collector current increases in proportion to the driving voltage. The collector current at these voltages is given by the expression: $I = \sigma CV/\varepsilon_0$. Gerdien condenser measurements have some sources of errors caused either due to non-fulfillment of some ideal conditions or due to improper adjustment of some important controlling factors such as potential, air flow rate etc and these factors have to be considered in the estimation of conductivities.

In case the humidity is very high, like during fog, it does not directly influence the proper function of the condenser. Apart from this, cobwebs, fragments of feathers, and accumulated dust, may, become conducting in humid air. In such a case the conductivity values increase drastically and reach unrealistic magnitudes. Similar effects can be observed when very small insects enter the inner cylinder. Hence, it is necessary to maintain the area surrounding the condenser free from the insects, cobwebs, dust, leaves etc.

The electric field inside the Gerdien condenser is assumed to be independent of axial distance. This is true only for an infinitely long condenser. For a condenser of finite length, the electric field distribution gets modified at the ends. This is called the 'edge effect'. At the ends of the condenser, some field lines originating at the outer surface of the driving electrode terminate at the collector. The field lines which are straight in the interior and perpendicular to the capacitor electrode are distorted at the edge. Because of this, some ions which are not present in the volume of air being sampled are also pulled in. Thus, the internal field does not end at the capacitor edge, but extends outside the region. Consequently, the ions experience an electric field even before entering the capacitor. Difficulties related to the edge effect can be reduced when the outer electrode is closer to the earth's potential. However, since it is easier to maintain the inner tube at a high level of insulation required for electrometer usage, the accelerating potential is most commonly applied to the outer tube and an electrometer is connected to inner cylinder. The diverging of the electric field around the entrance of the outer electrode may also be avoided with the aid of a grounded entrance cylinder. This and various other modifications suggested to homogenize the field at the capacitor inlet do not remove the field disturbance completely. Moreover, such arrangements introduce another source of error; the losses due to diffusion of ions to devices introduced to remove edge effects. Any such device is, therefore, not introduced in our condenser.

The atmospheric electrical conductivity of both positive and negative polarities is simultaneously measured with two Gerdien condensers housed separately in a single unit as in Fig. 1, locally fabricated at Pune, India. The apparatus consists of two identical cylindrical tubes of 10-cm diameter and 41 cm length joined by a U-shaped tube. The air is sucked through them with a single fan fixed at the end of U-shaped tube. The flow rate in each tube is about 19 liters per second. The inner co-axial aluminum electrodes in both the tubes are of 1-cm diameter and 20 cm length, and are fixed well inside the outer electrodes with Teflon insulation. The outer electrodes are shielded from any external electric field by two coaxial cylinders of 11 cm diameter and 35 cm length separated from the outer electrodes with Bakelite rings. Opposite but equal potentials of ± 35 V are applied to the

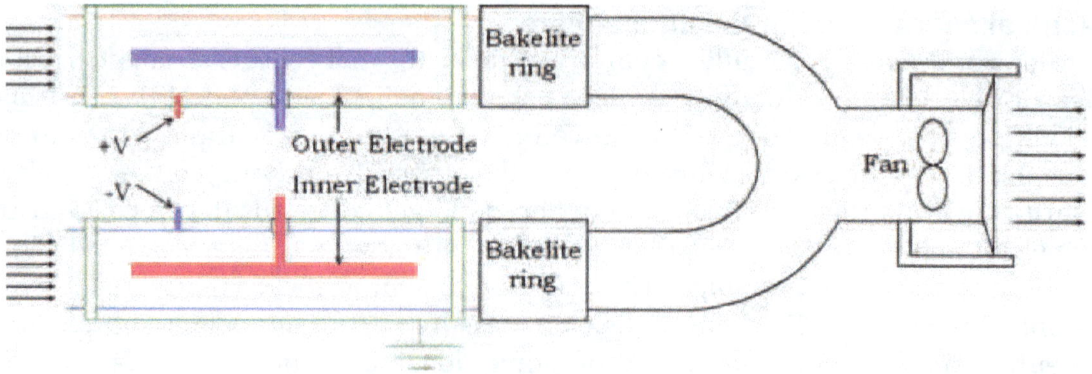

Fig. 1. Schematic representation of the Gerdien Condenser

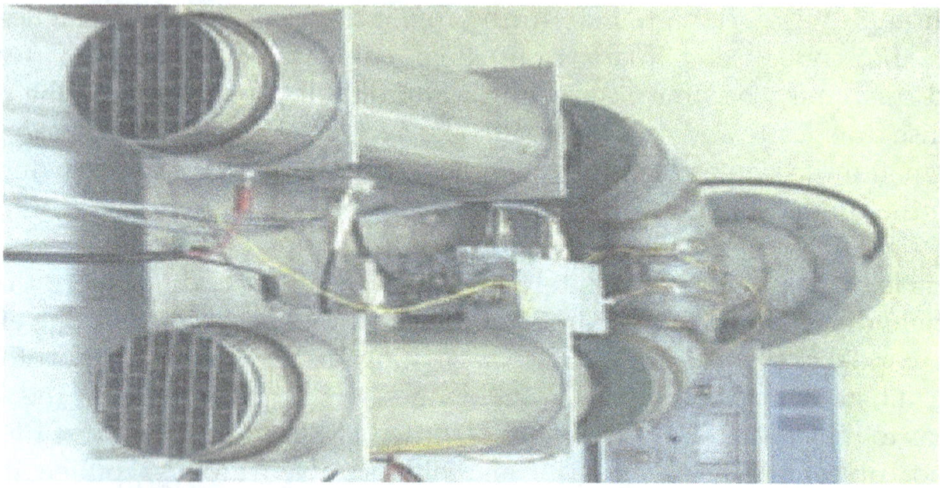

Plate 1. Gerdien Condenser

outer electrodes of the two condensers. The critical mobility of the instrument is greater than 2.93 x $10^{-4}m^2V^{-1}s^{-1}$ (Dhanorkar and Kamra, 1993) and is capable of resolving the values of conductivity as small as 3 x 10^{-16} $\Omega^{-1}m^{-1}$ (Dhanorkar and Kamra, 1992). To avoid leakages along the insulators at higher humidity, Teflon is used to insulate all high impedance points in the apparatus because it maintains good insulation even at high humidity. The signals from the two condensers are amplified separately with two electrometer amplifiers AD 549J, and are carried through a separate coaxial cable and then recorded on a data logger card, which has a voltage range of ±15V with 10 mV resolutions. Though each sensor is scanned every second, the data logger card records averaged values of each parameter for every one-minute and then stores the values on a computer. Then it is averaged for any desired length of time with the help of an external program. In our case we have averaged this for every hour. The card has the capability of storing data at the maximum of 100 samples per second. In our case we made 60 samples per second and then it was averaged for a minute or for any desired length of time with the help of an external program. The diurnal and / or seasonal variations of radioactivity and the electrical conductivity of the atmosphere, and their dependence on meteorological parameters are presented for a continental station Pune (12 °N, 76 °E), India.

2.6 Meteorological parameters

The meteorological parameters such as temperature, pressure, wind speed and wind direction near the surface of the earth are obtained by automatic weather station installed near the observatory. Relative humidity is estimated from the wet and dry bulb thermometers.

3. Simple ion aerosol model for the estimation of electrical conductivity

There have been no reports on modeling study of the electrical conductivity in the lower part of the troposphere, in particular near the surface. Modeling for this region requires, ionic aerosols, in addition to the molecular ions. Based on this, a Simple Ion-Aerosol Model is proposed and the schematic diagram is shown in Fig. 2. It involves primary ion pair production rate due to surface radioactivity and cosmic rays, the small ion densities ($N_\pm$) and the aerosol number density. For the model prediction of equilibrium ion density and conductivity, it is necessary to use all the four loss processes of ions involved. The various recombination coefficients that enter into the model are: α_i due to the loss of oppositely charged small ions, α_a due to oppositely charged aerosol ions, α_s between small ions and aerosol ions, β due to attachment of small ions of similar polarities with aerosols.

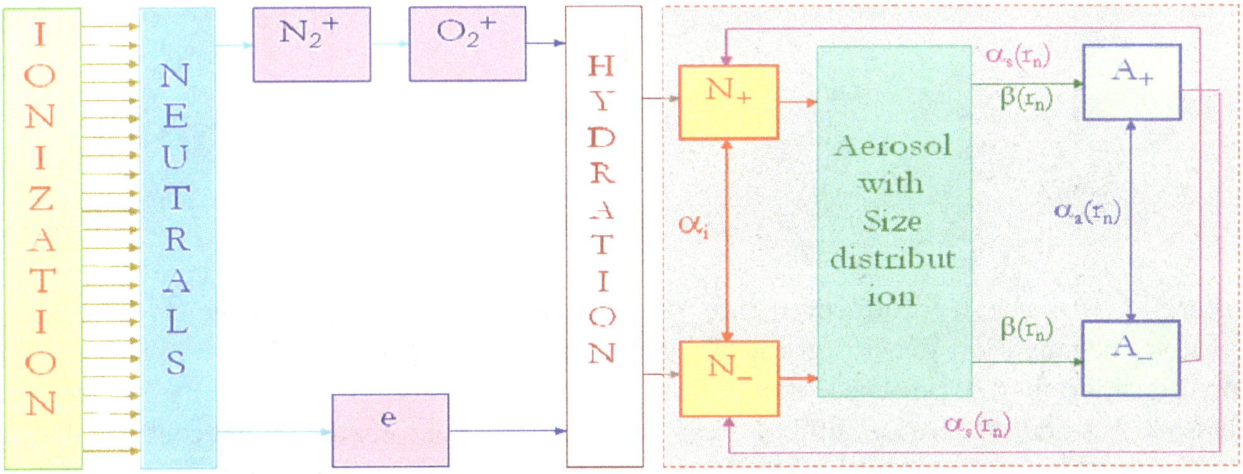

Fig. 2. Schematic representation of ion aerosol model

The temperature, pressure, relative humidity, ionization rate and aerosol number density along with the different attachment/recombination coefficients are the input parameters for the model. With these inputs the electrical conductivity is estimated. However, the model gives the total conductivity and does not make any distinction between positive and negative ion conductivities.

3.1 Theoretical description of the Model

The polar conductivity of air is defined as

$$\sigma = neb \tag{1}$$

where n is the concentration of ions of mobility b and having an elementary charge e. In the absence of aerosol particles, polar conductivity is mainly due to the high-mobility small ions. However, in the presence of aerosol particles, small ions get attached to the aerosol particles. In the Simple Ion - Aerosol Model, the detailed reaction paths for the formation of individual cluster ions are not considered. The total number densities of positive and negative ions are assumed to be equal from the charge neutrality criterion. The equilibrium ion densities are computed from the equations of continuity where the effect of transport is neglected. These equations include aforementioned four types of loss processes. The two values of β (for attachment of positive and negative ions with neutral aerosols) and α_s (for recombination of positive and negative ions with oppositely charged aerosols) are assumed to be equal.

The ion-aerosol interaction and involves primary ion pair production rate due to surface radioactivity and cosmic rays, the small ion densities ($N_\pm$) and the aerosol number density. For the model prediction of equilibrium ion density, it is necessary to use all the four loss processes of ions involved. The various recombination coefficients that enter into the model are: a) α_i - due to the loss of oppositely charged small ions, b) α_a - due to oppositely charged aerosol ions, c) α_s - is between small ions and aerosol ions and d) β - due to attachment of small ions of similar polarities with aerosols.

The equations of continuity (i) for small ions in the absence of aerosols and in the presence of aerosols and (ii) the charged aerosols are given by:

$$\frac{dN_0}{dt} = q - \alpha_i N_0^2 \tag{2}$$

$$\frac{dN_\pm}{dt} = q - \alpha_i N_\pm^2 - \beta Z N_\pm - \alpha_s A_\pm N_\mp \tag{3}$$

$$\frac{dA_\pm}{dt} = \beta Z N_\pm - \alpha_s A_\pm N_\mp - \alpha_a A_\pm^2 \tag{4}$$

where q is the ion production rate due to radioactivity and cosmic rays, N_0 is positive/negative ion density in the absence of aerosols, $N_\pm \, and \, N_\mp$ are positive/negative ion densities in the presence of aerosols, $A_\pm$ is positively/negatively charged aerosol density, Z is neutral aerosol number density, β is the aerosol-ion attachment coefficient, α_i is the ion-ion recombination coefficient, α_s is the charged aerosol-ion recombination coefficient, α_a is the charged aerosol-aerosol recombination coefficient.

Under steady state conditions Eqs.(2), (3) and (4) reduce to:

$$q - \alpha_i N_0^2 = 0 \quad \Rightarrow \quad N_0 = \left(\frac{q}{\alpha_i}\right)^{\frac{1}{2}} \tag{5}$$

$$q - \alpha_i N_\pm^2 - \beta Z N_\pm - \alpha_s A_\pm N_\mp = 0 \tag{6}$$

$$\beta Z N_\pm - \alpha_s A_\pm N_\mp - \alpha_a A_\pm^2 = 0 \tag{7}$$

Solving Eqs. (5), (6) and (7) for $N_\pm$ with $A_\pm = N_0 - N_\pm$ we get:

$$N_\pm = \frac{(\alpha_a N_0 + \beta Z) \pm \sqrt{(\alpha_a N_0 + \beta Z)^2 - (\alpha_a - \alpha_i)(q + \alpha_a N_0^2)}}{(\alpha_a - \alpha_i)} \tag{8}$$

Solving Eqs. (6) and (7) simultaneously, one can get:

$$\beta Z = \left(\frac{q - \alpha_i N_{\pm}^2 + \alpha_a A_{\pm}^2}{2N_{\pm}} \right) \tag{9}$$

$$\alpha_s = \left(\frac{q - \alpha_i N_{\pm}^2 + \alpha_a A_{\pm}^2}{2A_{\pm} N_{\mp}} \right) \tag{10}$$

The ion depletion ΔN, i.e., loss of molecular ions in presence of aerosols is given by $\Delta N = N_0 - N_{\pm}$. Further, this ion depletion is equal to the charged aerosol concentration, $A_{\pm}$. Hence, the fractional depletion, η of small ions is used and computed as follows:

$$\eta = \left(\frac{\Delta N}{N_0} \right) = \left(\frac{A_{\pm}}{N_0} \right) \tag{11}$$

Using Eqs. (11) and (5), we can write Eqs.(9) and (10), respectively, as

$$\beta Z = \left(\frac{N_0 \eta [\alpha_i (2 - \eta) + \alpha_a \eta]}{2(1 - \eta)} \right) \tag{12}$$

$$\alpha_s = \left(\frac{\alpha_i (2 - \eta) - \alpha_a \eta}{2(1 - \eta)} \right) \tag{13}$$

The conductivity of the atmosphere in the absence and presence of aerosols, respectively, are given by:

$$\sigma_0 = N_0 e b_{\pm} \;\; \& \;\; \sigma_{\pm} = N_{\pm} e b_{\pm} \tag{14}$$

where e is the charge of an electron, $b_{\pm}$ is the ionic mobility, σ_0 is the conductivity of the atmosphere in the absence of aerosols and $\sigma_{\pm}$ is the conductivity of the atmosphere in the presence of aerosols.

The ionic mobility ($b_{\pm}$) is expressed in terms of the reduced mobility (b_0) at normal temperature (T_0) and pressure (P_0), and the ambient temperature (T) as well as pressure (P) as follows:

$$b_{\pm} = \frac{b_o P_o T}{T_0 P} \tag{15}$$

The reduction in conductivity (Δσ) due to depletion of small ions is given by:

$$\Delta \sigma = \sigma_0 - \sigma_{\pm} \tag{16}$$

The value of Δσ is computed in terms of different model parameters from Eqs. (5)-(16) and by following the method of Gringel et al. (1978), we get:

$$\Delta \sigma = \frac{\left[(2\alpha_i - \alpha_s)\sigma_0 + \beta Z e b_{\pm} \right] \mp \sqrt{\left[(2\alpha_i - \alpha_s)\sigma_0 + \beta Z b_{\pm} \right]^2 + 4(\alpha_s - \alpha_i)\sigma_0 \beta Z e b_{\pm}}}{2(\alpha_i - \alpha_s)} \tag{17}$$

3.2 Computation of conductivity

Modeling the conductivity of the atmosphere requires a knowledge of recombination coefficients α_i, α_a and α_s. The value of α_i is estimated from the parametric formula (Smith and Adams, 1982; Datta et al., 1987). From the theoretical considerations, Hoppel (1985) has shown that, for singly charged aerosols, the relative magnitudes of α_a and α_s are such that $\alpha_a \leq \alpha_s \leq \alpha_i$. Initially, with a suitable assumed value of α_a, the value of η is computed from Eq.(11), then the value of α_s is computed using Eq.(13). It is noted that, in this step, α_s becomes negative if the assumed value of α_a is unrealistically large. In the present computations $\alpha_a = 10^{-7}$ cm^{3}s^{-1} is found to be suitable. In presence of aerosols and depending on the magnitude of α_i, α_s and βZ, the realistic values of ΔN are obtained which lies between zero and N_0, and satisfies the condition $0 < N_\pm < N_0$. The values of βZ and α_s are obtained by solving Eqs.(9) and (12) simultaneously for βZ and Eqs.(10) and (13) for α_s through iterative procedure until the difference between the values gets minimized. These values are substituted in Eq.(8) to obtain $N_\pm$. By knowing the experimental values of q, Z, T, P and different attachment and recombination coefficients computed from the ion-aerosol balance equations, it is possible to estimate the values of N_0 and $N_\pm$ and hence, the values of σ_o, $\sigma_\pm$ and $\Delta\sigma_\pm$.

4. Results and discussion

In an experimental programme at Pune (18^{0}32′N, 73^{0}51′E, 559 m above mean sea level), India measurements of radon and its progeny, aerosol number density and atmospheric conductivities are measured. Using the ionization rate from radioactivity, ambient aerosol density and meteorological parameters such as temperature, pressure and relative humidity, the small ion concentration and hence the conductivity of the atmosphere is estimated from an Ion-aerosol model. Model computed conductivity of the atmosphere is used to validate the measured conductivity from a Gerdien condenser setup. Nagaraja et al. (2006) have described the experimental method for the measurement of ground level atmospheric electrical conductivity near the earth's surface and its validation using a simple Ion-aerosol model. The present model work is an extension of the earlier work for the estimation of percent reduction in conductivity for assumed higher ambient aerosol levels, which may be the result of increased pollution due to vehicular exhaust, industrial effluents, anthropogenic activity etc. Thus, monitoring of ground level atmospheric conductivity can be useful in air pollution studies (Nagaraja Kamsali et al., 2009).

The diurnal variation of radon, its progeny, aerosols, small ion concentration and conductivity for a typical day are described. Even though, the observations differ quantitatively, they show similar trends of variations over a day. Variability in the data of any of these parameters can be associated with different meteorological conditions. Moreover, simultaneous measurements of the diurnal variations of radon, its progeny, aerosols, conductivity, dry and wet bulb temperature, and wind speed and direction at the same observatory confirm that the day chosen is in no way electrically or meteorologically abnormal from other fair weather days. Thus, the observations described below can be reasonably generalized for other fair weather days as well. The results of collocated measurements of the parameters such as concentrations of radon and its progeny, ionization rate, aerosols, electrical conductivity and meteorological parameters such as temperature, relative humidity are examined for a typical day and are taken as representative.

4.1 Variation of Ionization rate due to radon and its progeny

The diurnal variations of activities of radon and its progeny near the earth's surface are depicted in Fig. 3(a). The radon concentration varies from 2.1 to 23.0 Bq m^{-3} with a median of 6.8 Bq m^{-3} showing a significant diurnal variation, whereas values of radon progeny vary from 0.5 to 6.4 Bq m^{-3} with a median of 2.0 Bq m^{-3}. It is seen from the figure that the concentrations are maxima during the early morning hours, generally between 0500 and 0700 hrs of Indian Standard Time (IST). The concentration decreases after sunrise, attaining minima during the afternoon, 1000 to 1800 hrs of IST. The variation of meteorological parameters such as temperature and relative humidity are shown in Fig. 3(b). It is observed that the concentration of radon and its progeny follow the trend of the relative humidity, in general. This is due to the fact that as temperature increases, the saturation vapour pressure increases so that the given air can take more water vapour. Consequently the relative humidity will decrease. The raise of temperature causes increased vertical mixing and raising of aerosols to the higher altitudes resulting in lower concentration of radon at the ground level (Hoppel et al., 1986). When the temperature decreases and relative humidity increases, the vertical mixing and raising of aerosols to the higher altitude reduces. As a consequence, the aerosol to which radon is attached, will be present at higher concentrations during night and in the early morning hours at ground level. This results in the increase of radon concentrations near the surface of the earth (Porstendörfer, 1994).

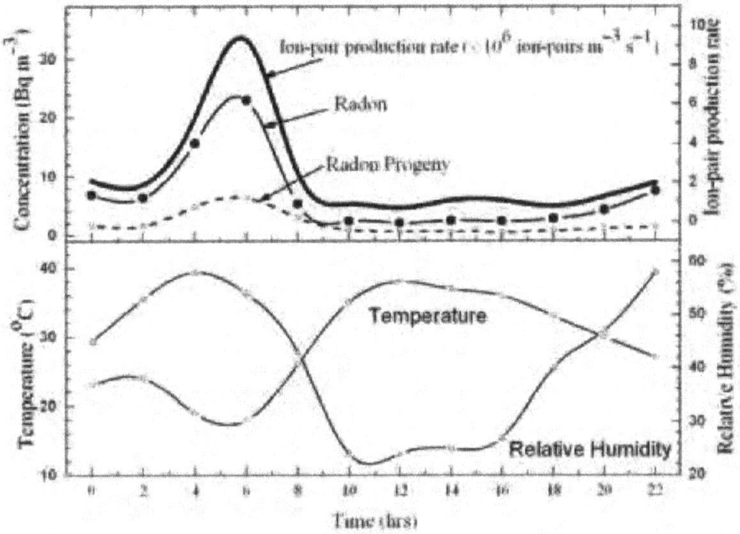

Fig. 3. Diurnal variation of a) radon, progeny and ion-pair production rate b) temperature and relative humidity

The ionization due to radioactivity also exhibits the diurnal variation as that of concentration of radon and its progeny and is shown in Fig. 3(a). At a height of 1 m above the surface ionization due to radioactive gases and their short-lived daughter products is predominantly caused by alpha particles. The rate of ionization due to radioactivity also exhibits the diurnal variation as that of concentration of radon and its progeny. The radon exhaled from the surface causes ionization of the atmosphere (Israel, 1970). It also shows a maximum in the early morning hours and a minimum in the afternoon (Dhanorkar and Kamra, 1994). The ion-pair production rate varies between 0.7 to 9.3 ion-pairs cm^{-3}s^{-1} with a mean of 2.4 ion-pairs cm^{-3}s^{-1}. At night, ionization rates close to the ground are enhanced because of the accumulation of radioactive emanations from ground under temperature inversions and due to their lesser dispersion because of low winds (Hoppel et al., 1986). It also shows that the concentration of radon and its daughter products is low at noon when air is unstable and is highest during the night and early morning hours, when mixing is low. The radioactive emanations from the ground are trapped below inversions, and their accumulation causes an increase in ionization in the lower stable atmosphere during the nighttime. Large rates of ionization, produced due to these radioactive emanations in the presence of large aerosol concentration at night, produce large concentrations of ions of all categories. The accumulation of radioactive emanations will increase through the night until morning, when there are rapid changes in the stability of the atmosphere. In the morning, the vertical mixing caused due to increased eddy diffusivity will dilute the concentration of aerosols and radioactive emanations close to the ground. This may explain the occurrence of large concentrations at nighttime as compared to daytime (Dhanorkar and Kamra, 1993).

4.2 Variation of aerosol concentration and its size distribution

Figure 4(a) represents two hourly values of the aerosol concentration obtained for each of the eight size categories in the range of 13 to 750 nm. It shows a diurnal variation with the concentration showing a minimum during the early morning around 0600 hours IST and an increasing trend towards noon-hours. Early in the morning, due to anthropogenic activity and also due to the onset of convection resulting from the increase in atmospheric temperature, aerosols are pushed into the atmosphere resulting in an increased aerosol concentration (Israelsson et al., 1994; Sasikumar et al., 1995).

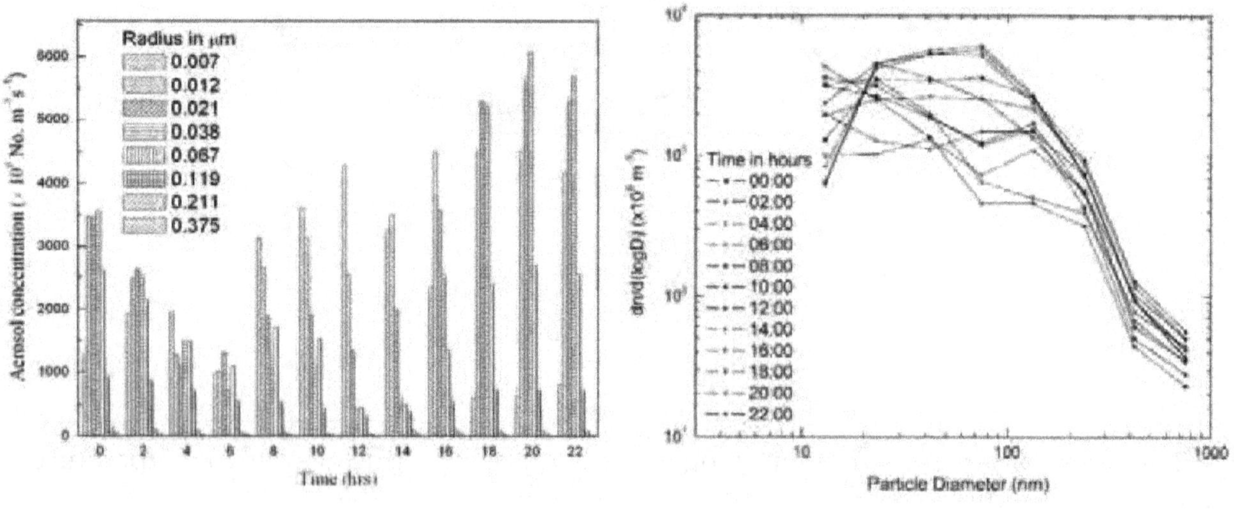

Fig. 4. Variation of aerosols a) over a day and b) its size distribution

The size distribution of aerosols is plotted in Fig. 4(b) are generally bimodal with their maxima at 75 and 23 nm during night (1800 to 0400 hrs). After 0400 hour with slight shift in maxima in the accumulation mode towards the higher size. During daytime, particularly in the afternoon, the shift in maxima in the accumulation mode to a higher diameter of 133 nm is distinct and the maxima in the nucleation mode seem to shift to smaller diameter. As a result of increase in the concentration of small particles, the size distribution curves during daytime are mostly open ended at the small particle side (Murugavel and Kamra, 1999). The magnitude of the peaks varies with time.

4.3 Variation of small ion concentration

The diurnal variation of estimated small ion concentration for a period of one week along with its mean values is shown in Fig. 5(a). Relatively a large day-to-day difference in the concentrations of ions that may arise from the difference in the accumulations of radioactive emanations and aerosol particles due to varying degrees of the lower atmospheric stability and/or due to varying intensities of advection. It is observed that the concentration show maxima in the early morning around 0600 hrs IST, it decreases after sunrise, and attains minimum in the afternoon. Radioactive emanations from ground are trapped below inversions, and their accumulations cause an increase in ionization and consequently increase in concentration in the lower atmosphere during the nighttime (Hoppel et al., 1986). Increased vertical mixing caused by increase in temperature and rising of aerosols to the higher altitudes results in the lower value of ionization rate at the ground level (Nagaraja et al., 2006). The concentration varies between 436 and 1663 ions cm^{-3} with a mean of 734 ions cm^{-3}.

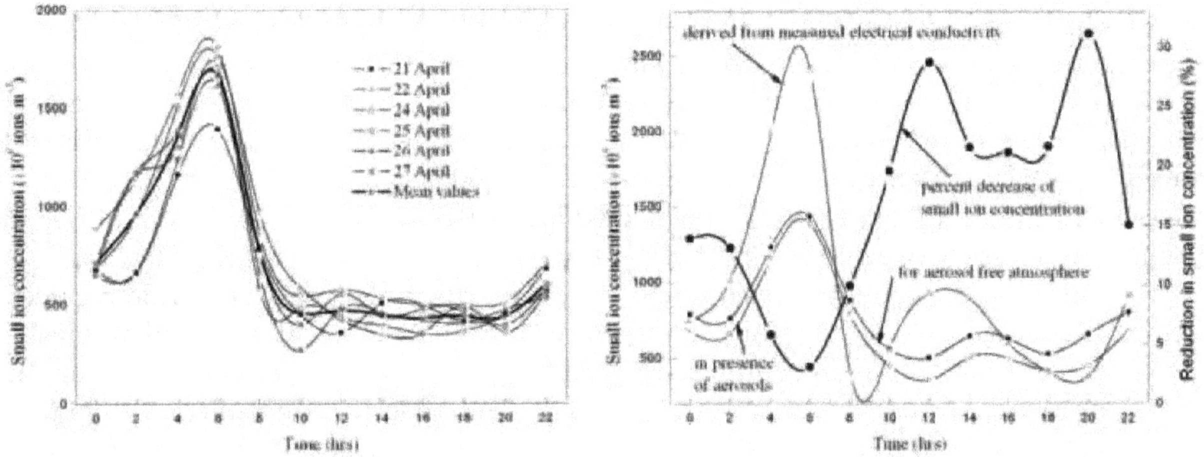

Fig. 5. Variation of a) small ion concentration b) small ion concentration in presence and absence of aerosols

The estimated small ion concentration shows a positive correlation with ionization rate due to surface radioactivity near to the Earth's surface. There is a correlation of 98.4% and suggests that the theoretical approach is best fitted and the correlation obtained indicates the correctness of ion-aerosol interaction considerations (Nagaraja Kamsali et al., 2009). Also shown in the figure is the estimated concentration of ions in the absence and in presence of aerosols using Eqs. (5) and (8), respectively. There is a fair agreement between theoretically estimated small ion concentration and experimentally derived values during daytime. However, there is a considerable difference during the morning hours. This difference is due to the large variability in measured conductivity. The percentage decrease in small ion concentration due to presence of aerosols is also shown in Fig. 5(b). The decrease is more during the noon-time when the aerosol concentration is high and less in the early morning when aerosol concentrations are minima. The maximum reduction in the estimated small ion concentration observed at around 2000 hrs coincides with the observed maximum concentration in aerosols. It is also seen that the existence of aerosols reduces the concentration and the percent reduction of small ion concentration is found to be minimum during early morning hours and starts increasing after sun rise and remains more or less same in the afternoon till sunset and reduces after words. This clearly shows that aerosols will have a control over the small ion concentration and hence the electrical structure of the atmosphere. A good correlation of 87.3% observed between the small ion concentration against derived values of small ion concentration from the measured electrical conductivity and suggests that the ion-aerosol balance equations considered are good enough to explain the interactions and are comparable with the experimental measurements of Dhanorkar and Kamra (1993).

4.4 Variation of conductivity

The diurnal behaviour of conductivity is found to be different at sites having different levels of natural radioactivity. Diurnal variations of experimental positive and negative ion conductivities for a continental station are shown in Fig. 6(a). The diurnal variation of conductivity of both polarities indicates that the pattern is representative of the regions. The positive and negative conductivities are approximately equal and their diurnal variations are generally mirror images to each other. It is observed that the conductivity of both polarities show maxima in the early morning hours between 0400 and 0700 hrs of IST, it decreases after sunrise, and attains minimum in the afternoon. This is due to the fact that the raise in temperature causes increased vertical mixing and rising of aerosols to the higher altitudes, which results in the lower value of ionization rate, and hence conductivity at the ground level (Dhanorkar et al., 1989; Nagaraja et al., 2006). The positive conductivity varies between 0.3 to 1.8 x 10^{-14} $\Omega^{-1}m^{-1}$ with a median of 0.8 x 10^{-14} $\Omega^{-1}m^{-1}$ and the negative conductivity varies between 0.2 to 2.0 x 10^{-14} $\Omega^{-1}m^{-1}$ with a median of 0.7 x 10^{-14} $\Omega^{-1}m^{-1}$.
During nighttime the atmosphere is relatively calm with low winds and little convective motion. The radon exhaled from the soil therefore accumulates near the ground and leads to increased ionization and higher conductivity. After sunrise, due to human activity and also due to the onset of convection resulting from the increase in atmospheric temperature, aerosols are pushed into the atmosphere. This causes a conversion of small ions, that are responsible for conductivity, into large ions through attachment, and an increase in the destruction of small ions through recombination with large ions of opposite polarity. The onset of circulation also removes radon from near the ground to higher altitude regions. These factors contribute to the observed reduction in conductivity in the afternoon (Subbaramu and Vohra, 1969). In the evening, with decreasing ground temperatures and also the anthropogenic activity, the aerosols that had been pushed to higher altitudes begin to settle down, and a greater fraction of small ions is lost through attachment. Conductivity therefore again starts raising after sunset. Finally, at nightfall, the aerosols settle down, and the conductivity recovers its normal night time high values. Since the major contribution to ionization at the surface comes from surface radioactivity and radon/thoron emanating from the soil, one would expect the diurnal variation pattern of conductivity to reflect that of the radon concentration near the surface. Radon and its short-lived daughters are understood to show a diurnal variation in concentration with a maximum early in the morning and a minimum in the afternoon (Nagaraja et al., 2003a). This is ascribed to the variations in the vertical mixing of air near the surface, which is controlled by atmospheric stability. As convection builds up with solar heating of the ground, radioactive gases are also transported upward, where they add to the ambient ionization. This leads to a reduction in the concentration of these gases, and consequently of ionization, near the surface (Nagaraja et al., 2003b). Therefore one should expect higher conductivities during

the period of atmospheric stability, with the values decreasing with the onset of turbulence (Sasikumar et al., 1995; Prasad et al., 2005). Wilkening and Romero (1981) reported the values of ionization rate, positive conductivity and negative conductivity, respectively, as 1.5 $cm^{-3}s^{-1}$, 1.4 x 10^{-14} $\Omega^{-1}m^{-1}$ and 1.4 x 10^{-14} $\Omega^{-1}m^{-1}$, for free atmosphere. These are of the same order as our measurements reported in this paper. The conductivity values presented here are of the same order of Dhanorkar and Kamra (1993). The total conductivity is the algebraic sum of the experimentally measured values of polar conductivities i.e., positive and negative conductivity. The variation of total conductivity is shown in Fig. 6(b). It varies between 0.6 to 3.9 x 10^{-14} $\Omega^{-1}m^{-1}$ with a median of 1.5 x 10^{-14} $\Omega^{-1}m^{-1}$. The increase of conductivity in the early morning hours is mainly because of the ionization produced by radioactive substances present in the atmosphere. The correlation coefficient between the ionization rate, obtained from the radioactivity, and the measured electrical conductivity is found to be 98% for a typical day and shows that the conductivity at the ground is mainly because of the ionization due to radioactivity near the surface.

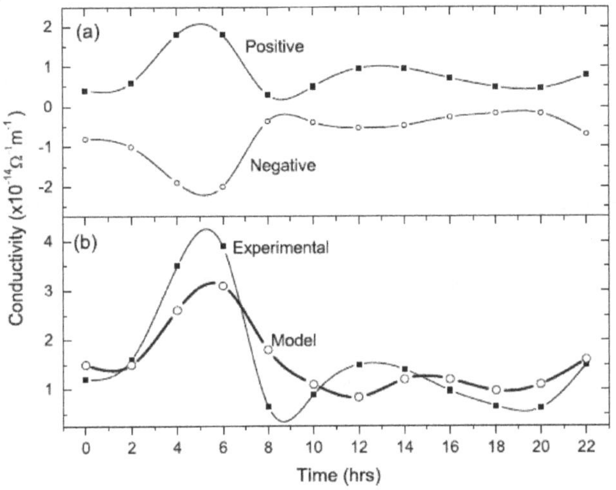

Fig. 6. Variation of a) positive and negative conductivity b) experimental and model conductivity

The experimental conductivity fluctuations in Fig. 6 is due to the similar variations in q which in turn is affected by the variations in meteorological parameters such as temperature, humidity, wind speed etc. The model of this study has reproduced well this trend of the diurnal variation in conductivity. It is seen that there is no exact one to one match between the experimental and model conductivities at certain points. However, there is a correlation coefficient of more than 90% between experimental and model derived conductivity values, which is an indication of correctness of the model results. In this study, the standard deviation in the measured ionization rates and aerosol concentration are both less than 10%. It is observed that the variations in the model conductivity values for the 10% variations in q and/or Z show variations of about 7% or less. It is evident that the simple Ion-aerosol model of this study has adequately reproduced the experimental conductivity.

4.5 Variation of conductivity under enhanced aerosols conditions

The conductivity of air is entirely due to small ions and depends upon the number of small ions present in the atmosphere. The conductivity decreases if the small ion concentration reduces. The small ion concentration is altered by the presence of aerosols due to attachment with aerosols, apart from the usual loss due to mutual recombination between the ions of opposite polarity. Hence, the conductivity alters due to the presence of aerosols in the atmosphere. Variation of conductivity due to enhanced (from the background) aerosols is plotted in Fig. 7(a). Three curves indicate that the conductivity decreases as the aerosol concentration increases. Fig. 7(b) shows the percent reduction in conductivity due to the presence of aerosols derived from the model for normal condition. In the early morning hours the concentration of aerosols is less and the conductivity is more (Houghton, 1985). Convective currents in the morning carry aerosol particles up from ground surface and the aerosols change the high mobility small ions into low mobility large ions. Thus the conductivity decreases after sunrise (Dhanorkar et al., 1989). As the day advances the aerosol concentration increases due to anthropogenic activities and conductivity decreases. The percent decrease in conductivity shows a diurnal variation with a minimum in the early morning hours when the aerosol concentration is minimum and reaches a maximum (30 - 32 %) in the afternoon when aerosol concentration also shows maximum. Due to pollution the concentration of aerosols increases and leads to decrease in conductivity of the atmosphere near the surface (Retalis et al., 1991; Dhanorkar and Kamra, 1993). Thus the measurement of atmospheric conductivity can be used as an index of pollution in the atmosphere. Also estimated from the Ion-Aerosol model is the reduction in small ion concentration and hence the conductivity of the atmosphere for varying levels of aerosols and as seen in Fig. 7(b). The results reveal that, for an increase of background aerosols by three-fold the percent reduction in conductivity is 7%, and for an increase of aerosols by six-fold the reduction in conductivity is 10% from the measured ambient aerosol level.

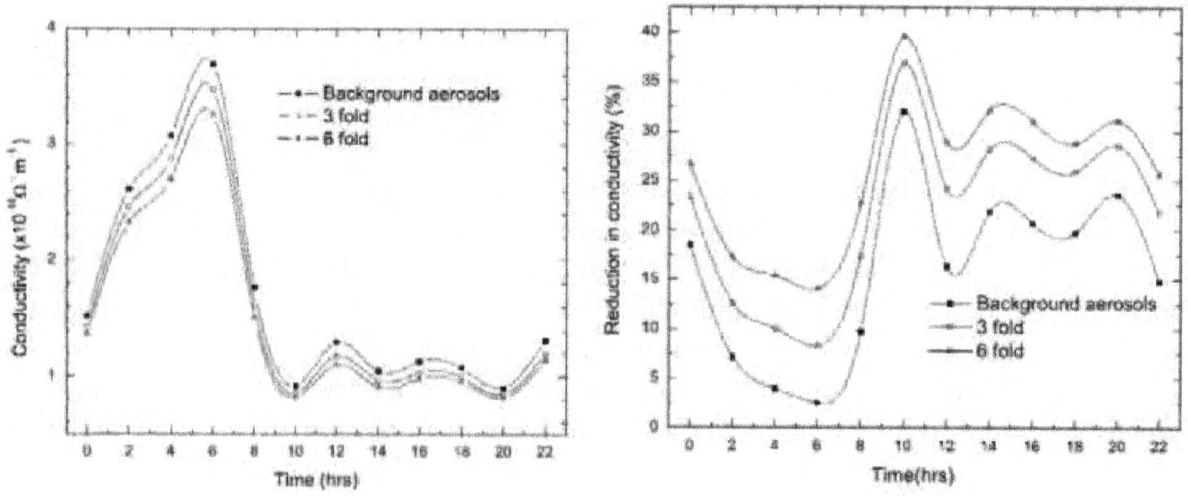

Fig. 7. Variation of conductivity a) in presence of aerosols and enhanced aerosols conditions b) percent decrease in conductivity for enhanced aerosol conditions

The diurnal, weekly and annual patterns of atmospheric electrical conductivity seem to be well correlated with the local aerosol air pollution patterns. The long-term decreasing trend seems to be affected by the air pollution build-up over metropolitan areas. Convective currents in the morning carry aerosol particles up into atmosphere; change the high mobility small ions into low mobility large ions. Thus the conductivity decreases after sunrise. As the day advances the aerosol concentration increases due to anthropogenic activities and conductivity decreases. The percent decrease in conductivity shows a diurnal variation with a minimum in the early morning hours when the aerosol concentration is minimum and reaches a maximum 32% for a typical day in the afternoon when aerosols also show maximum for background aerosols. Due to pollution the concentration of aerosols increases and leads to decrease in conductivity of the atmosphere near the surface. This gives an idea that the measurement of conductivity can be used as an index of pollution measurement. The result also reveals that, for an increase of aerosols by 3-fold the percent reduction in conductivity is 7%, and for an increase of aerosols by 6-fold the reduction in conductivity is 10% from the pollution free atmosphere.

5. Summary

The electrical conductivity of the atmosphere is mainly due to the radioactive substances in the earth's crust and their emanation near the surface. The radiations emitted by the radioactive substances ionize the atmospheric air and leads to conductivity. The conductivity gets altered because of the existence of aerosols in the atmosphere. A simple Ion aerosol model has been developed to deduce the conductivity theoretically by using ionization rate, aerosol concentration, and different attachment and recombination coefficients for the loss of small ions as input to the model. The model agrees well with the experimentally measured conductivity. In addition to this, it is seen from the model that the percent reduction in conductivity is 7% and 10% for an increase of ambient background aerosols by three-fold and six-fold, respectively.

6. Acknowledgements

The work was proposed under the financial support from the Department of Science and Technology (DST), Government of India. The authors (Nagaraja and Prasad) are thankful to the funding agency, and to Dr. A. K. Kamra and Dr. S. D. Pawar, Indian Institute of Tropical Meteorology, Pune, India for the fabrication of the Gerdien condenser and permitting us to carry out the experiments in Pune.

7. References

Chalmers, J.A. (1967). *Atmospheric Electricity*. Pergamon press, Oxford, London.

Cobb, W.E. (1973). Oceanic aerosol levels deduced from measurements of the electrical conductivity of the atmosphere, *Journal of Atmospheric Science*, 30, 101-106.

Cobb, W.E. & Wells, H.J. (1970). The electrical conductivity of oceanic air and its correlation to global atmospheric pollution. *Journal of Atmospheric Science*, 27, 814-819.

Datta, J., Chakravarty, S.C. & Mitra, A.P. (1987). A model for the cosmic ray produced ionization in the middle atmosphere. *Indian Journal of Radio and Space Physics*, 16, 257-266.

Dhanorkar, S. & Kamra, A. K. (1992). Relation between electrical conductivity and small ions in the presence of intermediate and large ions in the atmosphere, *Journal of Geophysical Research*, 97, 20345-20360.

Dhanorkar, S. & Kamra, A. K. (1993). Diurnal and seasonal variations of the small-, intermediate-, and large-ion concentrations and their contributions to polar conductivity, *Journal of Geophysical Research*, 98, 14895-14908.

Dhanorkar, S. & Kamra, A. K. (1994). Diurnal variation of ionization rate close to ground, *Journal of Geophysical Research*, 99, 18523-18526.

Dhanorkar, S. & Kamra, A. K. (1997). Calculation of electrical conductivity from ion-aerosol balance equations. *Journal of Geophysical Research*, 102, 30147-30159.

Dhanorkar, S., Deshpande, C. G. & Kamra, A. K. (1989). Observations of some atmospheric electrical parameters in the surface layer, *Atmospheric Environment*, 23, 839-841.

Gringel, W., Kaselan, K.H. & Muhleisen, R. (1978) Recombination rates of small ions and their attachment to aerosol particles. *Pure and Applied Geophysics,* 116, 1101–1113.

Hoppel, W.A. (1985). Ion-aerosol attachment coefficients, ion depletion, and the charge distribution on aerosols. *Journal of Geophysical Research*, 90, 5917-5923.

Hoppel, W.A., Anderson, R.V. & Willet, J.C. Atmospheric electricity in the planetary boundary layer, In: *The earth's electrical environment,* National Academy Press, Washington, D.C., USA, pp. 149-165, 1986.

Houghton, H.G. *Physical Meteorology*. MIT Publications, Cambridge, Massachusetts, England, 1–452, 1985.

Israel, H., *Atmospheric Electricity*, Vol.1, Israel Program for Scientific Translations, Jerusalem, 1-317, 1970.

Israelsson, S., Knudsen, E. & Anisimov, S.V. (1994). Vertical profiles of electrical conductivity in the lowermost part of the turbulent boundary layer over flat ground. *Journal of Atmospheric and Terrestrial Physics*, 56, 1545–1550.

Kamra, A.K. & Deshpande, C.G. (1995). Possible secular change and land-to ocean extension of air pollution from measurements of electrical conductivity over the Bay of Bengal. *Journal of Geophysical Research*, 100, 7105–7110, 1995.

Liu, B.Y.H. E. & Pui, D.Y.H. (1975). On the performance of the electrical aerosol analyzer. *Journal of Aerosol Science*, 6, 249-264.

Mani, A. & Huddar, B.B. (1972). Studies of the surface aerosols and their effect on atmospheric electric parameters. *Pure and Applied Geophysics,* 100, 154–166.

Misaki, M. & Takeuti, T. (1970). The extension of air pollution from land over ocean as related in the atmospheric electrical conductivity. *Journal of Meteorological Society of Japan*, 48, 263–269, 1970.

Misaki, M. (1964). Mobility spectrums of large ions in the New Mexico semidesert. *Journal of Geophysical Research*, 69, 3309–3318.

Morita, Y., Ishikawa, H. & Kanazawa, J. (1972). Atmospheric electrical conductivity measurements in the Pacific Ocean, exploring the background level of global pollution. *Journal of Meteorological Society of Japan*, 497–500, 1972.

Murugavel, P. & Kamra, A.K. (1999). Changes in the concentration and size-distribution of the sub-micron particles associated with the sea-and land-breezes at a coastal station. *Current Science*, 76, 994-997.

Nagaraja, K., Prasad, B. S. N., Madhava, M. S. & Paramesh, L., (2003a). Concentration of radon and its progeny near the surface of the earth at a continental station Pune (18 N, 74 E), *Indian Journal of Pure and Applied Physics*, 41, 562-569.

Nagaraja K., Prasad, B.S.N., Madhava, M.S., Chandrashekara, M.S., Paramesh, L., Sannappa, J., Pawar, S.D., Murugavel, P. & Kamra, A.K. (2003b). Radon and its short-lived progeny: Variations near the ground. *Radiation Measurement*, 36, 413-417.

Nagaraja K., Prasad, B.S.N., Srinivas, N. & Madhava, M.S. (2006). Electrical conductivity near the Earth's surface: Ion-aerosol model. *Journal of Meteorological Society of Japan*, 68, 757-768.

Nagaraja Kamsali., B.S.N. Prasad. & Jayati Datta. (2009). Atmospheric electrical conductivity measurements and modeling for application to air pollution studies. *Advances in Space Research*, 44, 1067-1078.

Porstendörfer, J. (1994). Properties and behaviour of radon and thoron and their decay products in the air, *Journal of Aerosol Science*, 25, 219-263.

Prasad, B.S.N., Nagaraja, K., Chandrashekara, M.S., Paramesh, L. & Madhava, M.S. (2005). Diurnal and seasonal variations of radioactivity and electrical conductivity near the surface for a continental location Mysore, India. *Atmospheric Research*, 76, 65-77.

Prasad, B.S.N., Srinivas, N. & Chandramma, S. (1991). A simplified ion-aerosol model for balloon measurements of ion conductivity and aerosol concentration. *Indian Journal of Pure and Applied Physics*, 20, 304-306.

Raghavayya, M. (1998). Modification of Kusnetz method for estimation of radon progeny concentration in air, *Radiation Protection and Environment*, 21, 127-132.

Retalis, D., Pitta, A. & Psallidas, P. (1991). The conductivity of the air and other electrical parameters in relation to meteorological elements and air pollution in Athens, *Meteorological Atmospheric Physics*, 46, 197-204.

Rosen, J.M. & Hofmann, D.J. (1981). Balloon-borne measurements of electrical conductivity, mobility and the recombination coefficients. *Journal of Geophysical Research*, 86, 7406–7410.

Sasikumar, V., Sampath, S., Muralidas, S. & Vijaykumar, K. (1995). Atmospheric electrical conductivity variations over different environments. *Geophysics Journal International*, 122, 89–96.

Smith, D. & Adams, N.G. (1982). Ionic recombination in the stratosphere. *Geophysical Research Letter*, 9, 1085-1087.

Srinivas, N. & Prasad, B.S.N. (1993). Seasonal and latitudinal variations of stratospheric small ion density and conductivity. *Indian Journal of Pure and Applied Physics*, 22, 122–127.

Srinivas, N. & Prasad, B.S.N. (1996). A detailed model study of stratospheric small ion density and conductivity. *Indian Journal of Pure and Applied Physics*, 25, 255-262.

Srinivas, N., Prasad, B.S.N. & Nagaraja, K. (2001). An ion-aerosol model study for the stratospheric conductivity under enhanced aerosol condition. *Indian Journal of Pure and Applied Physics*, 30, 31-35.

Subbaramu, M.C. & Vohra, K.G. (1969). Investigations on radioactive equilibrium in the lower atmosphere between radon and its short-lived decay products. *Tellus*, 21, 395–403.

Wilkening, M. & Romero, V. (1981). ^{222}Rn and atmospheric electrical parameters in the Carlsbad caverns. *Journal of Geophysical Research*, 86, 9911–9916.

5

Imprecise Uncertainty Modelling of Air Pollutant PM_{10}

Danni Guo[1], Renkuan Guo[2], Christien Thiart[2] and Yanhong Cui[2]
[1]Climate Change and Bioadapation Division, South African National Biodiversity Institute, Cape Town
[2]Department of Statistical Sciences, University of Cape Town, Cape Town, South Africa

1. Introduction

Particulate matter (PM) refers to solid particles and liquid droplets found in air. Many manmade and natural sources produce PM directly, or produce pollutants that react in the atmosphere to form PM. The resultant solid and liquid particles come in a wide range of sizes, and particles that are 10 micrometers or less in diameter (PM_{10}) can be inhaled into and accumulate in the respiratory system and are believed to pose health risks (Environmental Protection Agency, 2010). Particulate matter is one of the six primary air pollutants the Environmental Protection Agency (EPA) regulates, due to exposure to high outdoor PM_{10} concentrations causes increased disease and death (Environmental Protection Agency, 2010). Therefore, PM_{10} concentrations, amongst many other air pollutants, are sampled and measured in various places in California, United States.

The general trend of PM air pollutant concentrations in the air in California are on the decrease, but it continues to be monitored and observed. The California standards for annual PM_{10} concentrations is that the annual arithmetic mean is 20 $\mu g/m^3$, and the national standard is 50 $\mu g/m^3$ before 2006 (California Environmental Protection Agency Air Resources Board, 2010, Environmental Protection Agency, 2010). The State of California sets very high standards for their air quality, and air pollutants are carefully monitored.

However, in reality, it is too costly in terms of time, finance, and manpower to keep all the 213 sites to be monitoring and recording. In Fig. 1, a complete map of all 213 sample locations for PM_{10} are shown. However, one must note that these sample sites are never all used at any given year, PM_{10} samples are taken at different locations each year. At best, a maximum of 102 PM_{10} samples are collected during some years, and at worst, 61 PM_{10} samples are collected at that year. Therefore comparisons of PM_{10} between years are difficult, due to missing data at sample sites. It is difficult to construct kriging maps in terms of actual observations annually since the air pollutants were measured in different locations each year although the site design originally planned was quite delicate statistically.

Each year, approximately 40% of the 213 sites were actually observed. We call a site that does not have a recorded PM_{10} value as "missing value", and since there are no patterns so that serious problems would twist the kriging map constructions. In Fig. 2, this is clearly demonstrated. In 1989, there are 61 PM_{10} samples collected (29% of 213 locations), and in 2000, there are 94 PM_{10} samples collected (44%).

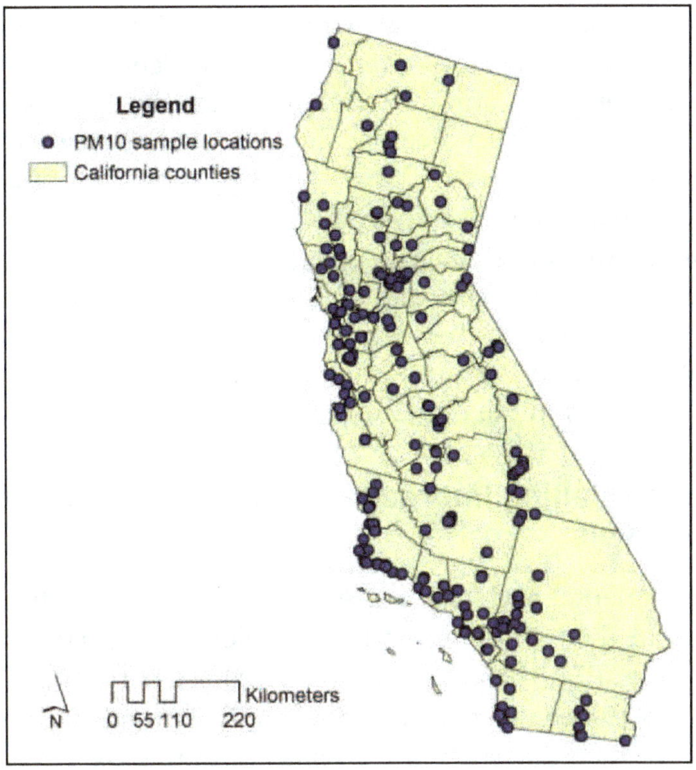

Fig. 1. Complete 213 Observational Sites in the California State

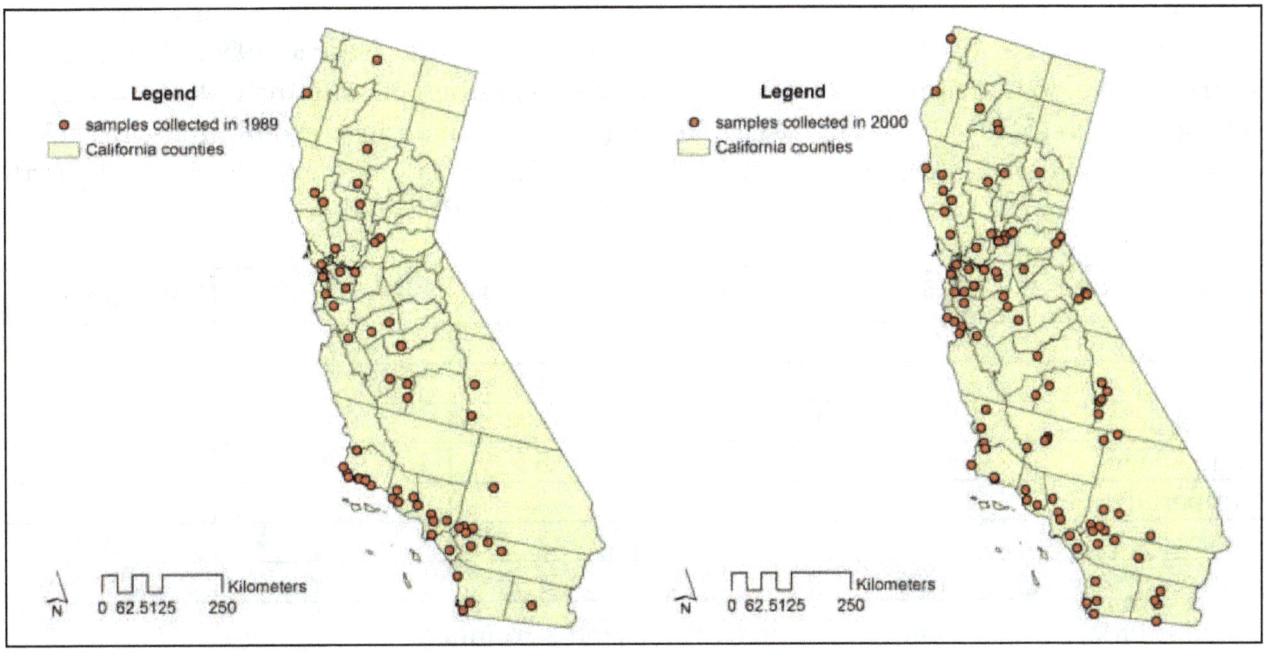

Fig. 2. PM_{10} Samples Collected in California in 1989 (61 sites) and 2000 (94 sites)

The data scarcity brings in a series of (five) fundamental issues into the spatial-temporal modelling and prediction practices for California PM_{10} data, namely:

1. The necessity to recognize the impreciseness in analyzing the spatial-temporal pattern in terms of California PM_{10} records, which inevitably acts the solidness of a geo-statistical analysis;

2. Which theoretical foundations are appropriate for modelling impreciseness uncertainty;
3. How to fill up the "missing value" sites so that the "complete" records are available, which is either an original annual average from the original observations (40%) recorded on the site or a or predicted value by "neighbourhood sites" (60%), i.e., to facilitate spatial-temporal imprecise PM_{10} value by interpolations and extrapolations;
4. How to estimate the parameters of uncertain processes (temporal patterns), particularly the rate of change parameter α_i, $i = 1,2,\cdots,213$;
5. Create annual kiging maps (19 maps) under spatially isotropy and stationarity assumptions so that the changes between annual maps can be analyzed by kriging map difference between 2007 and 1989 and kriging map of location rate of change.

These issues will be addressed in the remaining sections sequentially.

2. The necessity of modelling impreciseness in California PM_{10} spatial-temporal analysis

Impreciseness is a fundamental and intrinsic feature in the PM_{10} spatial-temporal modelling, due to the observational data shortage and incompleteness. Spatially, there are 213 sites involved, and temporally, PM_{10} observations were collected from 1989 to 2007, over a 19-year period. During the 19-year period, there are only two sites (Site 2125 and Site 2804) having complete 19 year records. There are 16 sites having only 1 record (8%) and 70 sites having 10 or above records. To have a statistically significant time-series analysis, 50 data points are minimal requirement for each site, so classical time-series analysis (probabilistic analysis) cannot be performed. In order to have a quick overall evaluation of PM_{10} records on each site, we borrow the statistical quality control idea here (Electric, 1956, Montgomery, 2001). But we do not carry on traditional 6-sigma rule, rather, classify the PM_{10} records into four groups: 1-$(5,20]$, 2-$(20,35]$, 3-$(35,50]$, 4-$(50,160]$. These four-group limits in Table 1 reflect the national standard, (50 $\mu g/m^3$) and California state standard (20 $\mu g/m^3$) respectively. For example, 1 - $(5,20]$ is for a location whose PM_{10} fall in 5 to 20 ($\mu g/m^3$).

County name	1-(5,20]	2-(20,35]	3-(35,50]	4-(50,160]	No. of Sites
Los Angeles			3	7	10
Kern	1	2	1	5	9
Riverside	3		3	3	9
San Diego			5	3	8
Imperial	3			4	7
Lake		3			3
Inyo	2	2	4		8

Table 1. PM_{10} Hazard level evaluation over selected 7 counties

One must be aware that the classification is not in absolute sense, rather, additional rules are adding (similar to quality control chart pattern analysis (Electric, 1956):
(1) if a single point, then, classify the site hazard level according to which group it falls in;
(2) if a sequence of records, some of them, particularly early points may fall in higher (or lower) hazard level, but if last three points fall in a lower (or higher) hazard level, the later level would be chosen for the site.

The additional rule 1 can attribute to expert's knowledge confirmation, while the additional rule 2 can be regarded as an expert's decision based on trend pattern.

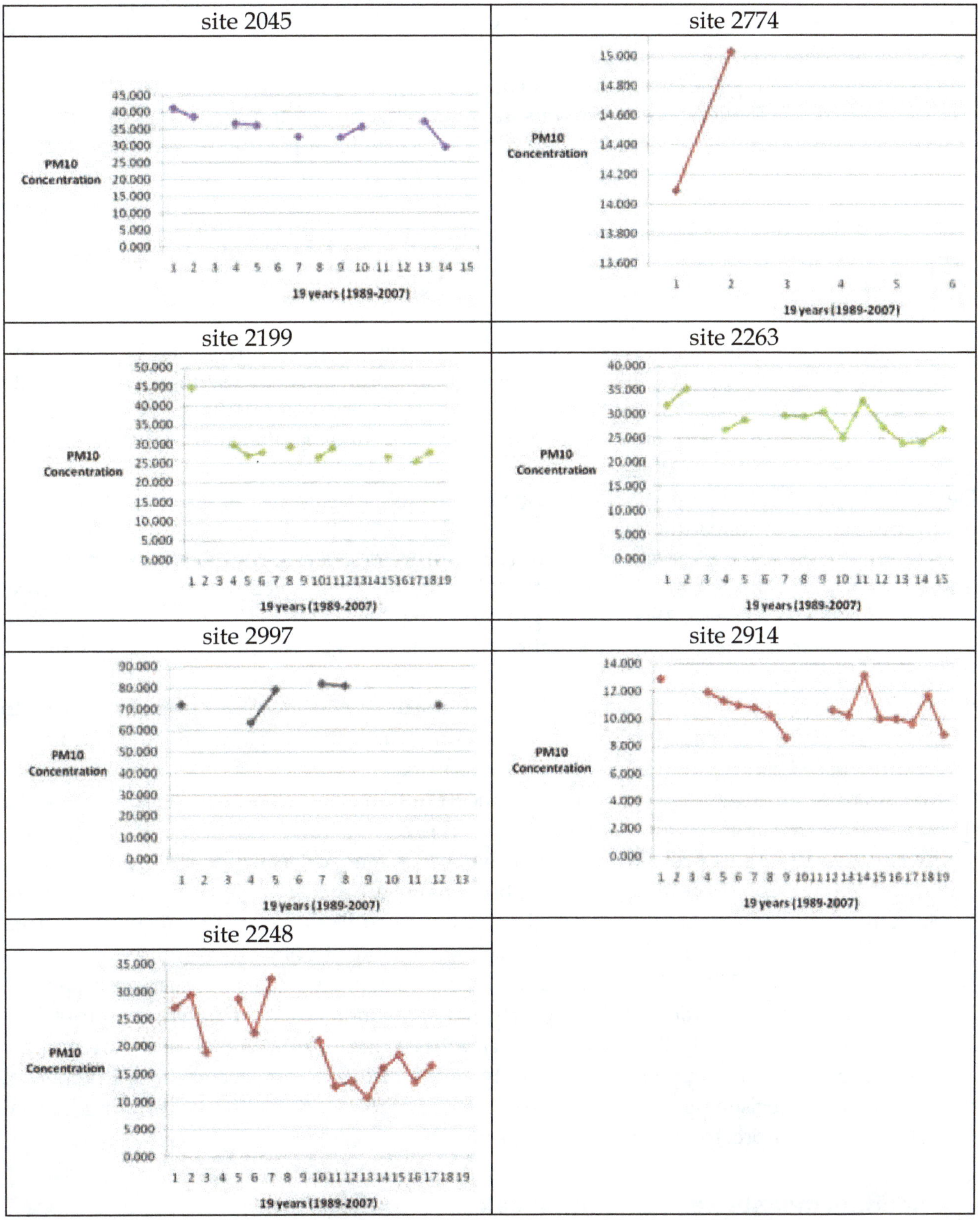

Fig. 3. The 7 sites from the selected 7 counties with original PM_{10} data plots and the hazard level classifications

Fig. 3 shows the classifications of a seven sites from the selected 7 counties in Table 1, each county one site is picked up for illustration purpose. The red coloured plot means the hazard level 1 $(5,20]$; the green coloured plot means the hazard level 2$(20,35]$; the purple coloured plot means the hazard level 3 $(35,50]$; and the black coloured plot means the hazard level 4 $(50,160]$.

It is evident that facing the impreciseness caused by incomplete data recording, one has to rely on expert's knowledge to compensate the inadequacy and accuracy in collected observational data. Impreciseness is referred to a term with a connotation specified by an uncertain measure or an uncertainty distribution for each of the actual or hypothetical members of an uncertainty population (i.e., collection of expert's knowledge). An uncertain process is a repeating process whose outcomes follow no describable deterministic pattern, but follow an uncertainty distribution, such that the uncertain measure of the occurrence of each outcome can be only approximated or calculated.

The uncertainty modelling without a measure specification will not have an rigorous mathematical foundations and consequently the modelling exercise is baseless and blindness. In other words, measure specification is the prerequisite to spatial-temporal data collection and analysis. For example, without Kolmogrov's (1950) three axioms of probability measure, randomness is not defined and thus statistical data analysis and inference has no foundation at all.

Definition 2.1: Impreciseness is an intrinsic property of a variable or an expert's knowledge being specified by an uncertain measure.

It is therefore inevitably to seek appropriate form of uncertainty theory to meet the impreciseness challenges. In the theoretical basket, interval uncertainty theory (Moore, 1966), fuzzy theory (Zadeh, 1965, 1978), grey theory (Deng, 1984), rough set theory (1982), upper and lower provisions (or expectations) (Walley, 1991), or Liu's uncertainty theory (2007, 2010) may be chosen.

While imprecise probability theory (Utikin and Gurov, 1998) may be a typical answer to address the observational data inaccuracy and inadequacy. However the imprecise probability based spatial modelling requires too heavy assumptions. Just as Utikin and Gurov (2000) commented, "the probabilistic uncertainty model makes sense if the following three premises are satisfied: (i) an event is defined precisely; (ii) a large amount of statistical samples is available; (iii) probabilistic repetitiveness is embedded in the collected samples. This implies that the probabilistic assumption may be unreasonable in a wide scope of cases." Guo et al. (2007) and Guo (2010) did attempt to address the spatial uncertainty from the fuzzy logic and later Liu's (2007) credibility theory view of point.

Nevertheless, Liu's (2007, 2010) uncertainty theory is the only one built on an axiomatic uncertain measure foundation and fully justified with mathematical rigor. Therefore it is logical to engage Liu's (2007, 2010, 2011) uncertainty theory for guiding us to understand the intrinsic character of imprecise uncertainty and facilitate an accurate mathematical definition of impreciseness in order to establish the foundations for uncertainty spatial modelling under imprecise uncertainty environments.

3. Uncertain measure and uncertain calculus foundations

Uncertainty theory was founded by Liu in 2007 and refined in 2010, 2011. Nowadays uncertainty theory has become a branch of mathematics.

A key concept in uncertainty theory is the uncertain measure, which is a set function defined on a sigma-algebra generated from a non-empty set. Formally, let Ξ be a nonempty set (space), and $\mathfrak{A}(\Xi)$ the σ-algebra on Ξ. Each element, let us say, $A \subset \Xi, A \in \mathfrak{A}(\Xi)$ is called an uncertain event. A number denoted as $\lambda\{A\}$, $0 \le \lambda\{A\} \le 1$, is assigned to event $A \in \mathfrak{A}(\Xi)$, which indicates the uncertain measuring grade with which event $A \in \mathfrak{A}(\Xi)$ occurs. The normal set function $\lambda\{A\}$ satisfies following axioms given by Liu (2011):

Axiom 1: (Normality) $\lambda\{\Xi\} = 1$.

Axiom 2: (Self-Duality) $\lambda\{\cdot\}$ is self-dual, i.e., for any $A \in \mathfrak{A}(\Xi)$, $\lambda\{A\} + \lambda\{A^c\} = 1$.

Axiom 3: (σ- Subadditivity) $\lambda\left\{\bigcup_{i=1}^{\infty} A_i\right\} \le \sum_{i=1}^{\infty} \lambda\{A_i\}$ for any countable event sequence $\{A_i\}$.

Axiom 4: (Product Measure) Let $(\Xi_k, \mathfrak{A}_{\Xi_k}, \lambda_k)$ be the kth uncertain space, $k = 1,2,\cdots,n$. Then product uncertain measure λ on the product measurable space $(\Xi, \mathfrak{A}_\Xi)$ is defined by

$$\lambda = \lambda_1 \wedge \lambda_2 \wedge \cdots \wedge \lambda_n = \min_{1 \le k \le n} \{\lambda_k\} \tag{1}$$

where

$$\Xi = \Xi_1 \times \Xi_2 \times \cdots \times \Xi_n = \prod_{k=1}^{n} \Xi_k \tag{2}$$

and

$$\mathfrak{A}_\Xi = \mathfrak{A}_{\Xi_1} \times \mathfrak{A}_{\Xi_2} \times \cdots \times \mathfrak{A}_{\Xi_n} = \prod_{k=1}^{n} \mathfrak{A}_{\Xi_k} \tag{3}$$

That is, for each product uncertain event $\Lambda \in \mathfrak{A}_\Xi$ (i.e, $\Lambda = \Lambda_1 \times \Lambda_2 \times \cdots \times \Lambda_n \in \mathfrak{A}_{\Xi_1} \times \mathfrak{A}_{\Xi_2} \times \cdots \times \mathfrak{A}_{\Xi_n} = \mathfrak{A}_\Xi$), the uncertain measure of the event Λ is

$$\lambda\{\Lambda\} = \begin{cases} \sup\limits_{A_1 \times \cdots \times A_n \subset \Lambda} \min\limits_{1 \le k \le n} \lambda\{\Lambda_k\} & \text{if } \sup\limits_{A_1 \times \cdots \times A_n \subset \Lambda} \min\limits_{1 \le k \le n} \lambda\{\Lambda_k\} > 0.5 \\ 1 - \sup\limits_{A_1 \times \cdots \times A_n \subset \Lambda^c} \min\limits_{1 \le k \le n} \lambda\{\Lambda_k\} & \text{if } \sup\limits_{A_1 \times \cdots \times A_n \subset \Lambda^c} \min\limits_{1 \le k \le n} \lambda\{\Lambda_k\} > 0.5 \\ 0.5 & \text{otherwise} \end{cases} \tag{4}$$

Definition 3.1: (Liu, 2007, 2010, 2011) A set function $\lambda : \mathfrak{A}(\Xi) \to [0,1]$ satisfies *Axioms 1-3* is called an uncertain measure. The triple $(\Xi, \mathfrak{A}(\Xi), \lambda)$ is called an uncertainty space.

Definition 3.2: (Liu, 2007, 2010, 2011) An uncertainty variable is a measurable function ξ from an uncertainty space $(\Xi, \mathfrak{A}(\Xi), \lambda)$ to the set of real numbers, i.e., for any Borel set B of real numbers, the set $\{\tau \in \Xi : \xi(\tau) \subset B \in \mathfrak{B}(\mathbb{R})\} \in \mathfrak{A}(\Xi)$, i.e., the pre-image of B is an event.

Remark 3.3: Parallel to revelation of the connotation of randomness in geostatistics, impreciseness occupies an fundamental position in geospatial-temporal uncertainty statistical analysis. In California PM_{10} spatial-temporal study, nearly 60% sites do not have "complete" temporal sequences so that in order to fill the "missing" observations, we have to engage expert's knowledge to pursue "complete sequences" (i.e., to have 19 PM_{10} values at each individual site), which is inevitably imprecise and incomplete. Impreciseness is referred to a term here with an intrinsic property governed by an uncertainty measure or an uncertainty distribution for each of the actual or hypothetical members of an uncertainty population (i.e., collection of expert's knowledge). An uncertainty process is a repeating process whose outcomes follow no describable deterministic pattern, but follow an uncertainty distribution, such that the uncertain measure of the occurrence of each outcome can be only approximated or calculated.

Remark 3.4: Impreciseness exists in engineering, business and research practices due to measurement imperfections, or due to more fundamental reasons, such as insufficient available information, ... , or due to a linguistic nature, because it is an unarguable fact that impreciseness exists intrinsically in expert's knowledge on the real world.

Definition 3.5: Let ξ be a uncertainty quantity of impreciseness on an uncertainty measure space $(\Xi, \mathfrak{A}(\Xi), \lambda)$. The uncertainty distribution of ξ is

$$\Psi_{\xi}(x) = \lambda\{\tau \in \Xi \mid \xi(\tau) \leq x\} \tag{5}$$

An imprecise variable ξ is an uncertainty variable and thus is a measurable mapping, i.e., $\xi : \mathbb{D} \to \mathbb{R}$, $\mathbb{D} \subseteq \mathbb{R}$. An observation of an imprecise variable is a real number, (or more broadly, a symbol, or an interval, or a real-valued vector, a statement, etc), which is a representative of the population or equivalently of an uncertainty distribution $\Psi_{\xi}(\cdot)$ under a given scheme comprising set and σ-algebra. The single value of a variable with impreciseness should not be understood as an isolated real number rather a representative or a realization from the uncertain population.

Definition 3.6: (Lipschitz condition) Let $f(x)$ be a real-valued function, $f : \mathbb{R} \to \mathbb{R}$. If for any $x, y \in \mathbb{R}^n$, there exists a positive constant $M > 0$, such that

$$|f(x) - f(y)| < M|x - y| \tag{6}$$

Definition 3.7: (Lipschitz continuity) Let $f : \mathbb{R}^m \to \mathbb{R}^m$

1. for $\forall B \subset \mathbb{R}^m$, B to open set, f is Lipschitz continuous on B if $\exists M > 0$ such

$$\|f(x) - f(y)\| < M\|x - y\|, \ \forall x, y \in B \tag{7}$$

where $\|\cdot\|$ is some metric (for example, Euclidean distance in $\mathbb{R}^m$), such

$$d(f(x), f(y)) < Md(x, y), \ \forall x, y \in B \tag{8}$$

2. for each $z \in \mathbb{R}^m$, f is Lipschitz continuous locally on the open ball B of center z radius $M > 0$ such

$$B_M(z) = \{y \in \mathbb{R}^m \mid d(y,z) < M\} \tag{9}$$

3. if f is Lipschitz continuous on the whole space $\mathbb{R}^m$, then the function is called globally Lipschitz continuous.

Remark 3.8: For continuity requirements, Lipschitz continuous function is stronger than that of the continuous function in Newton calculus but it is weaker than the differentiable function in Newton differentiability sense. In other words, Lipschitz-continuity does not warrant the first -order differentiability everywhere but it does mean nowhere differentiability. Lipschitz-continuity does not guarantee the existence of the first-order derivative everywhere, however, if exists somewhere, the value of the derivative is bounded since

$$\frac{|f(x) - f(y)|}{|x - y|} < M \tag{10}$$

by recalling the definition of the Newton derivative

$$\lim_{y \to x} \frac{f(x) - f(y)}{x - y} = f'(x) \tag{11}$$

Similar to the concept of stochastic process in probability theory, an uncertain process $\{\xi_t, t \geq 0\}$ is a family of uncertainty variables indexed by t and taking values in the state space $\mathbb{S} \subseteq \mathbb{R}$.

Definition 3.9: (Liu 2010, 2011) Let $\{C_t, t \geq 0\}$ be an uncertain process.

1. $C_0 = 0$ and all the trajectories of realizations are Lipschitz-continous;
2. $\{C_t, t \geq 0\}$ has stationary and independent increments;
3. every increment $C_{t+s} - C_s$ is a normal uncertainty variableb with expected value 0 and variance t^2, i.e., the uncertainty distribution of $C_{t+s} - C_s$ is

$$\Psi_{C_{t+s} - C_s}(z) = \left(1 + \exp\left(-\frac{xz}{\sqrt{3}t}\right)\right)^{-1} \tag{12}$$

Then $\{C_t, t \geq 0\}$ is called a canonical process.

Remark 3.10: Comparing to Brownian motion process $\{B_t, t \geq 0\}$ in probability theory, which is continuous almost everywhere and nowhere is differentiable, while Liu's canonical process $\{C_t, t \geq 0\}$ is Lipschitz-continuous and if $\{C_t, t \geq 0\}$ is differentiable somewhere, the derivative is bounded. Therefore $\{C_t, t \geq 0\}$ is smoother than $\{B_t, t \geq 0\}$.

Definition 3.11: (Liu, 2010, 2011) Suppose $\{C_t, t \geq 0\}$ is a canonical process, and f and g are some given functions, then

$$d\xi_t = f(t, \xi_t)dt + g(t, \xi_t)dC_t \tag{13}$$

is called an uncertain differential equation. A solution to the uncertain differential equation is the uncertain process $\{\xi_t, t \geq 0\}$ satisfying it for any $t > 0$.

Remark 3.12: Since dC_t and $d\xi_t$ are only meaningful under the umbrella of uncertain integral, i.e., the an uncertain differential equation is an alternative representation of

$$\xi_t = \xi_0 + \int_0^t f(s,\xi_s)ds + \int_0^t g(s,\xi_s)dC_s \tag{14}$$

Definition 3.13: The geometric canonical process $\{G_t, t \geq 0\}$ satisfies the uncertain differential equation

$$dG_t = \alpha G_t dt + \sigma G_t dC_t \tag{15}$$

has a solution

$$G_t = \exp(\alpha t + \sigma C_t) \tag{16}$$

where α can be called the drift coefficient and $\sigma > 0$ can be called the diffusion coefficient of the geometric canonical process $\{G_t, t \geq 0\}$ due to the roles played respectively.

4. Spatial interpolation and extrapolation via inverse distance approach

Statistically, spatial interpolation and extrapolation modeling is actually a kind of linear regression modeling exercises, say, kriging methodology. Considering the shortage of California PM_{10} data records, we will utilize a weighted linear combination approach, which was first proposed by Shepard (1968). The weights are the inverse distances between the missing value cell to the actual observed PM_{10} value cells. The weight construction is a deterministic method, which is neutral and does not link to any specific measure theory. It is widely used in spatial predictions and map constructions in geostatistics, but is not probability oriented, rather, molecular mechanics stimulated. A unique aspect of geostatistics is the use of regionalized variables which are variables that fall between random variables and completely deterministic variables. The weight of an observed PM_{10} value is inversely proportional to its distance from the estimated value. Let:

C_{ij} — The jth cell on the Site i, (i represent the actual site number), $i = 1,2,\cdots,213$. Note index j points to the cell where no PM_{10} value is recorded, i.e., missing value cell.

x_i — Longitude of site i

y_i — Latitude of site i

d_{ij} — The distance between site i and site j (where missing value j^{th} cell is located)

$$d_{ij} = \sqrt{\left(x_j - x_i\right)^2 + \left(y_j - y_i\right)^2}$$

w_{ij} — Weight, $w_{ij} = \begin{cases} 0 & \text{Cell } (i,j) \text{ has no } PM_{10} \text{ obs.} \\ 1/d_{ij} & \text{Cell } (i,j) \text{ has } PM_{10} \text{ obs.} \end{cases}$

Then the inverse distance formula is,

$$C_{ij} = \frac{w_{ij}}{\sum_{i=1}^{213} w_{ij}} z_{ij} \tag{17}$$

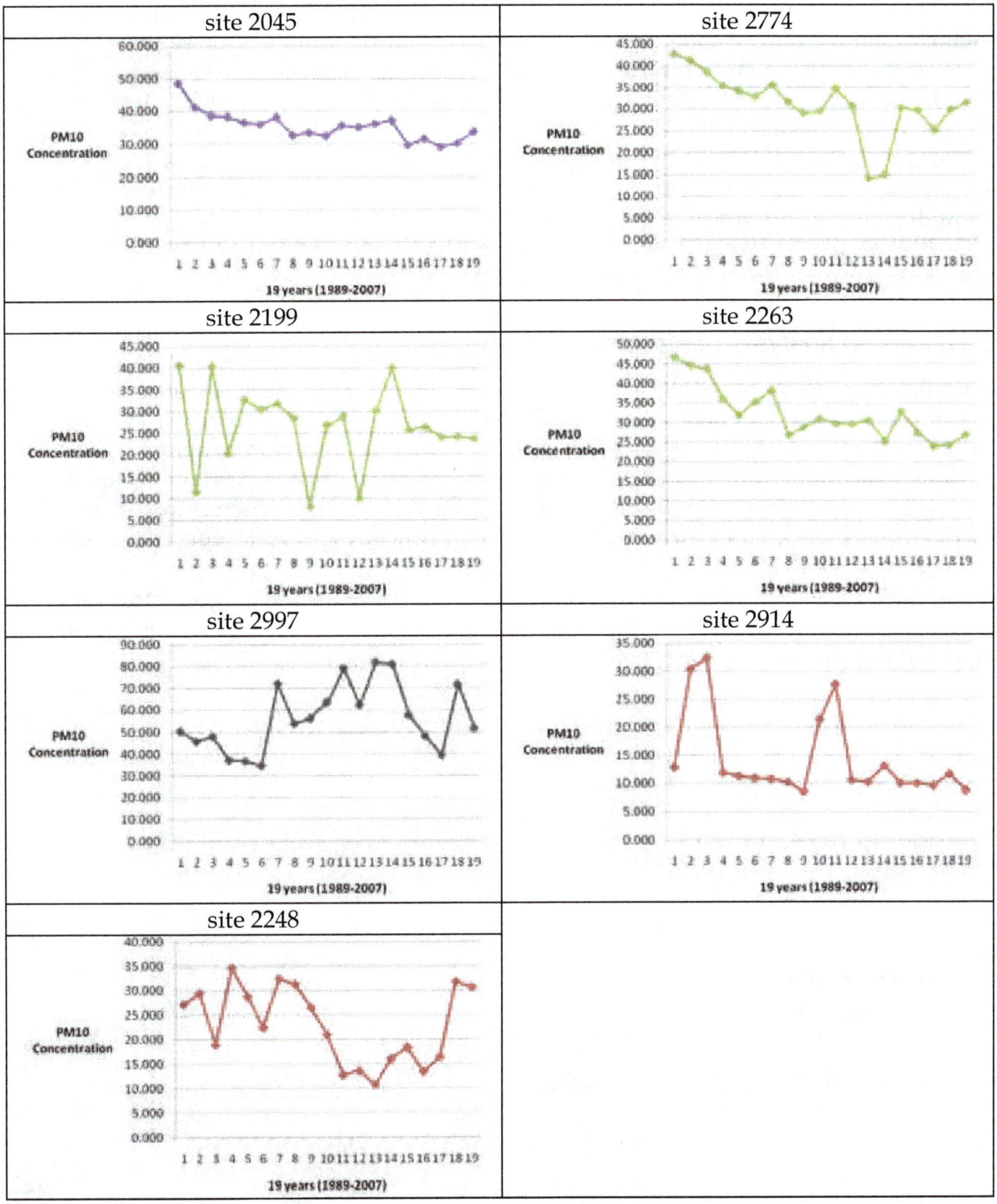

Fig. 4. The 7 sites from the selected 7 counties with completed 19 year observations of PM_{10}

We wrote a VBA Macro to facilitate the interpolations and the extrapolations to "fill" up the 2048 missing value cells in terms of the 1639 cells with PM_{10} values. With the interpolations and the extrapolations, every site has 19 PM_{10} values now. As to whether the inverse distance approach can facilitate highly accurate predictions for each cell without a observed PM_{10} value, we performed a re-interpolation and re- extrapolation scheme (by deleting a true PM_{10} record, then fill it by the remaining records one by one) to evaluate the mean square value for error evaluation, the calculated mean of sum of error squares is 59.885, which is statistically significant (asymptotically).
We plotted sites 2045, 2744, 2199, 2263, 2297, 2914, and 2248 (appeared in Fig. 3) respectively in Fig. 4. By comparing Fig. 3 and 4, it is obvious that only Site 2744 the hazard level changed (moving up to next higher hazard level), while the hazard level of other six sites are unchanged. This may give an justification of the inverse distance approach. Keep in mind, the aim of this article is investigate whether the PM_{10} level is changed over 1989 to 2007 19-year period. The change is not necessarily be accurate but reasonably calculated because of the impreciseness features of PM_{10} complete records.

5. Uncertain analysis of site temporal pattern

Once the interpolations and the extrapolations in terms of the inverse distance approach is completed, a "complete" data set is available, containing 4047 data records of 213 sites over 19 years. The next task is for a given site, how to model the uncertain temporal pattern. It is obvious that the "complete" data set contains impreciseness uncertainty due to the interpolations and the extrapolations. We are unsure that the impreciseness uncertainty is of random uncertainty, so that we still use uncertain measure theory to pursue the temporal uncertainty modelling.
Recall that the **Definition 3.13** in Section 3 facilitates a uncertain geometric canonical process, $\{G_t, t \geq 0\}$. Notice that $G_0 = 0$ may not fit the data reality so that we propose a modified uncertain geometric canonical process, $\{G_t^*, t \geq 0\}$ with $G_0 > 0$:

$$G_t^* = G_0 G_t = G_0 \exp(\alpha t + \sigma C_t) \tag{18}$$

Note that

$$\ln G_t^* = \ln G_0 + \alpha t + \sigma C_t \tag{19}$$

Let $y_t = \ln G_t^*$, $\alpha_0 = \ln G_0$, then we have

$$y_t = \alpha_0 + \alpha t + \sigma C_t, \, t = 1, 2, \cdots, 18 \tag{20}$$

Recall the relevant definitions in Section 3, we have

$$\mathrm{E}[C_t] = 0 \text{, and } V[C_t] = t^2 \tag{21}$$

But note that for $\forall s < t$,

$$\begin{aligned} \mathrm{E}[C_t C_s] &= \mathrm{E}[(C_s + (C_t - C_s))C_s] \\ &= \mathrm{E}[C_s^2] + \mathrm{E}[C_s(C_t - C_s)] \\ &= s^2 + \mathrm{E}[C_s(C_t - C_s)] \end{aligned} \tag{22}$$

Notice that the increment $C_t - C_s$ is independent of C_s, i.e., C_{t-s} is independent of C_s. Therefore,

$$\begin{aligned} E[C_{t-s}C_s] &= \int_{-\infty}^{\infty}\int_{-\infty}^{\infty} z_1 z_2 d\Phi_{C_{t-s},C_s}(z_1,z_2) \\ &= \int_{-\infty}^{\infty}\int_{-\infty}^{\infty} z_1 z_2 d\min\left\{\Psi_{C_{t-s}}(z_1),\Upsilon_{C_s}(z_2)\right\} \\ &= \int_{-\infty}^{\infty}\int_{-\infty}^{\infty} z_1 z_2 d\min\left\{\left(1+\exp\left(-\frac{\pi z_1^2}{\sqrt{3}(t-s)}\right)\right)^{-1},\left(1+\exp\left(-\frac{\pi z_2^2}{\sqrt{3}s}\right)\right)^{-1}\right\} \end{aligned} \tag{23}$$

since if ξ_1 and ξ_2 are independent uncertain variables with uncertainty distributions Ψ_{ξ_1} and Υ_{ξ_2} respectively, then the joint uncertainty distribution of (ξ_1,ξ_2) is $\Phi_{\xi_1,\xi_2}(z_1,z_2)=\min\left\{\Psi_{\xi_1}(z_1),\Upsilon_{\xi_2}(z_2)\right\}$. Hence we obtain the expression of $\sigma_{s,t}$:

$$\sigma_{s,t} = E[C_sC_t] = s^2 + \int_{-\infty}^{\infty}\int_{-\infty}^{\infty} z_1 z_2 d\min\left\{\left(1+\exp\left(-\frac{\pi z_1^2}{\sqrt{3}(t-s)}\right)\right)^{-1},\left(1+\exp\left(-\frac{\pi z_2^2}{\sqrt{3}s}\right)\right)^{-1}\right\} \tag{24}$$

Then the ith "variance-covariance" matrix for uncertain vector $(y_{i1},y_{i2},\cdots,y_{i19})'$

$$\Gamma_i = \sigma^2\left(\sigma_{j,k}^i\right)_{19\times 19} \tag{25}$$

where i is the site index, $j,k=1,2,\cdots,19$ are the entry pair in Γ_i matrix. Hence we have a regression model (Draper and Smith, 1981, Guo et al., 2010, Guo, 2010).

For the ith site, the regression model is

$$y_{it} = \alpha_{i0} + \alpha_i t + \sigma_i C_{i,t} \tag{26}$$

Then in terms of the weighted least square criterion we can define an objective function as

$$J_i(\alpha_{i0},\alpha_i) = \left(\underline{Y}_i - X\underline{\beta}_i\right)'\Gamma^{-1}\left(\underline{Y}_i - X\underline{\beta}_i\right) \tag{27}$$

where

$$\underline{Y}_i = \begin{bmatrix} y_{i1} \\ y_{i2} \\ \vdots \\ y_{i19} \end{bmatrix} \quad \underline{\beta} = \begin{bmatrix} \alpha_{i0} \\ \alpha_i \end{bmatrix} \quad X = \begin{bmatrix} 1 & 1 \\ 1 & 2 \\ \vdots & \vdots \\ 1 & 19 \end{bmatrix} \tag{28}$$

We further notice that

$$\begin{aligned} r_{ij} &= y_{i,j} - y_{i,j-1} \\ &= \alpha_i + \sigma_i\left(C_{ij} - C_{i,j-1}\right) = \alpha_i + \sigma_i \Delta C_{ij} \\ j &= 2,3,\cdots,19 \end{aligned} \tag{29}$$

Then it is reasonable to estimate α_i by

$$\hat{\alpha}_i = \frac{1}{18}\sum_{j=2}^{19} r_{ij} \tag{30}$$

Furthermore, we notice that

$$\hat{\sigma}_i = \sqrt{\frac{1}{18}\sum_{j=2}^{19}\left(r_{ij} - \hat{\alpha}_i\right)^2} \tag{31}$$

Also, we can evaluate

$$\sigma_{j,k}^{i} = j^2 + \int_{-\infty}^{\infty}\int_{-\infty}^{\infty} z_1 z_2 d \min\left\{\left(1+\exp\left(-\frac{\pi z_1}{\sqrt{3(k-j)}}\right)\right)^{-1}, \left(1+\exp\left(-\frac{\pi z_2}{\sqrt{3j}}\right)\right)^{-1}\right\} \tag{32}$$

in terms of numerical integration, Then an estimate for Γ_i matrix is obtained:

$$\hat{\Gamma}_i = \hat{\sigma}_i^2\left(\sigma_{j,k}^{i}\right)_{19\times 19} \tag{33}$$

Finally, we use the approximated objective function

$$\hat{J}_i(\alpha_{i0}, \alpha_i) = \left(\underline{Y}_i - X\underline{\beta}_i\right)' \hat{\Gamma}_i^{-1}\left(\underline{Y}_i - X\underline{\beta}_i\right) \tag{34}$$

to obtain a pair of estimates $(\tilde{\alpha}_{i0}, \tilde{\alpha}_i)$. Repeat this estimation process until all the 213 weighted least square estimate $(\tilde{\alpha}_{i0}, \tilde{\alpha}_i)$ are obtained.

Recall the definition of coefficient α_i so that the sign and the absolute value of α_i indicates the geometric change over the 19 years. Since the estimation procedure of α_i involves all the spatial-temporal information, it is reasonable to have them plotted in a kriging map to reveal the overall changes over 19-year period.

6. Kriging maps and time-change maps based on completed PM_{10} data

Kriging map presentation is vital for a geostatistian's visualization, and maps reveal hidden information or the whole picture. A sample statistic is typically condensing the wide-spread information into a numerical point. While, a kringing map is actually a map statistic (or a statistical map) which contains infinitely many information aggregated from limited "sample" information (i.e., observations). Kriging itself is not specifically probability oriented, it is another weighted linear combination prediction, but requires more mathematical assumptions. In fuzzy geostatistics, the fuzzy kriging scheme has also been developed (Bardossy et al., 1990).

Ordinary kriging (abbreviated as OK) is a linear predictor, see Cressie (1991) and Mase (2011). The formula is

$$Z(s_0) = \sum_{j=1}^{N} \lambda_j Z(s_j) \tag{35}$$

where s_j are spatial locations with observation $Z(s_j)$ available and the coefficients λ_j satisfy the OK linear equation system

$$\begin{cases} \sum_{j=1}^{N} \lambda_j \gamma\left(\varepsilon(s_j) - \varepsilon(s_i)\right) - \psi = \gamma\left(\varepsilon(s_0) - \varepsilon(s_i)\right), \; i = 1,2,\cdots,N \\ \sum_{j=1}^{N} \lambda_j = 1 \end{cases} \tag{36}$$

The OK system is generated under the assumptions of an additive spatial model

$$Z(s) = \mu(s) + \varepsilon(s) \tag{37}$$

where $\mu(s)$ is the basic (expected) spatial trend and $\varepsilon(s)$ is a Gaussian error $N\left(0,\sigma^2(s)\right)$, i.e., Gaussian variable with mean and variance

$$\mathrm{E}\left[\varepsilon(s)\right] = 0, \; V\left[\varepsilon(s)\right] = \sigma^2(s) \tag{38}$$

respectively. Accordingly, the variogram 2γ of the random error function $\varepsilon(\cdot)$ is just defined by

$$2\gamma(h) = \mathrm{E}\left[\left(\varepsilon(s+h) - \varepsilon(h)\right)^2\right] \tag{39}$$

where h is the separate vector between two spatial point $s+h$ and s under the isotropy assumption.

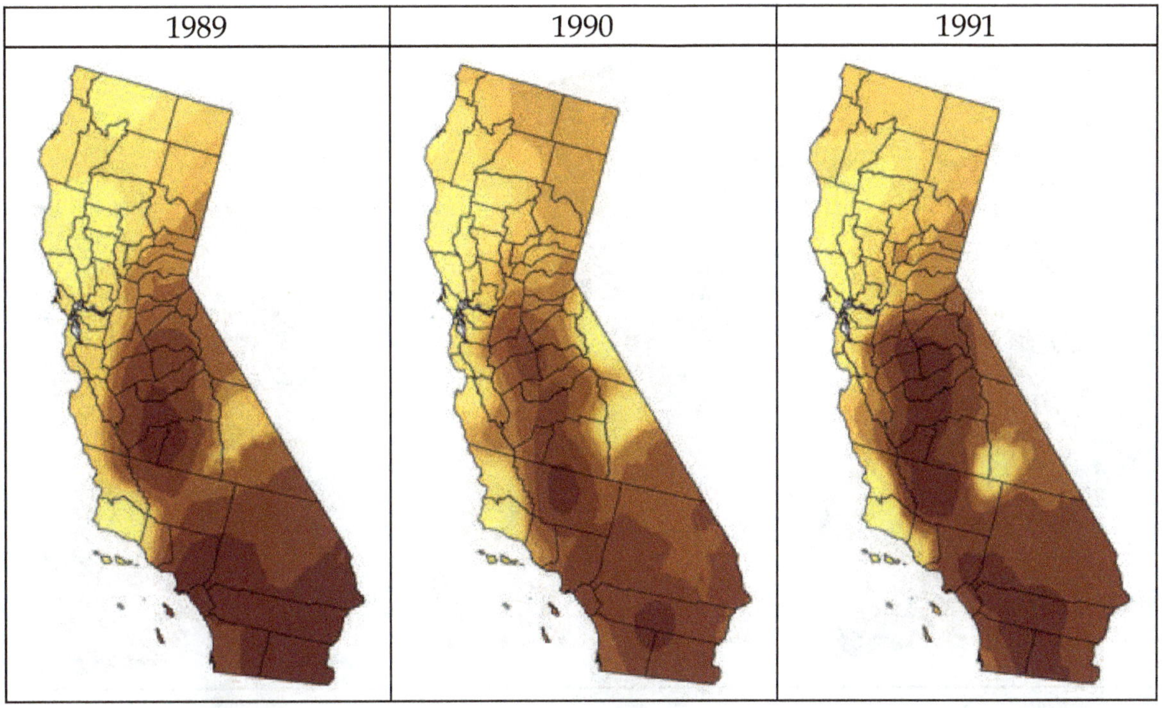

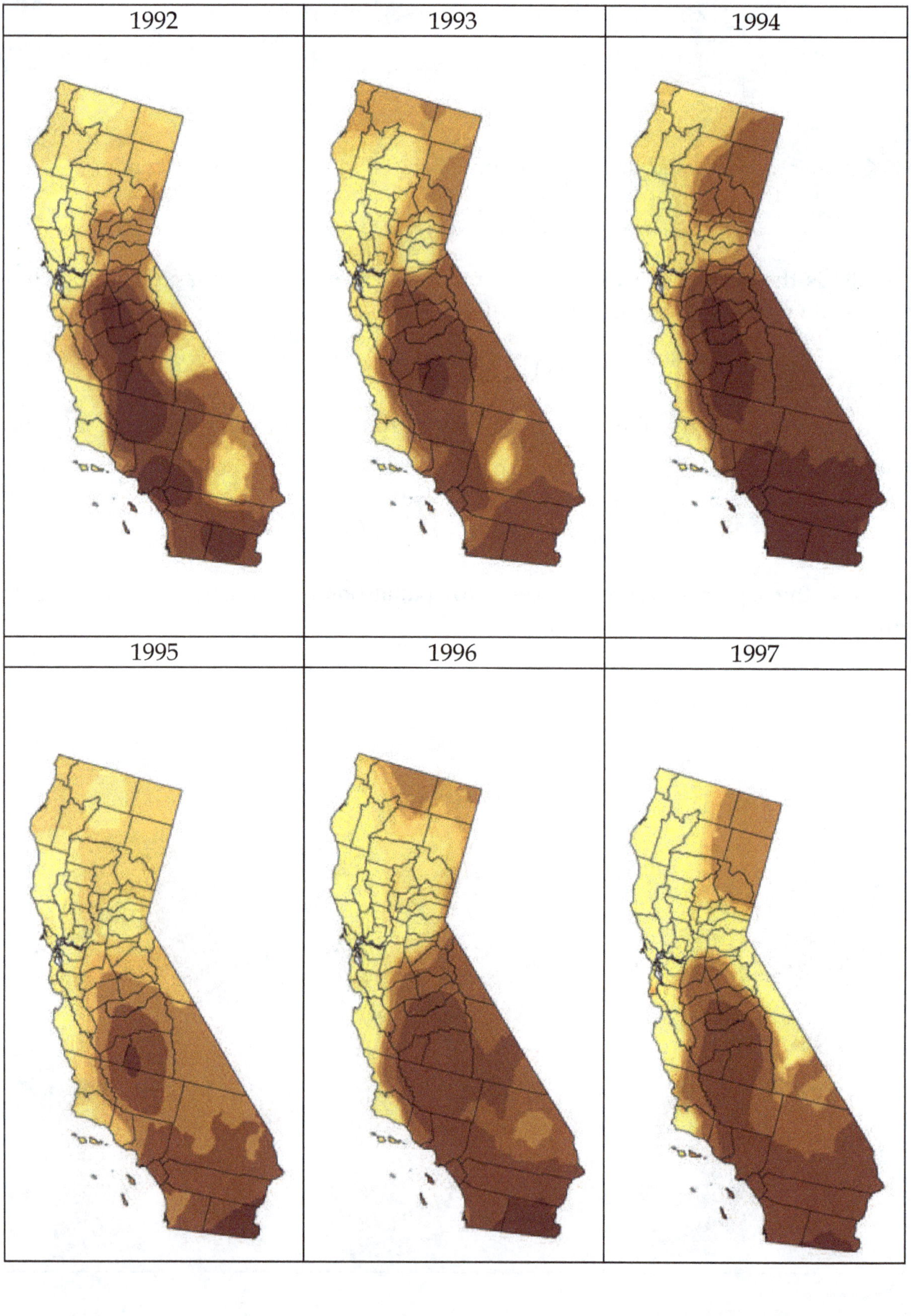
1992
1993
1994
1995
1996
1997

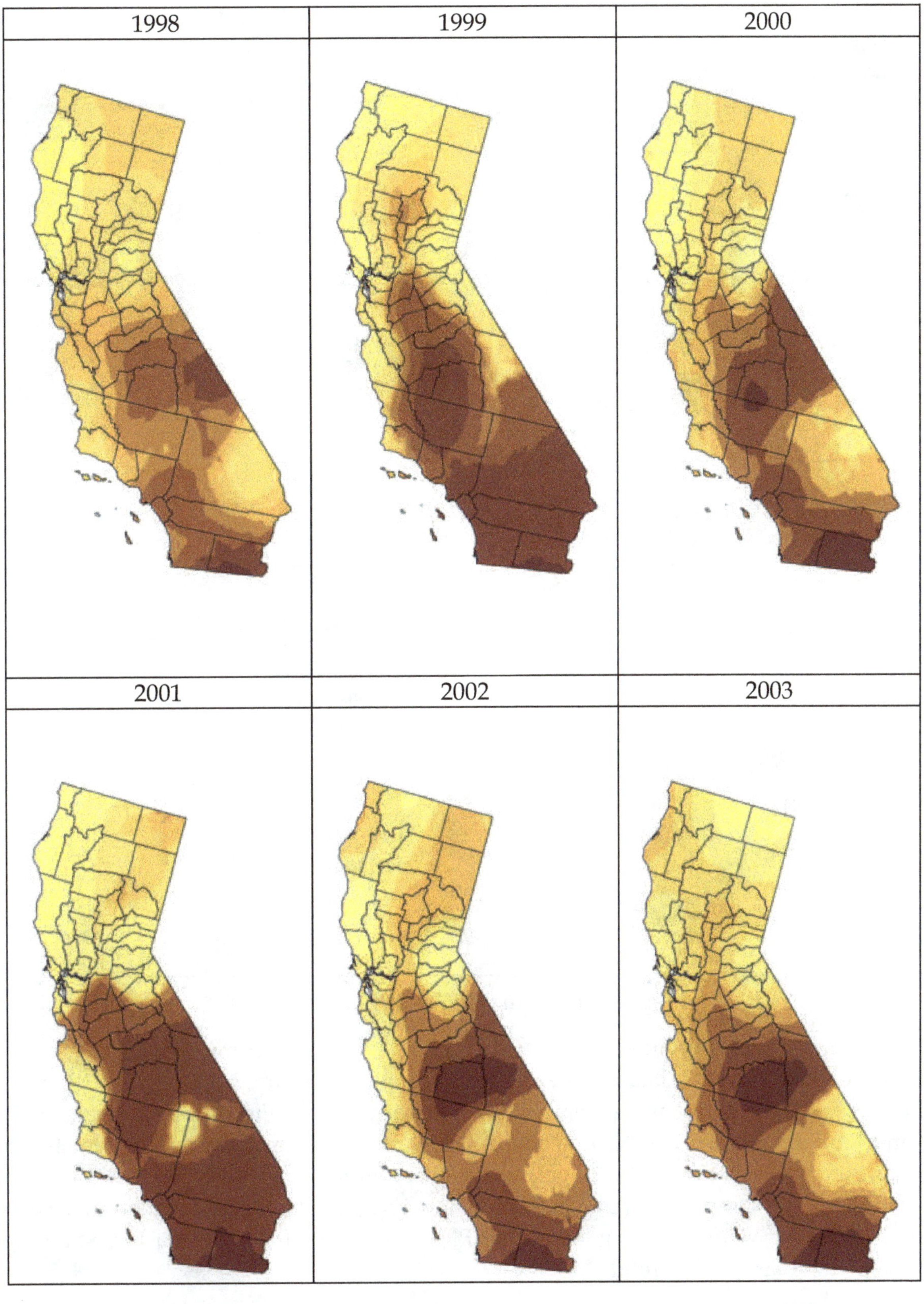
1998
1999
2000
2001
2002
2003

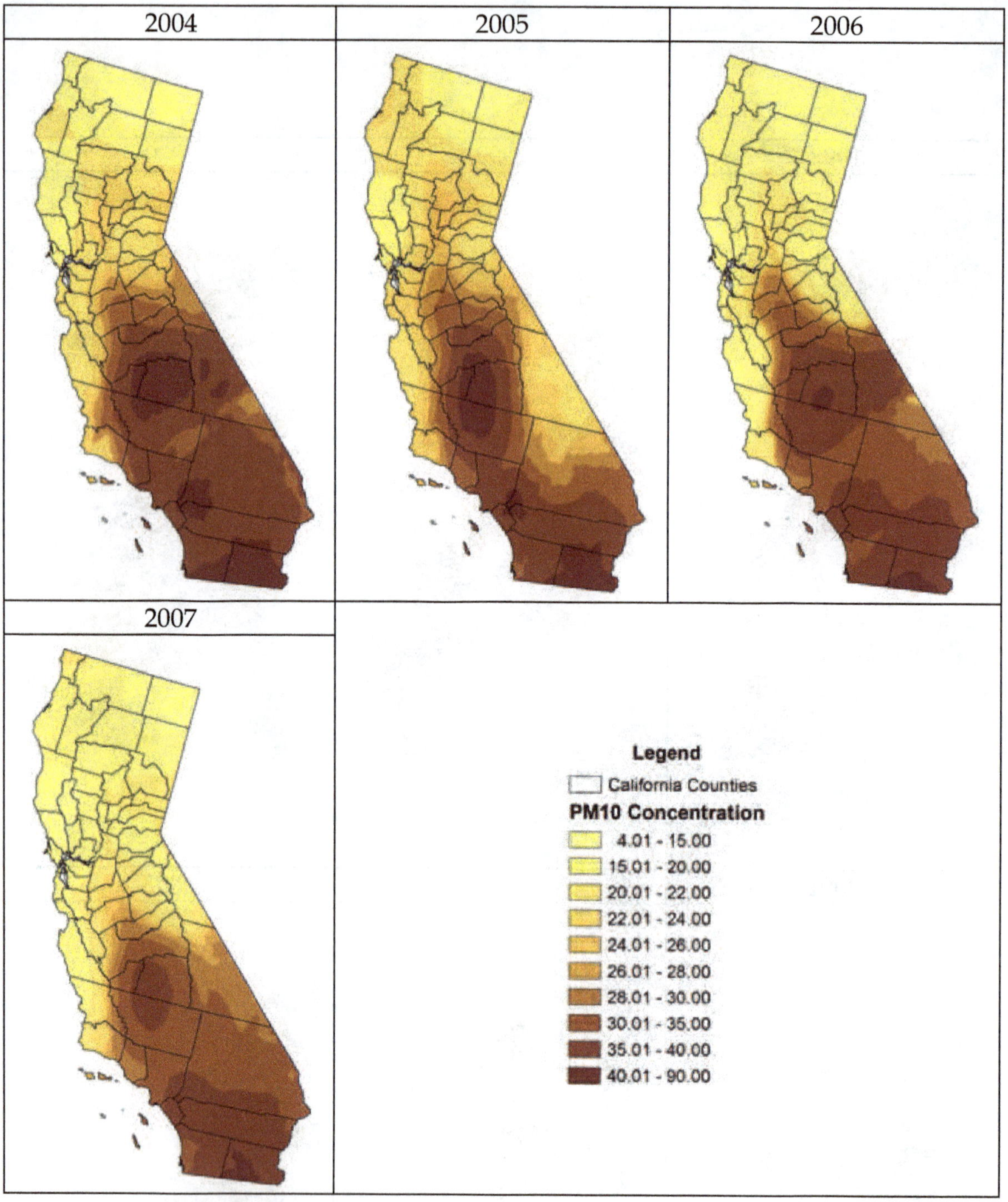

Fig. 5. Kriging Prediction Maps for PM_{10} in California 1989-2007.

The 213 observation sites now have 19-year PM_{10} values, a "complete" data set is now available, containing 4047 data records of 213 sites over 19 years, and then the 19 ordinary kriging pred4iction maps are generated for comparisons. In Fig. 5, all 19 years of PM_{10} concentration in California State are shown. It is very interesting to examine the change in PM_{10} concentrations through the 19 years, based upon the modelled complete 213 site data. In particular, 1998 shows to have an extremely low PM_{10} concentration. Although air quality is varied over the years, but in general, the PM_{10} concentration is decreasing, showing an improvement of air quality trend.

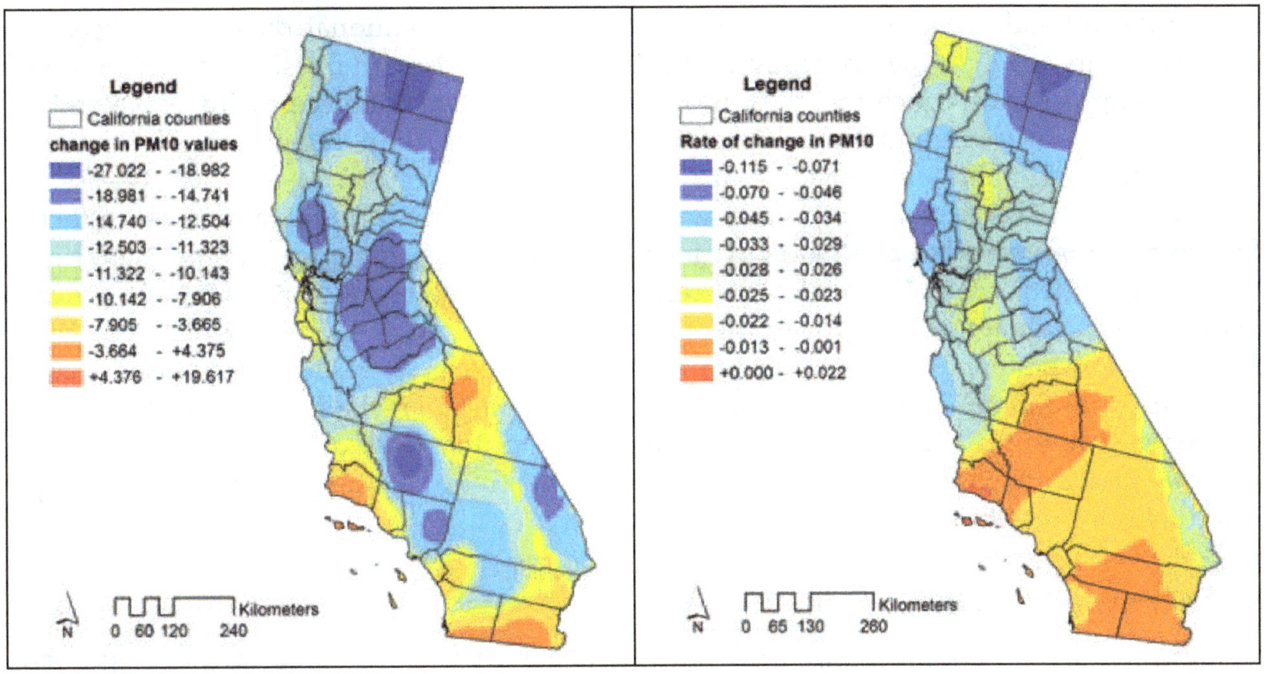

Fig. 6. Changes in PM_{10} values and the rate of change of PM_{10} in California between 1989 and 2007.

As one can clearly see from Fig. 6, that PM_{10} concentration has clearly decreased over the 19 years, and air quality has improved remarkably over the years. The blue and green colours show negative changes, and red shows positive changes or near positive changes. Counties such as San Diego, Inyo, Santa Barbara, Imperial, still show an increase in PM_{10} concentration in the air, and indicate bad air quality. While Kern, Modoc, Siskiyou counties show the most improvement in air quality. The left map in Fig. 6 is PM_{10} record difference between 2007 and 1989 at each location, in total 231 values, and then a difference map is constructed. It is obvious that the difference map only utilizes 1989 and 2007 two-year PM_{10} records, 1990, 1991, ..., 2006 seventeen years' information do not participate the change map construction. The right map in Fig. 6 show completed $\tilde{\alpha}_i$, $i = 1,2,\cdots,213$, the rate of change over 1989 to 2007 19-year period.

Note that the calculations of $\tilde{\alpha}_i$, $i = 1,2,\cdots,213$ involve all nineteen years by temporal regression, the dependent variable y are estimated form the actual PM_{10} observations cross over all the available locations. Therefore, the rate of change parameter α_i at each individual location contains all spatial-temporal information. It is reasonable to say the rate of change parameter $\tilde{\alpha}_i$ is an aggregate statistic for revealing the 19-year changes over 213 locations. $\tilde{\alpha}_i$ kriging map is thus different from 2007-1989 kriging maps. The positive sign of $\tilde{\alpha}_i$ indicates the increasing trend in PM_{10} concentration, while the negative sign f $\tilde{\alpha}_i$ indicates the decreasing trend in PM_{10} concentration. The absolute value of $\tilde{\alpha}_i$ reveals the magnitude of change of PM_{10} concentration. It is worth to report, among 213 locations, 193 locations have negative $\tilde{\alpha}_i$, while the negative $\tilde{\alpha}_i$ locations are 20 (9% approximately).

7. Discussion

Air quality and health is always a central issue to public concern on the quality of life. In this chapter, we examined PM_{10} levels over 19 years, from 1989 to 2007, in the California State.

Facing the difficult task of a lack of "complete" PM_{10} observational data, we utilised the inverse distance weight methodology to "fill" in the locations with missing values. By doing so, the impreciseness uncertainty is introduced, which is not necessarily explained by probability measure foundation. We noted the character of a regionalized variable in geostatistics and therefore engage Liu's (2010, 2011) uncertainty theory to address the impreciseness uncertainty. In this case, we developed a series of uncertain measure theory founded spatial-temporal methodology, including the inverse distance scheme, the kriging scheme, and the geometric canonical process based weighted regression analysis in order to extract the change information from the incomplete 1989-2007 PM_{10} records. The use of the rate of change parameter alpha is a new idea and it is an aggregate change index utilized all spatial-temporal data information available. It is far better than classical change treatments. However, due to the limitations of our ability, we are unable to demonstrate the detailed uncertain measure based spatial analysis model. In the future research, we plan to develop a more solid uncertain spatial prediction methodology.

8. Acknowledgements

I would like to thank the California Air Resources Board for providing the air quality data used in this paper. This study is supported financially by the National Research Foundation of South Africa (Ref. No. IFR2009090800013) and (Ref. No. IFR2011040400096).

9. References

Bardossy, A.; Bogardi, I. & Kelly, E. (1990). Kriging with imprecise (fuzzy) variograms, I: Theory. *Mathematical Geology*, Vol. 22, pp. 63–79.

California Environmental Protection Agency Air Resources Board. (2010). Ambient Air Quality Standards (AAQS) for Particulate Matter. (www.arb.ca.gov)

Cressie, N. (1991). *Statistics for Spatial Data.* Wiley-Interscience, John-Wiley & Sons Inc. New York.

Deng, J. L. (1984). *Grey dynamic modeling and its application in long-term prediction of food productions.* Exploration of Nature, Vol. 3, No. 3, pp. 7-43.

Draper, N. & Smith, H. (1981). *Applied Regression Analysis.* 2nd Edition. John Wiley & Sons, Inc. New York.

Electric, W. (1956). *Statistical Quality Control Handbook.* Western Elctric Corporation, Indianapolis.

Environmental Protection Agency (EPA). 2010. National Ambient Air Quality Standards (NAAQS). U.S. Environmental Protection Agency. (www.epa.gov)

Guo, D.; Guo, R. & Thiart, C. (2007). Predicting Air Pollution Using Fuzzy Membership Grade Kriging. *Journal of Computers, Environment and Urban Systems.* Editors: Andy P Jones and Iain Lake. Elsevier, Vol. 31, No. 1, pp. 33-51. ISSN: 0198-9715

Guo, D.; Guo, R. & Thiart, C. (2007). Credibility Measure Based Membership Grade Kriging. *International Journal of Uncertainty, Fuzziness and Knowledge-Based Systems.* Vol. 15, No. Supp. 2, (April 2007), pp. 53-66. B.D. Liu (Editor). ISSN 0218-4885

Guo, D. (2010). *Contributions to Spatial Uncertainty Modelling in GIS.* Lambert Academic Publishing (online.lap-publishing.com). ISBN 978-3-8433-7388-3

Guo, R.; Cui, Y.H. & Guo, D. (2010). Uncertainty Statistics. (Submitted to Journal of Uncertainty Systems, under review)

Guo, R.; Cui, Y.H. & Guo, D. (2010). Uncertainty Linear Regression Models. (Submitted to Journal of Uncertainty Systems, under review)

Guo, R.; Guo, D. & Thiart, C. (2010). Liu's Uncertainty Normal Distribution. *Proceedings of the First International Conference on Uncertainty Theory*, August 11-19, 2010, Urumchi & Kashi, China, pp 191-207, Editors: Dan A. Ralescu, Jin Peng, and Renkuan Guo. International Consortium for Uncertainty Theory. ISSN 2079-5238

Kolmogorov, A.N. (1950) *Foundations of the Theory of Probability*. Translated by Nathan Morrison. Chelsea, New York.

Liu, B.D. (2007). *Uncertainty Theory: An Introduction to Its Axiomatic Foundations*. 2nd Edition. Springer-Verlag Heidelberg, Berlin.

Liu, B.D. (2010). *Uncertainty Theory: A Branch of Mathematics of Modelling Human Uncertainty*. Springer-Verlag, Berlin.

Liu, B.D. (2011). *Uncertainty Theory*, 4th Edition, 17 February, 2011 drafted version.

Liu, S.F. & Lin, Y. (2006). *Grey Information*. Springer-Verlag, London.

Mase, S. (2011). GeoStatistics and Kriging Predictors, In: *International Encyclopedia of Statistical Science*. Editor: Miodrag Lovric, 1st Edition, 2011, LVIII, pp. 609-612, Springer.

Montgomery, D.C. (2001). *Introduction to Statistical Quality Control*. 4th Edition. John Wiley & Sons, Now York.

Moore, R.E. (1966). *Interval Analysis*. Prentice-Hall, Englewood Cliff, NJ. ISBN 0-13-476853-1

Pawlak, Z. (1982). Rough Sets. *International Journal of Computer and Information Sciences*, Vol. 11, pp. 341-356.

Shepard, D. (1968). A two-dimensional interpolation function for irregularly-spaced data. *Proceedings of the 1968 ACM National Conference*, pp. 517–524.

Utkin, L.V. & Gurov, S.V. (1998). New reliability models on the basis of the theory of imprecise probabilities. *Proceedings of the 5th International Conference on Soft Computing and Information Intelligent Systems*, Vol. 2, pp. 656-659.

Utkin, L.V. & Gurov, S.V. (2000). New Reliability Models Based on Imprecise Probabilities. *Advanced Signal Processing Technology, Soft Computing*. Fuzzy Logic Systems Institute (FLSI) Soft Computing Series - Vol. 1, pp. 110-139, Charles Hsu (editor). Publisher, World Scientific. November 2000. ISBN 9789812792105

Walley, P. (1991). *Statistical Reasoning with Imprecise Probabilities*. London: Chapman and Hall. ISBN 0412286602

Zadeh, L. A. (1965). Fuzzy sets. *Information and Control*, Vol. 8, pp. 338-353.

Zadeh, L. A. (1978). Fuzzy sets as a basis for a theory of possibility. *Fuzzy Sets and Systems*, Vol. 1, pp. 3-28.

6

New Approaches for Urban and Regional Air Pollution Modelling and Management

Salvador Enrique Puliafito, David Allende, Rafael Fernández, Fernando Castro and Pablo Cremades
Grupo de Estudios Atmosféricos y Ambientales (GEAA)
Universidad Tecnológica Nacional – Facultad Regional Mendoza
Consejo Nacional de Investigaciones Científicas y Técnicas (CONICET)
Argentina

1. Introduction

Air pollution is a complex problem that plays a key role in human well-being, environment and climate change. Since cities are, by nature, concentrations of humans, materials and activities, air pollution is clearly a typical phenomenon associated with urban centres and industrialized regions (Fenger, 1999; de Leeuw et al., 2001). Since approximately half the population of the world lives in medium to large cities, it is essential to evaluate the air quality levels of the atmosphere in order to assess the possible health impact from exposition to pollutants (World Health Organization [WHO], 2002; Brunekreef & Holgate, 2002). Additionally, air pollution is not only a human health problem: the effects of pollution in ecosystems and materials are well identified and documented (Fowler et al., 2009); economic costs can also be associated with poor air quality, and with political/governmental measures taken in order to prevent or reduce pollution (Muller & Mendelsohn (2007)).

The simplest technique for evaluating patterns of local-scale urban air pollution concentration involves the interpolation of ambient concentrations from existing monitoring networks (Ballesta et al., 2008; Ferretti et al., 2008). However, the measured data from these stations are not necessarily representative of areas beyond their immediate vicinity, since concentrations of pollutants in urban areas may greatly vary on spatial scales that range from tens to hundreds of metres. At the same time, the temporal behaviour of primary and secondary pollutants changes considerably between day and night due to solar radiation, so that daily average measurements become unsatisfactory in determining or explaining high pollution episodes.

Air Quality Models (AQMs) are mathematical tools that simulate the physical and chemical processes that involve air pollutant dispersion and reaction in the atmosphere. Furthermore, they improve the limitations of monitoring networks by providing prediction of the temporal and spatial distribution of actual pollution levels. Modelling studies, in combination with air quality monitoring, are then essential and complementary tools for long and short term air pollution control strategies. A well calibrated model is a unique tool that allows the representation of the atmospheric dynamics and chemistry. Thus, AQMs have become a valid instrument for environmental managers in many activities, such as *a)* setting emission control regulations, *b)* testing the compliance of actual pollution levels, *c)* predicting the impact of new facilities on human health, *d)* selecting the best location for monitoring stations, and *e)* assessing the impact of different emissions scenarios on Global Climate Change.

This Chapter presents an overview of several techniques and available codes applied to urban modelling, plus describing three study cases from different approaches that we have selected according to the magnitude of the simulated problem.

2. Air Quality Models (AQMs)

The major elements of an AQM are depicted schematically in Fig. 1.

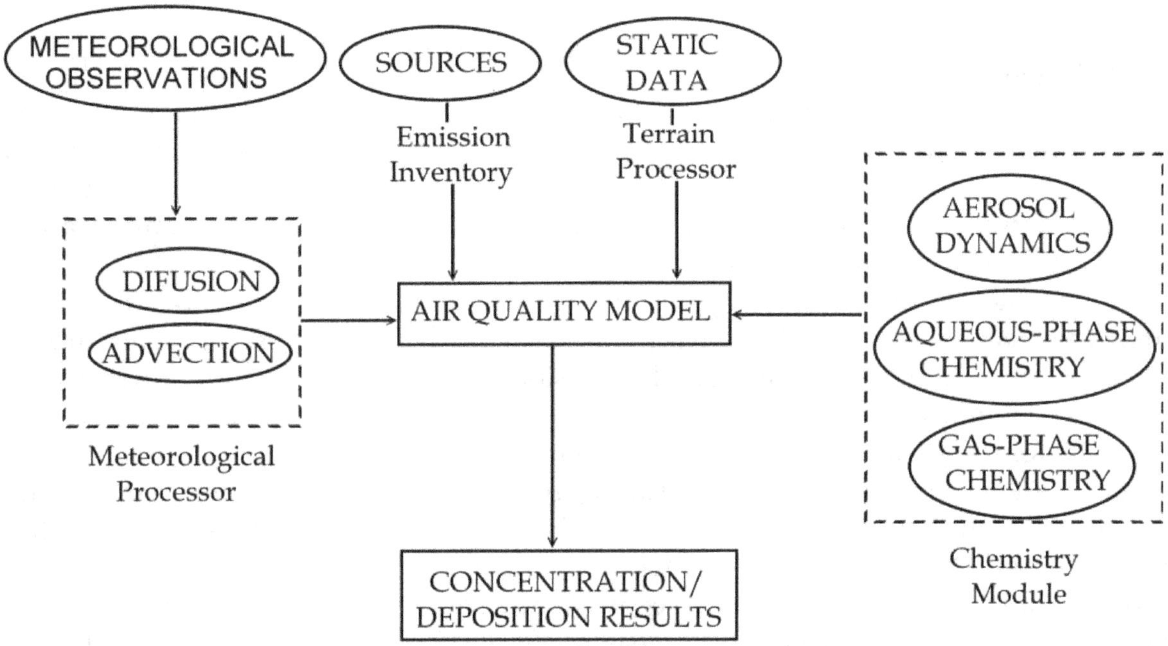

Fig. 1. Main components of an Air Quality Model and their interactions.

Basically, an air quality study implies the estimation of some pollutants concentration in a region of interest, during a finite period of time. The characteristics of each specific problem will define the physical and chemical processes involved, and consequently, the best model to use. The main criteria for choosing adequate software are: the dimension of the area under study, the number of pollution sources, the chemical species involved and the time scale of the episode.

2.1 Air quality and spatial scales

In a geophysical context, it is possible to define several superimposed scales for the air pollution phenomena, extending from local to global scales with several inputs and feedbacks at every level. A pollutant released from traffic or industries in a city may have a significant impact on the city itself or on a farther region, according to the properties and atmospheric lifetime of the pollutant. Consequently, local scale problems, such as the evaluation of the maximum ground level impact of primary pollutants released from traffic or industrial sources, cannot be treated similarly to long range problems due to the influence of regional scale processes, such as chemical transformation or deposition, which need not to be taken into account on a smaller scale. Therefore, following Zannetti (1990), five types of scales can be distinguished, depending on the distance from the emitting source: *a)* near-field phenomena (< 1 km from the source); *b)* short range transport (< 10 km from the source); *c)* intermediate range transport (between 10 and 100 km from the source); *d)* long range transport (> 100 km from the source); and *e)* global effects.

Recently, Air Quality Models have been extended to consider the interactions between processes at different scales, by including nesting capabilities to work with different grid resolutions simultaneously, or by integrating several models designed for a specific problem. Examples of the first approach are CMAQ (Binkowski & Roselle, 2003) and WRF/Chem (Grell et al., 2005) models. The BRUTAL model (Oxley et al., 2009) and the modelling system proposed by Jiménez-Guerrero et al. (2008) are examples of the second approach.

2.2 Emission inventories

Emission inventories have long been fundamental tools for air quality management. Emission estimates are important for developing emission control strategies, determining applicability to permit and control programs, ascertaining the effects of sources and appropriate mitigation strategies; and a number of other related applications by an array of users, including federal, state, and local agencies, consultants and industries.

Emission inventories are also a key input for AQMs and therefore must match their requirements, namely the spatial and temporal distribution of emissions in accordance with the model setup. Spatial and temporal emissions distribution of individual chemical compounds may not be available in emission inventories compiled at national, regional or local levels for regulatory purposes. Consequently, besides emission estimation itself, emission inventory preparation is a critical stage in air quality modelling. Many authors have clearly shown that emission input to AQMs is one of the main sources of uncertainty (Russell & Dennis, 2000; Hanna et al., 2001).

Data from source-specific emission tests or continuous emission monitors are usually preferred for estimating pollutant releases because those data provide the best representation of emissions from tested sources. However, test data from individual sources are not always available, and may not even reflect the variability of actual emissions over time. Thus, emission factors are frequently the best or only method available for estimating emissions, in spite of their limitations. The general methodology to estimate emissions from emission factors has the form:

$$E = A \times EF \times \left(1 - \frac{ER}{100}\right) \tag{1}$$

in which E represents the emissions (mass of pollutant), A is the activity rate in hours/year, EF is the emissions factor and ER defines the overall emission reduction efficiency, %. The emission factors are usually expressed as the weight of pollutant divided by a unit weight, volume, distance, or activity duration that releases the pollutant (e.g., kilograms of particulate emitted per megagram of coal burned). Such factors facilitate estimation of emissions from various sources of air pollution. In most cases, these factors are simply averages of all acceptable quality data available, and are generally assumed to be representative of long-term averages for all facilities in the source category (i.e., a population average).

The estimations of primary pollutants from numerous types of sources are generally performed by following well established methodologies, such as the Intergovernmental Panel on Climate Change (IPCC) Guidelines for National Greenhouse Inventories (IPCC, 2006), the EMEP Air Pollutant Emission Inventory Guidebook of the European Environmental Agency (EMEP/EEA, 2009), or the U.S. Environmental Protection Agency (U.S. EPA) AP-42 Compilation of Air Pollutant Emission Factors (U.S. EPA, 2010).

Even when the general methodology of estimation is similar, emissions from a specific type of sources have their own characteristics and must be treated by AQMs in a different way. Emissions from point sources are estimated for individual sources. Besides the amount of pollutants emitted (usually in g/s) and a precise geographic location, a few physical parameters are used in AQMs to characterise the emission, such as stack height and diameter, emission escape velocity and exit temperature.

Area (e.g., due to residential heating) or mobile sources are not treated individually and their emissions can be estimated by following one of two major approaches, namely top-down and bottom-up. The first one uses aggregated activity data and general *EF* to calculate the amount of emissions from all sources, often giving little spatial and temporal detail. Bottom-up approaches use detailed activity and source-specific data to calculate the emissions with high temporal and spatial resolution. Top-down approach is usually chosen because it needs a reduced amount of input parameters compared to bottom-up methods, therefore being more cost-efficient and easier to implement.

Emissions from area and mobiles sources are incorporated into an AQM as a regular grid with a resolution that depends on the modelling setup and the scope of the study. In order to spatially distribute emissions obtained through a top-down method, a series of simplified approaches of spatial disaggregation based on surrogate data (e.g. population density, principal road network density) and Geographical Information System (GIS) tools are generally used (Puliafito et al, 2003; Tuia et al, 2007). Fig. 2 shows a scheme for the spatial distribution of traffic emissions obtained from different approaches.

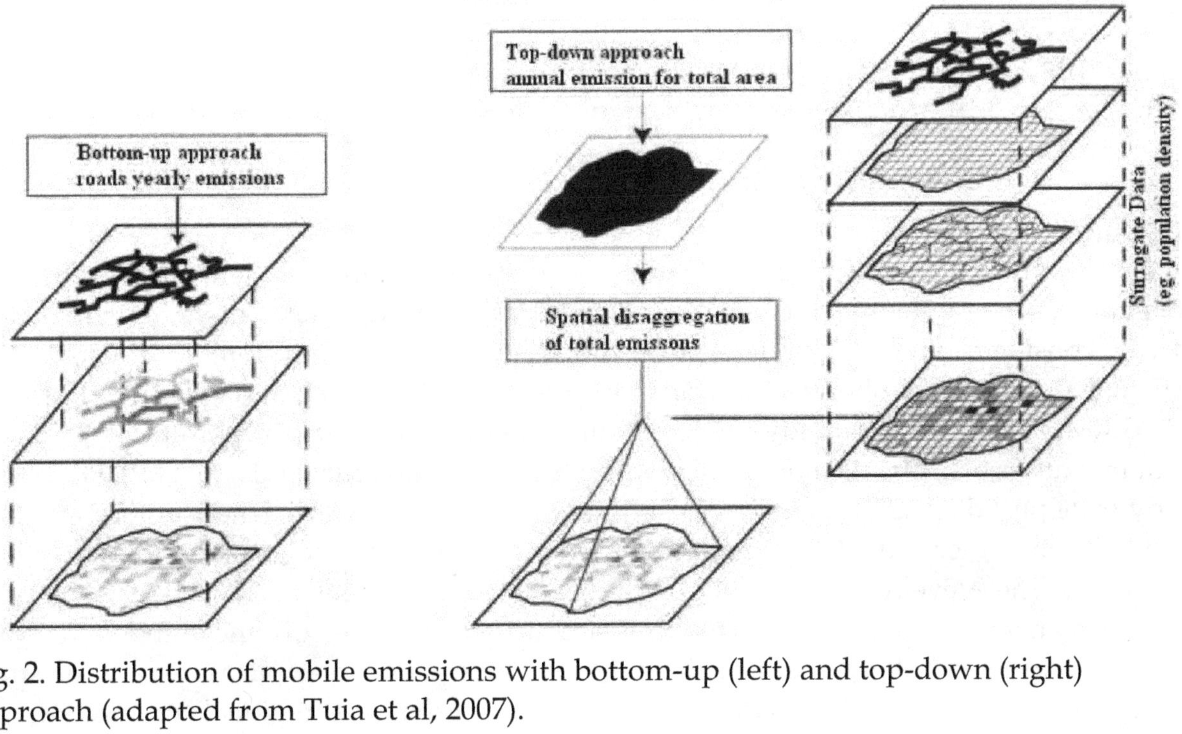

Fig. 2. Distribution of mobile emissions with bottom-up (left) and top-down (right) approach (adapted from Tuia et al, 2007).

Due to the intrinsic complexity of some sources, emission models are generally used in order to obtain total emissions and to evaluate different emission scenarios. For example, mobile emission can be estimated by using several models like COPERT (Ntziachristos et al, 2009), MOVES (U.S. EPA, 2009a) and IVE (Davis et al., 2005).

Temporal distribution of emission is another issue that needs to be addressed in order to generate acceptable inputs to AQMs. Time-resolved emissions are needed to properly simulate air quality through deterministic Eulerian models like WRF/Chem. Generally, emissions are expressed in annual or daily bases and temporal profiles are used to break them down to the temporal resolution expected by AQMs (usually 1 hour). These profiles are obtained by evaluating the temporal activity patterns from each type of source.

Emission inventories are built and reported for a variety of compounds or chemical classes such as CO, NO_X, NMVOC, PM_{10}, and SO_2. However, photochemical mechanisms, included in some AQMs like WRF/Chem, contain a simplified set of equations that use representative "model species" to simulate atmospheric chemistry (Dodge, 2000). Consequently, when chemical transformations are going to be considered in the AQMs, "inventory species" must be converted to "model species" through the use of speciation profiles. Among the most popular chemical mechanisms are CBIV (Gery et al., 1989), SAPRC-99 (Carter et al., 2000) and RADM-2 (Stockwell et al., 1990). Most of the speciation profiles are derived from North-American references (SPECIATE database, U.S. EPA, 2009b).

As described above, compiling and adapting emission inventories to meet the requirements of AQMs are demanding tasks. In order to simplify this process, a series of emission processing systems have been developed. These are capable of processing national or local inventories and global emission databases, as GFED (van der Werf et al., 2010), GEIA (GEIA/ACCENT, 2005) or RETRO (Schultz et al., 2008). For biogenic or mobile sources, some processing systems can be used to directly estimate the emissions. SMOKE is one of them and has been designed to create gridded, speciated and hourly disaggregated emissions inputs for a variety of AQMs in the United States. Adaptations of this model to non U.S. cases have been made for Spain (Borge et al., 2008) and for Europe (Bieser et al., 2010). Another example of emission pre-processor is the *Prep_chem_sources* code (Freitas et al., 2010) designed to work mainly with WRF/Chem, but adaptable to other AQMs.

2.3 Selection of the proper AQM

The complexity of the study case (meteorological effects, terrain interactions) and the scale and significance of potential effects (sensitivity of the receiving environment, human health) must be considered in order to determine which modelling tool is more appropriate for a particular problem. For short-range transport, models with a Gaussian approach like ISC3 and AERMOD, are used to study the interaction of emissions and concentrations of criteria pollutants in urban streets and surrounding urban neighbourhoods. CALPUFF, a non steady-state puff dispersion model, is best fitted for urban scales, with meteorological inputs provided by either on-site measurements and/or a separate meteorological model simulation. The development of the new WRF/Chem model (WRF with Chemistry) constitutes an adaptable and useful tool intended to perform the "on-line" modelling of the chemistry and meteorology over a wide range of scales. Its main applications are the study of secondary pollutants and formation of aerosols in an urban and regional context.

2.3.1 Short-range transport models

ISC3 (Industrial Source Complex; U.S. EPA, 1995) is a steady-state Gaussian plume model appropriate for assessing gas and particle pollutant concentrations from a wide variety of sources associated with an industrial zone. The model includes point, area, line, and volume sources; it can account for settling and dry deposition of particles, building downwash, and includes a limited terrain adjustment. ISC3 operates in both long-term and short-term modes (ISCLT3, ISCST3). The U.S. EPA, recently changed the model status to "alternative model" for regulatory applications, and it suggests using the AERMOD model instead.

AERMOD (American Meteorology Society/EPA Regulatory Model; Cimorelli et al., 2003) is a steady-state plume model. The concentration distribution function differs for the Stable Boundary Layer (SBL) and the Convective Boundary Layer (CBL): a vertical and horizontal Gaussian concentration distribution is considered into the SBL, while a vertical bi-Gaussian probability density function is included for the CBL. The model includes a boundary-layer similarity theory to define turbulence, including a variable altitude treatment for emissions released at different heights. Dispersion coefficients are treated as a continuum rather than as a discrete set of stability classes, including a non-Gaussian representation for unstable conditions, like those observed close to a stack under convective conditions. Variation of turbulence with height allows a better treatment of dispersion for different release heights. The model also uses a simple approach to incorporate terrain interactions. The modelling system consists of one main program (AERMOD) and two pre-processors (AERMET and AERMAP). The main purpose of AERMET is to incorporate all meteorological observations (from surface and upper air stations) and to calculate boundary layer parameters to be used by AERMOD. AERMAP is a terrain pre-processor used to create receptor grids and to generate gridded terrain data from several Digital Elevation Models datasets.

In general, ISC3 and AERMOD do not require significant computational resources, they are easy to use and have simple meteorological requirements. ISC3 is principally designed for calculating impacts on flat terrain regions. The more advanced AERMOD is designed for use with a complex terrain. In any case, Gaussian-plume models are best used for near-field applications where the steady-state meteorology assumption is most likely to apply.

2.3.2 CALPUFF

The CALPUFF modelling system (Scire et al., 2000a) is a multi-layer, multi-species non-steady-state Gaussian puff dispersion model capable of simulating time and space effects of varying meteorological conditions on pollutant transport. The model accounts for a variety of effects such as spatial variability of meteorological conditions, causality effects, dry deposition and dispersion over different types of land surfaces, plume fumigation, low wind-speed dispersion, primary pollutant transformation and wet removal. Recently, an improved chemical module has been included but it is still under evaluation. CALPUFF has various algorithms to include the use of turbulence-based dispersion coefficients derived from a similarity theory or observations. The system has three components: CALMET, a meteorological model that generates three dimensional hourly gridded fields of winds and temperatures, and two dimensional parameters such as mixing height, surface characteristics and dispersion parameters. CALPUFF is the transport and dispersion model that computes the advection of "puffs" containing the material emitted from modelled sources, using the fields generated from the meteorological model. CALPOST is the post-processing tool designed to summarize the results of the simulations (hourly concentrations or deposition fluxes) of CALPUFF.

The model is usually recommended by the U. S. EPA to simulate the effects of pollutant dispersion in long range transport, typically between 50 and 200 km, but contains algorithms which apply to much shorter distances (U.S. EPA, 2008). Advanced models like CALPUFF are more sophisticated than Gaussian models and are aimed at producing more realistic results. However, the potential benefits of using this advanced tool must be compared to the costs of producing more detailed meteorology, complete source inventories and terrain datasets, configuring and running the code.

2.3.4 WRF/Chem

The modelling of the chemical processes in AQMs is usually performed after and independently of the meteorological modelling. This is often called "off-line" integration of the chemical mechanisms, and computes the reactivity differential equations into an external grid containing the information of mass transport and meteorological fields. The development of the novel model WRF/Chem (WRF with Chemistry) allows performing a coupled modelling of the chemistry and the meteorology into a unique coordinate system (Grell et al., 2005; Wang et al., 2010). In this way, a wide range of chemical and physical parameterizations can be used without the need of interpolating them into different spatial and temporal domains. Of all WRF dynamic cores, the mass coordinate version, named Advanced Research WRF (ARW; Wang et al., 2009) possesses the ideal properties to perform the "on-line" modelling of the atmospheric chemistry. The WRF/Chem model has a modular structure which allows considering a variety of coupled physical-chemistry processes such as: transport, deposition, emission, homogeneous and heterogeneous chemistry, aerosol interaction, photolysis, long-wave and short-wave radiative transfer, etc. (Peckham et al., 2010). WRF/Chem application in different fields can explain past episodes, evaluate the potential effects of emission reduction strategies and perform air quality forecasting, always considering the interaction between high resolution chemistry and 3-D meteorology.

Besides the meteorological physical parameterizations (see Section 2.4.3), WRF/Chem model offers a wide range of chemical parameterizations which must be selected in order to perform a forecasting simulation. These include: *i)* the chemical mechanism and the chemical speciation; *ii)* the type of photolysis rate constant computation; *iii)* the aerosol optical properties, *iv)* the biogenic emissions; *v)* the inclusion of biomass and/or plume rise options; *vi)* the dust, gas and aerosol initial and boundary conditions (Grell et al., 2005; Peckham et al., 2010).

One of the most user-dependent features of WRF/Chem is the inclusion of pollutant emission inventories: because of its wide applicability, a time-dependent and high-resolution local inventory must be constructed and adapted by the user to fit the area of interest. By default, the WRF/Chem global configuration includes the implementation of the RETRO (Schultz, 2008) and EDGAR (Oliver & Berdowsky, 2001) emission inventories, which is performed by using the *Prep_chem_sources* application (see Section 2.2). This methodology is limited to very low resolution databases ($\sim 1^{\circ} \times 1^{\circ} \approx 100$ km × 100 km) with monthly average values of pollutants. Then, the spatial and temporal resolution must be improved if a local or regional air quality forecasting system is to be implemented. Only in the case of spatial domains located inside the U. S., the National Emissions Inventory (NEI) can be used to include high resolution pollutant emissions by means of the *emiss_v3* application. For all other countries, a detailed emission inventory must be developed as described in section 2.2.

2.4 Meteorological data

Meteorological data is a critical input for AQMs, as it is necessary to obtain accurate description of winds, turbulence fields and radiation in order to correctly describe transport, dispersion, deposition and chemical reactions of a released pollutant (Seaman, 2003; Gilliland et al., 2008; Schürmann et al., 2009; Demuzere et al., 2009; Pearce et al., 2011). The meteorological data requirements for local scale models (i.e., steady-state Gaussian, Puff-models) and more complex models vary considerably.

2.4.1 Meteorological data for Gaussian models

Steady-state Gaussian plume models, like ISCST3 (U.S. EPA, 1995) and AERMOD (Cimorelli et al., 2003), need data only from a single station, since they assume that meteorological conditions do not vary throughout the domain up to the top of the boundary layer. Although both codes can be used with screening options generally available from the model website, this is only suggested to generate first conservative guess estimates of ground level concentrations. The most appropriate practice is the use of existing data sets available in the study area providing enough accuracy to meet the local environmental authority criteria. National and local weather services and private consultants usually produce site-specific meteorological data sets and provide advice on its applicability for specific modelling studies. Also, the existing data are supplemented by new measurements from meteorological automatic stations collected on the site of interest. This approach is fully covered in U.S. EPA (2000).

2.4.2 Meteorological data for advanced models

More advanced models (both puff and grid models), allow meteorological conditions to vary across the modelling domain and up through the atmosphere, requiring thus more complex meteorological data. An example is the CALPUFF modelling system (Scire et al., 2000a). Since there are no meteorological measurements at every point of the domain, these models use pre-processed data for analysis. The CALMET meteorological model (Scire et al., 2000b) is a diagnostic model developed as a component of the CALPUFF modelling system for its use in air quality applications. CALMET in its basic form is designed to produce hourly fields of three-dimensional winds and various micro-meteorological variables based on the input of routinely available surface and upper air meteorological observations.

2.4.3 Mesoscale meteorological models: WRF

All prognostic models contain realistic dynamic and physical formulations, and can produce the most realistic meteorological simulations for regions where data are sparse or non existent. If local meteorological data is unavailable, the use of a prognostic modelling system could be a sensible option as part of a regulatory assessment.

The Weather Research and Forecasting Model, WRF (Skamarock et al., 2008) is now the most commonly used prognostic meteorological model. The WRF model is a state-of-the-art model developed and maintained by NCAR, NOAA/NCEP and other research centers (Michalakes et al., 2005). It is an open source code originally conceived as a next-generation mesoscale model oriented to cover both atmospheric research studies and operational forecasting. The model can be run in a nested way with the outer domain on a regional scale, covering distances usually in the order of 500–1,000 km. All domains are initialised using analyses from global models, like the NCEP Global Forecasting System (NCEP-GFS), which describe the three-dimensional fields of temperature, wind speed and direction, and moisture in global 1.0 x 1.0 degree grids prepared operationally every six hours for the whole world.

The WRF model can be easily configured to select and define the size and resolution of the computational domain. Moreover, it allows introducing "nested" domains, where smaller or child domains with higher resolution are inserted into bigger or parent domains. The resolution ratio between child and parent domains is usually a factor of three, which is a compromise between improving the spatial resolution and increasing the computing time. In order to configure a WRF simulation, it is necessary to define the modeled domain and its properties, as well as to determine the initial and boundary conditions for the selected case. This is usually performed by the WRF Pre-processing System (WPS). This tool allows setting up the model by including: *a)* static databases like terrain elevation (with different spatial resolutions), Land Use Land Cover (LULC) fields and surface levels; and *b)* initial values for boundary meteorological conditions (with temporal resolution), the number of vertical levels, among others.

Before the forward forecasting is performed, a number of physical and dynamic parameterization must be selected. This includes a wide range of options from simple and efficient modules (used for operational purposes) to sophisticated and computationally expensive routines (for atmospheric research purposes). The physical parameterizations are classified in: *i)* micro-physic schemes; *ii)* long-wave and short-wave radiative modules; *iii)* surface layer models; *iv)* land surface models; *v)* Planetary Boundary Layer (PBL) schemes; and *vi)* cumulus parameterizations; among others.

2.4.4 CALWRF

When modelling a large domain with complex geography or when there is a lack of meteorological data, it is possible to use the outputs of any mesoscale meteorological model as input to the AQM. That is the case of CALWRF, an external tool developed to perform a CALPUFF simulation including WRF output data. CALWRF is basically a data filter and format converter code, which produces an intermediate three dimensional data file which is used by CALMET as a "first-guess" meteorological field. Then, CALMET merges these initial and boundary conditions with the terrain elevation data and land-use, and CALPUFF is run in its usual diagnostic mode. This approach not only increases the horizontal resolution of the meteorological fields, but also reduces the computational time that running the WRF/Chem model with a greater resolution would require.

2.5 Geophysical data

Terrain features around a pollutant source can significantly affect the pattern of dispersion. Steady-state Gaussian models like ISC3 contain limited algorithms to include terrain effects. Advanced models like CALPUFF contain more sophisticated procedures for modelling the effects of terrain, with a correspondingly greater effort required by the user to specify the static data. Since terrain data will be required for every receptor on the grid, there are several pre-processing tools that extract and format the Digital Elevation Model (DEM) data. The most common global data sets are: the United States Geological Service (USGS) GTOPO30 with a horizontal grid spacing of 30 arc-seconds (approximately 1km); USGS SRTM30, with the same horizontal grid spacing, but covering the globe only from 60° N latitude to 56° S latitude, with a seamless and uniform representation; and SRTM3 data with a horizontal grid spacing of 3 arc-seconds (about 90 m). Advanced models also need surface parameters, generally as gridded fields, to compute properly the dispersion of pollutants. These include surface roughness length, albedo, Bowen ratio, soil heat flux parameter, vegetation leaf area index and anthropogenic heat flux. Land Use and Land Cover (LULC) data are also available from the USGS, at the 1:250,000 scale, or in some cases at the 1:100,000 scale. The USGS Global Land Cover Characterization (GLCC) Database (GLCC; U.S. Geological Survey (USGS), 2010) is developed in a continent basis for land use, while land cover maps are classified into 37 categories, with a spatial resolution of 1 km.

2.6 Model evaluation

Evaluation of an AQM is the process of assessing its performance in simulating spatial-temporal features embedded in the air quality observations. The Atmospheric Modelling and Analysis Division of U.S. EPA classifies the different aspects of model evaluation under four general categories: operational, dynamic, diagnostic, and probabilistic: *a)* Operational performance evaluation is accomplished by comparing model simulated values, against observed data. The concept applies not only to air quality parameters, but also to meteorological variables in the case of advanced coupled models like WRF; *b)* Dynamic evaluation focuses on assessing the AQM response to changes in emissions and meteorology, which is central to applications in air quality management. This type of analysis can help in determining the key factors governing air pollution; *c)* Diagnostic evaluation investigates the processes and input drivers that affect model performance in order to find possible improvements to the algorithms; and *d)* Probabilistic evaluation generally requires a set of tools that helps addressing the model response to statistically varying inputs, without having to run the simulation for every case, which is very time consuming.

When evaluating air quality management strategies, policy-makers need information about relative risk and likelihood of success of different options. In these cases, a range of values reflecting the model uncertainties is more important than the model best guess, or actual output. End users are more likely to work with operational and dynamic evaluation tools, while the other two categories of evaluation are more related to model development.

The kind of data needed for verifying model output, will depend on the model itself and the user's needs. For models with meteorological pre-processors, like CALMET, or coupled meteorological/chemical models like WRF/Chem, atmospheric variables observation in some points of the domain would be required in order to validate results. Observations can be made at ground level or with a vertical profile, in the case of three dimensional simulations. In the case of chemical species concentration, monitoring stations could supply data needed to check model results. Some ground or satellite instruments can also provide vertical profile for chemical species (Martin, 2008). In any case, a consistent procedure should be applied in order to evaluate the model performance.

The most usual practice is to use the information content shown between the observed and the model-predicted values. In this respect, Willmott (1982) and Seigneur et al. (2000) propose some statistical performance measures. Table 1 summarizes these commonly used statistical tools for analysing and comparing model results versus measurements.

Indicator	Formula	Range	Ideal value	Comments
Correlation coefficient	$r = \frac{\sum_{i=1}^{N}\left[(C_{oi} - \bar{C}_o)(C_{pi} - \bar{C}_p)\right]}{\sqrt{\sigma_o \sigma_p}}$	[0,1]	1.0	C_o and C_p are observed and predicted concentrations.
Mean Bias	$MBE = \frac{1}{N}\sum_{i=1}^{N}(C_{pi} - C_{oi})$	n.a.	0.0	C_{pi} and C_{oi} are observed and predicted concentrations for N cases.
Fractional Bias	$FB = \frac{\bar{C}_o - \bar{C}_p}{0.5(\bar{C}_o + \bar{C}_p)}$	[-2,2]	0.0	$\bar{C}_O$ and $\bar{C}_P$ are the averages of observed and predicted concentrations.
Root mean square error	$RMSE = \sqrt{\frac{1}{N}\sum_{i=1}^{N}(C_{oi} - C_{pi})^2}$	n.a.	0.0	
Normalized mean square error	$NMSE = \frac{\overline{(C_o - C_p)^2}}{\overline{C_o C_p}}$	n.a.	0.0	
Index of agreement	$d = 1 - \frac{\sum_{i=1}^{N}(C_{pi} - C_{oi})^2}{\sum_{i=1}^{N}\left(\lvert C_{pi} - \bar{C}_o\rvert + \lvert C_{oi} - \bar{C}_p\rvert\right)^2}$	[0,1]	1.0	
Geometric Mean	$MG = \exp\left(\ln\overline{(C_o)} - \ln\overline{(C_p)}\right)$	> 0	1.0	
Geometric variance	$VG = \exp\left[\overline{\left(\ln(C_o) - \ln(C_p)\right)^2}\right]$	> 0	1.0	

Table 1. Statistical indicators for the evaluation of AQMs.

3. Air pollution modelling in Argentina: results of several study cases

The following study cases were selected with the aim of describing the simulation process, from the selection of the appropriate model until the evaluation of final results.

3.1 Urban and regional modelling in Mendoza: inventories and monitoring

The extended or Gran Mendoza urban area is located in the north-central part of the Province of Mendoza (32.8° S; 68.8° W), Argentina, in a region of foothills and high plains, on the eastern side of the Andes Range. The metropolitan population is around 1,000,000 inhabitants, making this city the fourth largest metropolitan area in the country. The area is located in an arid to semi-arid zone of low rainfall: 120–400 mm/yr, mean 230 mm/yr, which occurs especially during the summer months (November to March).The closeness of the Andes Mountains has a strong influence on local meteorology and, therefore, on air quality.

Air quality in the area under study is affected by intensive and intermediate industrial activities, emissions due to transportation, residential sources and, to a lesser degree, agriculture and animal husbandry (Puliafito et al., 2003). Food and wine industry are very low energy consumers, but alloys, cement and petrochemical industries are relevant to energy consumption and their contribution to total emissions is important. Transportation is the second most important source of pollutants in the emission inventory, but the main concern in the downtown area. The emissions by the residential sector are mainly due to the use of natural gas for heating. Animal husbandry is concentrated in the South East of the Province and its contribution to total emission is low.

3.1.1 Modelling approaches

The impact of all anthropogenic sources in the air quality of the urban centre was evaluated using two different approaches: *a)* to reproduce the pollutant dispersion in the metropolitan area, we chose CALPUFF modelling system, setting the modelling domain in the urban area of Gran Mendoza; and *b)* to evaluate the impact of regional circulation in tropospheric chemistry, we used WRF/Chem in a greater modelling domain.

3.1.2 Urban pollutant dispersion

The modelling domain covers an area of 41×36 km^2 (32.8° S to 33.1° S and 68.8° W to 69.0° W) including the urban zone and part of the Andes. Because of the complex topography, detailed terrain features were incorporated using the USGS global 3 arc-sec SRTM3 data (~ 90m resolution). The LULC data was obtained from the Global Land Cover Characterization (GLCC) database captured by a 1-km resolution. Finer features of land use were obtained using the soil classification from the EcoAtlas Program from Mendoza's Rural Development Institute (IDR), which includes the interpretation and classification of Landsat images during the year 2006. Surface and upper meteorological data was obtained from the National Weather Service at the local airport station El Plumerillo (32.78° S, 68.78° W, 704 m above sea level) located at 6 km North-East of the urban center. Fig. 3 shows a view of the modelling domain. Industrial sources are located in two industrial areas in the periphery of the city. The position and emission rates for the industrial stacks are known, measured and compiled by the local environmental authority. Twenty one fixed (point) sources at constant rates for the modelling period were considered. Emissions from residential and commercial sources were estimated using data of natural gas consumption in the city and spatially distributed according to the land-use maps of the urban center.

A top-down approach (macro scale) associated to GIS was used to calculate mobile source emissions and their spatial distribution. The COPERT III model (Ntziachristos et al., 2000) was selected to estimate road transport emissions because of the similarities between the Argentinean and the European fleets (D'Angiola et al., 2010). Emission factors for Natural Gas Vehicles (NGV) were derived from local studies (ARPEL, 2005). The vehicle fleet was arranged into 4 vehicle classes and 28 technology categories depending on fuel type, vehicle size, fuel delivery system and exhaust control system; 19 of these categories, belong to light duty vehicles and passenger cars, 6 to heavy duty vehicles and 3 to urban buses.

Information from the National Vehicle Registration Directory (DNRPA) and the Automobile Manufacturers Association (ADEFA) was used to determine the fleet composition. To derive

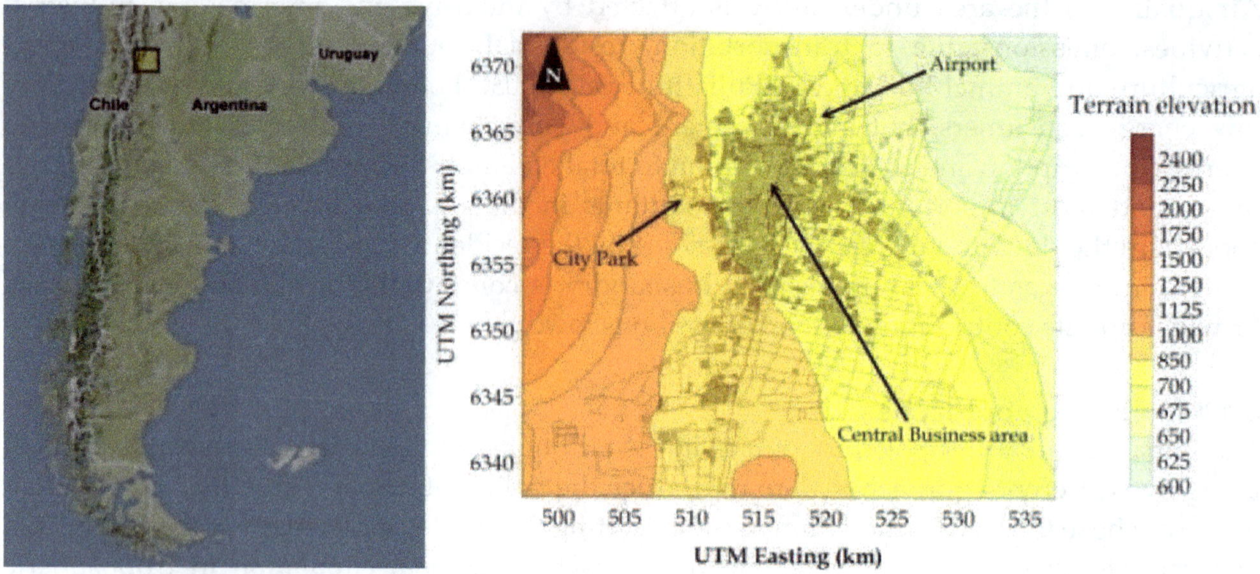

Fig. 3. Geographical location of the City of Mendoza (left) and terrain elevations (metres a.s.l.) in the modelling domain (right).

emission rates, the model considers different average speeds for different types of driving conditions. Three driving conditions were associated with specific road hierarchies: *a)* Highways: roads characterized as main intra-county, suburban areas or inter-regional highways connecting main town poles in the metropolitan area, with high traffic, no traffic lights and high average speed (70-100 km/h); *b)* Primary roads: roads that are main streets connecting important urban districts, with a high vehicle density, with most intersections regulated by traffic lights, and a low average speed (20-30 km/h); *c)* Secondary roads: roads that are mainly residential streets with a low vehicle density, very few or no traffic lights regulating intersections, although, in some cases, with the presence of speed limiters and a low–medium average speed (25-35 km/h). Average speed on each road hierarchy and annual average kilometres traveled (VKT) were estimated from information collected on a set of vehicles equipped with a Global Positioning Satellite (GPS) unit. The total amount of emissions estimated for highways, primary roads and secondary roads was distributed in the road network proportionally to the length of segments. Then, these emissions were geographically distributed into a gridded map with cells of 500 m x 500 m (Fig. 4, left).

Fig. 4 (right) depicts the daily mean ambient PM_{10} concentration modeled with CALPUFF. It shows concentrations of (70−80 µg/m^3) in downtown area, mainly produced by mobile sources. These values reduce towards the periphery of the city. The south-western area of the modelling domain is greatly influenced by the industrial complex located in the south western area of the city. PM_{10} emissions from dust sources were estimated according to OAQPS (1977) and modeled as area sources in CALPUFF. The emissions were incorporated as a discontinuous process when wind speeds were above a threshold of 6.8 m/s. Fig. 5 shows a time series comparison of the modeled results versus the PM_{10} measurements at the monitoring station in downtown for June 2009. Despite some peak values due to high traffic density the results show a good correlation between modelled concentration and measured data. A more detailed analysis of pollutant distribution in the metropolitan area of Mendoza can be found in Puliafito & Allende (2007) and Puliafito et al. (2011).

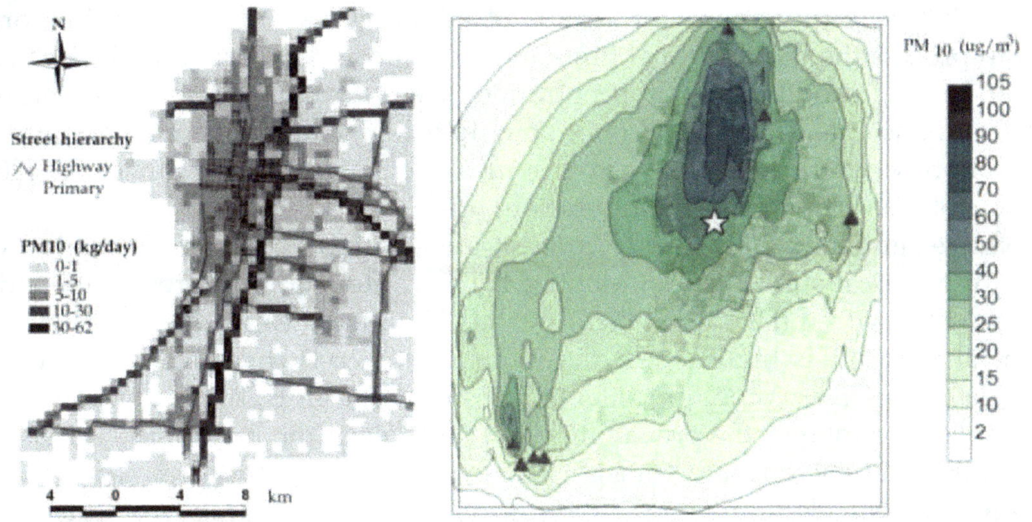

Fig. 4. Left panel: distribution of PM_{10} emission from mobile sources (kg/day). Right panel: spatial distribution of PM_{10} daily mean concentration values. Black triangles show the location of the main industrial sources. The white star indicates the location of a monitoring site.

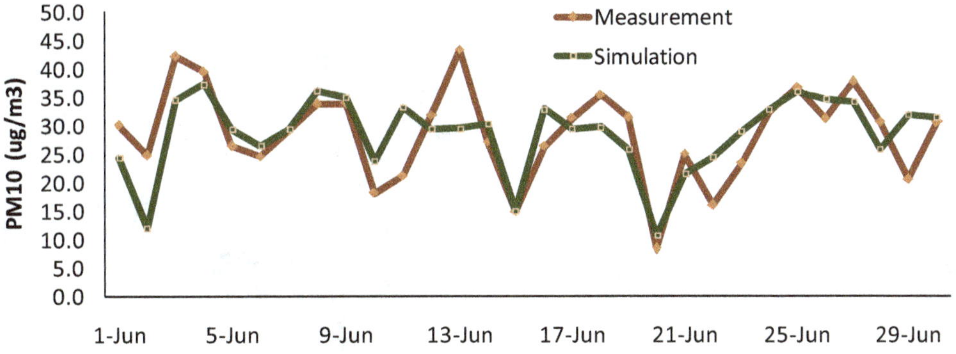

Fig. 5. Comparison between measured and simulated PM_{10} concentrations at the measurement site, indicated in Fig. 4.

3.1.3 Regional modelling

The mountain-valley circulation dominates the medium-range-transport in the area, so not only urban but also regional air pollution levels are influenced by these synoptic scale weather patterns. Then, in order to study the temporal and spatial distribution of primary and secondary pollutants in the northern region of the Province of Mendoza, the WRF/Chem model was used. Three nested domains with 28 vertical altitude levels were defined to downscale the chemical and physical properties of the atmosphere. The smaller domain covers the northern region of Mendoza Province (4 km resolution, 200 km N/S x 160 km E/W), including the urban center.

In order to capture the complexity of the terrain features in the study area, the WRF Preprocessing System (WPS) was modified to introduce the Shuttle Radar Topography Mission (SRTM3) data, increasing standard resolution over 10 times. Also, LULC maps developed by several local institutions and agencies were adapted to the WPS module using a GIS system (Fernandez et al., 2010). The emission inventory for Mendoza was developed to meet WRF/Chem requirements by means of adapting the *emiss_v3* routine, originally developed to process the U.S. National Emissions Inventory database (NEI; U.S. EPA, 2005). Boundary and initial conditions for concentrations were taken from global chemical model MOZART (Emmons et al., 2009). A 24 hour spin up was used for the inner domain. Default parameterization schemes were used in the model runs (see Peckham et al. (2010) for detailed descriptions of the model configuration).

Fig. 6 shows the simulated NO_2 surface concentration in p.p.m. during the night (left) and day (right) for a typical spring day in Mendoza. The contour lines represent the altitude of the calculated PBL within the WRF/Chem model. The model captures the behavior of the PBL height: at night, it is higher over the mountains (west) than over the valley, while during the day this effect is the opposite, surpassing 1,500 m right over the foothills. The NO_2 plume direction follows the prevailing wind patterns in the region, i.e., during the day pollutants emitted in the city are transported towards the South-West, while during the night the same air parcels return to the city in the opposite direction. Absolute values are greater during the night, mainly because of the smaller dilution volume limited by the PBL altitude. Here, the reader should recognize that, because of the local topography of Mendoza, the mountain-valley circulation is not exactly West-East, but shifted SW-NE.

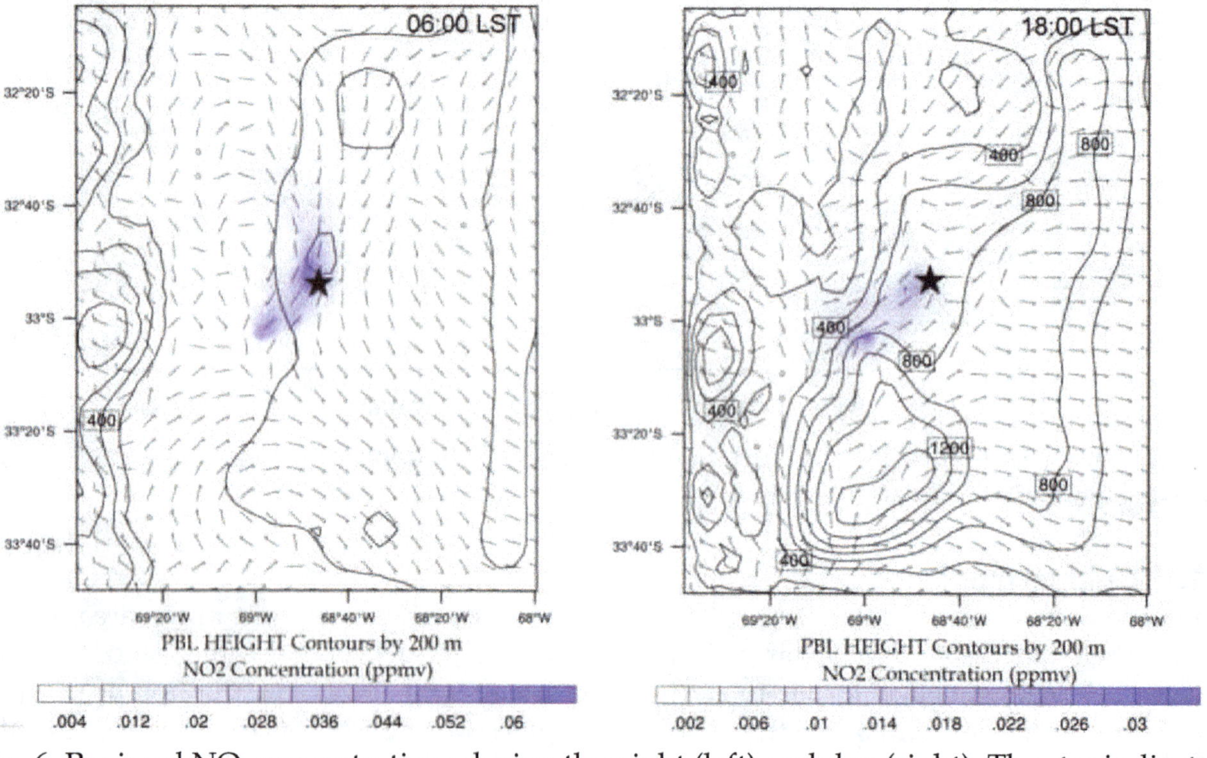

Fig. 6. Regional NO_2 concentrations during the night (left) and day (right). The star indicates the location of Mendoza City. The surface plots represent the altitude of the PBL in m. Wind barbs are superimposed in the figure.

3.2 Study case of Bahía Blanca

The district of Bahía Blanca (62.2° W; 38.5° S) is located on the southern Atlantic coast of the Province of Buenos Aires, in the boundary between the southern Pampas and the northern Patagonia. Bahía Blanca and its surroundings concentrate a population of 310,000 inhabitants in the urban area. The landscape is mainly flat with terrain elevations between

200 m and 500 m, except for the low mountain system of "Sierra de la Ventana", which reaches an altitude of 1,200 m. The area is located in a temperate zone with warm continental type weather. Its annual precipitation is about 600 mm, with high monthly variations. March, September, October and November are the months with higher probability of rainfall episodes. Winds are generally moderate, with predominant direction from North or Northwest. Mean wind speed is 5.3 m/s, with 3 % calms.

Bahía Blanca has an important seaport system handling intense industrial and commercial activities. The city has a highly developed oil refinery and petrochemical industries, producing ethanol, naphtha, gasoline, asphalt GLP, ethylene, PVC, polyethylene, urea, ammonia, chlorine and caustic soda. An important power station is located outside the industrial zone, SW of the urban centre, in the port area, which may operate with natural gas or heavy-oil, producing the main contribution of SO_2 emissions in the zone. Along the north-eastern shore of the bay, there are several small seaports dedicated to grains and containers. The ambient concentration of gases like NO_x, CO, NH_3 and SO_2 and concentrations of particulate matter have been measured by the local environmental authority since 1997.

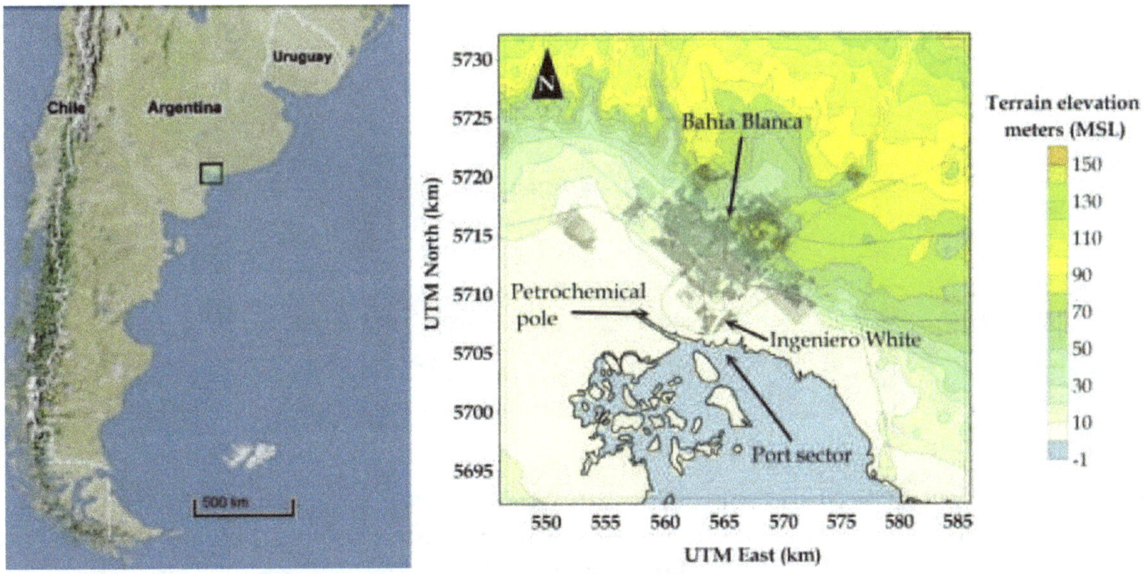

Fig. 7. Map of Argentina showing location of the modelling domain. To the right, a detail of the urban area of Bahia Blanca, major highways, the industrial complex, and the port area.

In order to complement the information provided by the monitoring data, pollutant dispersion was simulated with the CALPUFF modeling system. The modelling domain covered an area of 1,600 km^2 between 38.5° and 39.0° S latitude and 62.0° and 62.5 W longitude in a flat region with elevations up to 150 m, increasing to the NE. A view of the modelling domain is presented in Fig. 7. All industrial sources are located on a Petrochemical Pole, an Industrial Park and the port sector and were treated as point sources in CALPUFF.

Emissions of NO_x, SO_2 and NH_3 due to transportation (including road, railway and ships) were estimated following the top-down emission model COPERT III similarly to the Mendoza study case. Cargo oriented railway transportation emissions were estimated using specific fuel consumption factors and a number of train operations, while maritime activity emissions were calculated using a number of operations, port waiting time and bulk emission factors. Both of them were calculated with the simplest CORINAIR methodology (EMEP/EEA, 2009). The sum of all emissions from mobile sources was spatially distributed using Geographical Information System (GIS) tools in a regular grid of 500 m x 500 m, covering the whole modelling domain. The transportation emissions were included in the dispersion model as area sources. Residential and fugitive sources were also included in the simulation, estimated with natural gas consumption statistics, as in the Mendoza study case.

Meteorological data was obtained from the local airport station Comandante Espora (38° 42′ S, 62° 09′ W, 75 m above sea level) located 12 km East of the urban center. The model runs were conducted over one year, and five chemical species (NO_x, SO_2, NH_3, SO_4^{2-} and NO_3^-) were simulated. The model includes the MESOPUFF II scheme to incorporate chemical transformation of SO_2 and NO_x. Monthly background concentrations for ammonia and ozone were provided by the environmental authority monitoring data over one year period. Fig. 8 shows the contribution of each source type to the total emissions of CO, SO_2, NO_x, PM_{10} and NH_3 for Bahia Blanca (left) and the average 24-h concentrations for NO_x (right).

Finally, the 24 h concentrations obtained with CALPUFF were compared to the measured ones in the monitoring station. Fig. 9 shows the comparisons between the modelling results

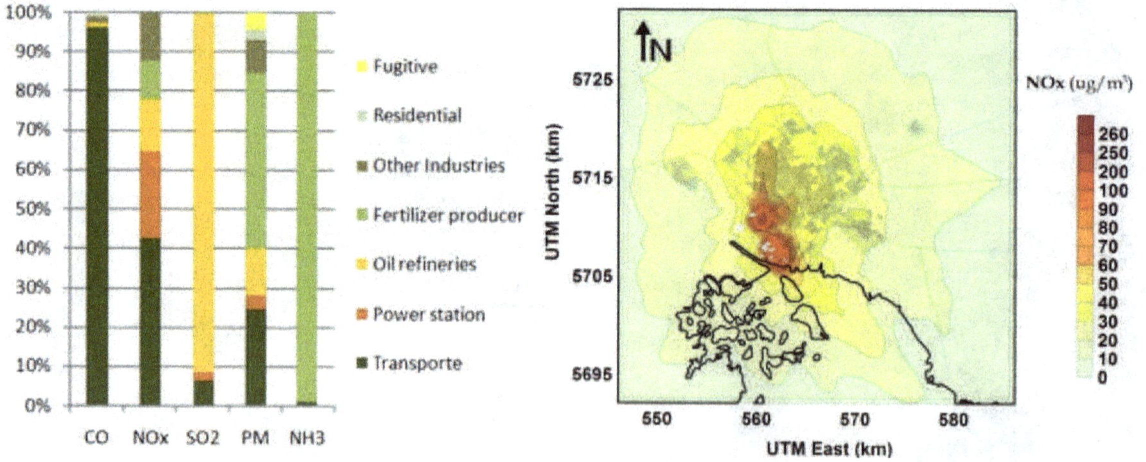

Fig. 8. Left: Emission distribution for all pollutant modelled. Industrial activities are the main contributor for all pollutants, except for transportation that is the most important CO emitter. Right: Iso-concentration plots simulated with CALPUFF for NOx. The white points represent industrial stacks.

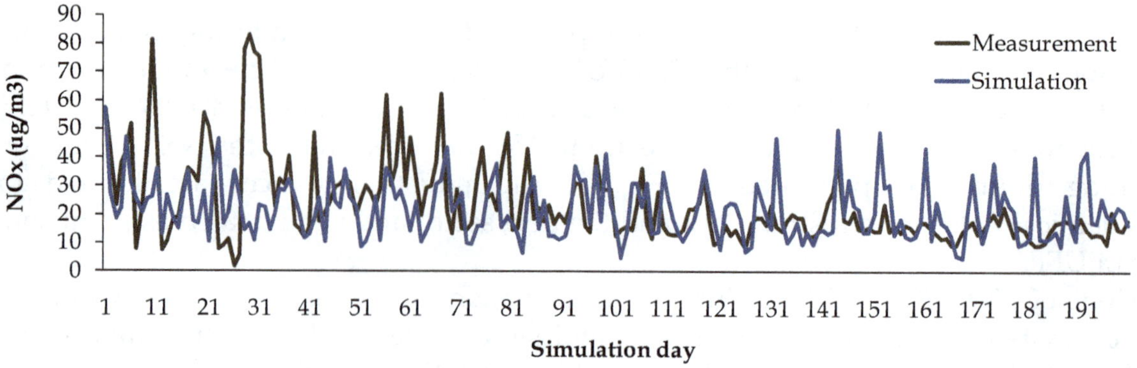

Fig. 9. Daily variation of NO_x concentration (µg/m³) at the monitoring point and values simulated with CALPUFF.

and the observations near the industrial complex. Although the model is unable to reproduce the concentration peaks exactly, the values of the statistical measures (MBE: -2.02; FB: -0.04; Willmott's d: 0.97) suggest that the dispersion simulation approaches the concentrations patterns with acceptable accuracy. The modelling results showed that the most polluted region is the town of Ingeniero White, located to the SE of the metropolitan area, and very close to the Petrochemical Pole (see detailed analysis in Puliafito & Allende (2007) and Allende et al. (2010a)). The fair agreement between concentrations measured by the environmental authority over one year and the simulated ones with CALPUFF validated the proposed emissions and dispersion models.

3.3 The Buenos Aires case

The City of Buenos Aires (34.6° S, 58.3° W) is the Capital and largest city of Argentina, and the second largest metropolitan area (around 13 million inhabitants), in South America after São Paulo, Brazil. It is located on the western shore of the Río de la Plata estuary, on the south-eastern coast of the South American continent. Buenos Aires has a humid subtropical climate (Köppen classification) with four distinct seasons, January being the warmest month and July the coldest. Relative humidity tends to be high throughout the whole year (~ 72%). Winds that blow clean air from Río de La Plata towards the city have an annual frequency of 58.0%, and there are 3.8% calms. At present, in spite of its extension and its high population density, Buenos Aires City has only a small air quality monitoring network, mainly dedicated to capture traffic pollution in 4 points of the city. Besides this fact, there have been several measurement campaigns at different sites in the city (Arkouli et al., 2010; Bogo et al., 1999; Mazzeo et al., 2005; Reich et al., 2006). Even though there are no large industries inside the city, a great industrial complex is located in the southeast. Also, four large partly gas and heavy-oil fired power stations are located close to the river shore.

For the air pollution evaluation in the area we used the CALPUFF model in a 25 km x 25 km modelling domain, extending from 34.5° S to 34.7° S and 58.3°S W to 58.6° W in a flat region with elevations up to 40 m increasing to the SW (Fig. 10). Surface meteorological observations were obtained from the local airport meteorological station in Aeroparque (34°34′ S, 58°30′ W). Since CALMET requires twice daily upper air meteorological

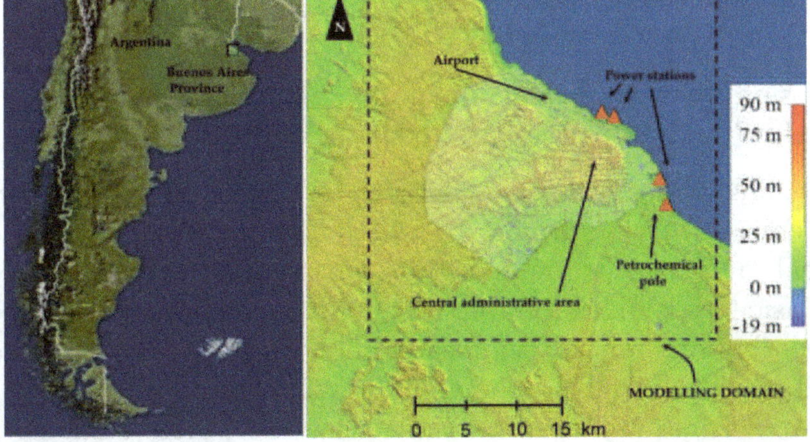

Fig. 10. Map of Argentina (left) and location of the modelling domain (right). Terrain elevations are in metres. In red are depicted major highways and the triangles indicate the location of the power stations.

observations for dispersion simulating, and since observations with such frequency are not available for this site, the required meteorological parameters were derived from the Weather Research and Forecasting model. The CALWRF off-line preprocessor was used to generate an intermediate three dimensional data file to be used by CALMET as a "first-guess" meteorological field. The WRF model was run with nested domains, reaching a resolution of 3 km, high enough to resolve the important meteorological features which can only be simulated by the prognostic WRF model, such as land and sea breezes.

Emissions from the power stations were estimated using emission factors of CORINAIR (EMEP/EEA, 2009) for most pollutants and IPCC (IPCC, 2006) for greenhouse gases, according to their thermal capacity and use of fuel. The emissions from oil refineries, chemical industries and storage tanks located very close to the city, near the SE border, were previously measured and included directly in the CALPUFF model. All these emissions were associated to 54 point sources. Temporal allocation of emissions was estimated using statistical data from the local Electricity Regulation Office and gross domestic product evolution for the industrial sector.

Emissions of NO_x, SO_2 and CO due to transportation were estimated following the top-down emission model COPERT III similar to the study cases of Mendoza and Bahía Blanca (Perez Gunella et al., 2009). Daily and monthly variation of road traffic in the city is often measured, with traffic rush hours during working days being from 08:00 to 09:00 a.m. and from 07:00 to 09:00 p.m. These emissions were geographically distributed into a gridded map with cells of 300 m x 300 m (Fig. 11, right).

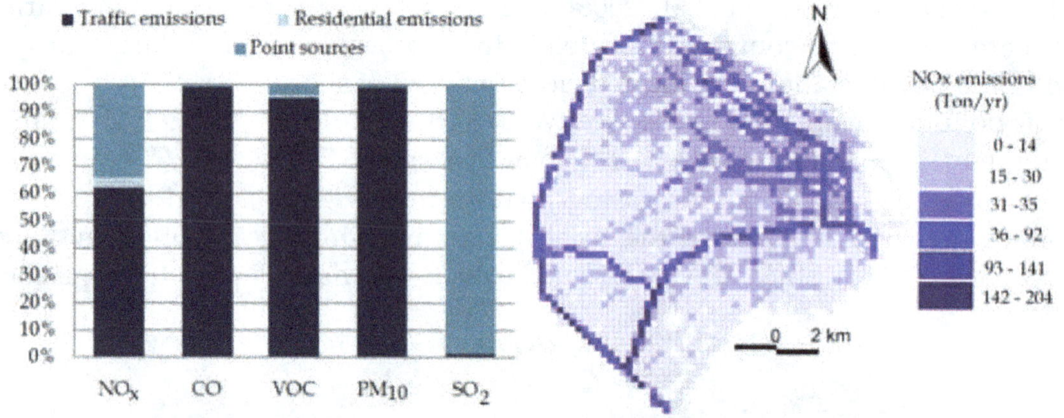

Fig. 11. Left: Emission contributions for all pollutants simulated in the study area. Right: Spatial allocation for NO_x traffic emissions in the city of Buenos Aires.

Residential emissions were associated with natural gas and LPG consumption, and estimated using emission factors from CORINAIR. Spatial allocation of the emissions was made proportional to the population density in the city districts and further weighted with the Unsatisfied Basic Needs index (NBI) to account for the population socioeconomic levels. The emission distribution from all sources is shown in Fig. 11 (left), and concentration patterns obtained with CALPUFF are shown in Fig. 12.

SO_2 concentration maxima are clearly located near the power stations. The road traffic appears to be responsible for high NO_x values near major roadways (Puliafito et al., 2010; Allende et al., 2010b). Monitoring values obtained in a toll gate located in the south of the city were compared to simulated ones. For CO, the comparison is shown in Fig. 13.

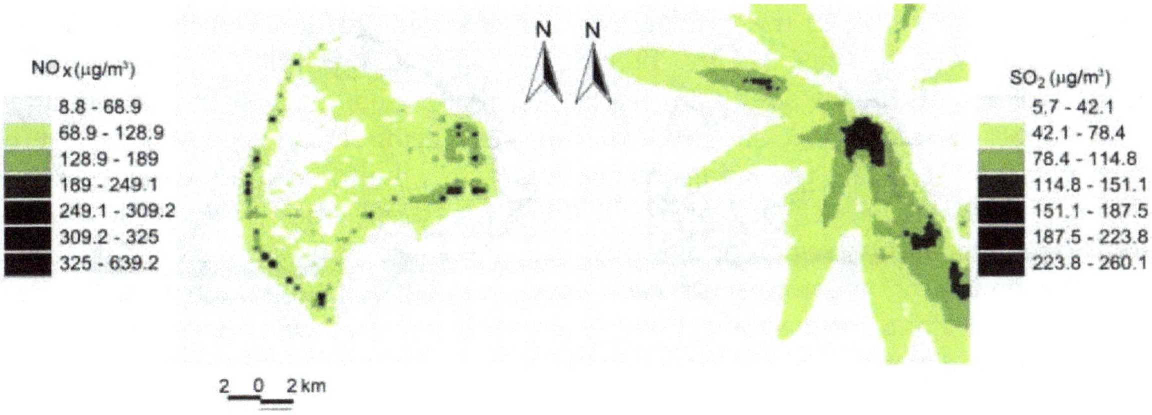

Fig. 12. 24 h average iso-concentration plots simulated with CALPUFF for NOx and SO_2.

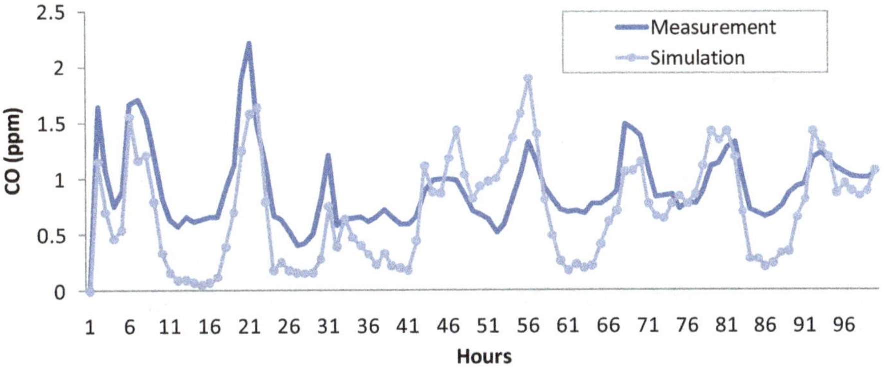

Fig. 13. Hourly variation of CO concentration (p.p.m.) at the monitoring point and values simulated with CALPUFF.

4. Summary and concluding remarks

The purpose of this chapter was to provide an overview of several modelling tools available for air quality evaluation, providing general and prescriptive but also flexible recommendations toward the adoption of best practice. Air Quality Models (AQMs) are mathematical tools that simulate the physical and chemical processes that involve air pollutants as they disperse and react in the atmosphere. The use of AQMs improve the limitations faced by the use of the monitoring network approach by providing prediction of the temporal and spatial distribution of actual pollution levels. Modelling studies, in combination with air quality monitoring are essential and complementary tools for long and short term air pollution control strategies. A well calibrated model is a unique tool that allows the representation of the atmosphere dynamics and chemistry, including real conditions and atmospheric disturbances.

From a theoretical standpoint, AQMs were divided in short-range Gaussian models (ISC3, AERMOD), advanced urban models (CALPUFF) and complex Chemical Transport Models coupled to mesoscale atmospheric models (WRF/Chem). The greatest differences between the model types are the meteorological input required, computer resources and user experience. For each of these models, we have discussed the configurations, data requirements, applicability and some physical and chemical formulations. We have identified several important aspects for selecting the proper AQM model: *a)* identification of the proper scale problem; *b)* the availability of meteorological data; *c)* the preparation of emission inventories; *d)* the inclusion of land use and topographical information; *e)* the integration of results in an adequate geographical information system; and finally *f)* a comparison of the simulated data to existing monitoring information, if any. From a practical point of view, and following the above scheme, the three study cases presented at the end of this chapter give practical implications for the selection of an AQM applied to different data availability, scale, topography and meteorology. We have described how to prepare the emission inventories for each case, and used local and regional AQMs to calculate the pollutant concentration. The acceptable accordance between the simulated concentrations and the monitored data confirms the proposed methodology described in this chapter.

Finally, we hope, this chapter encourages readers to use "state of the art" models since they provide the most realistic representation of the phenomena involved in the fate of pollutants released to the atmosphere.

4. Acknowledgements

We thank the authorities of the National Technological University of Argentina (UTN) and the Argentine Research Council (CONICET) for their support in our research activities. We further thank the Argentine National Weather Service (Servicio Meteorológico Nacional, SMN) for the availability of the meteorological data. This research was supported partially by grant PICT2005 # 23-32686 from the National Science Agency (Agencia Nacional de Promoción Científica y Tecnológica, ANPCyT).

5. References

Allende, D. G., Castro, F. H., & Puliafito, S. E. (2010a). Air Pollution Characterization and Modeling of an Industrial Intermediate City. *International Journal of Applied Environmental Sciences*, 5(2), 275-296.

Allende, D. G., Cremades, P., Puliafito, Enrique, Fernandez, R. P., & Perez Gunella, F. (2010b). Estimación de un Factor de Riesgo de Exposición a la Contaminación Urbana par ala población de la Ciudad de Buenos Aires. *Avances en Energías Renovables y Medio Ambiente*, 14, 127-134.

Arkouli, M., Ulke, A. G., Endlicher, W., Baumbach, G., Schultz, Eckart, Vogt, U., et al. (2010). Distribution and temporal behavior of particulate matter over the urban area of Buenos Aires. *Atmospheric Pollution Research*, 1(1), 1-8.

ARPEL. (2005). Medición de emisiones de vehículos en servicio en San Pablo, Santiago y Buenos Aires. *Informe Ambiental ARPEL* Número 25-2005.

Ballesta, P. P., Field, R. A., Fernandez-Patier, R., Madruga, D. G., Connolly, R., Caracena, A. B., et al. (2008). An approach for the evaluation of exposure patterns of urban populations to air pollution. *Atmospheric Environment*, 42(21), 5350-5364.

Bieser,J., Aulinger,A., Matthias,V., Quante, M., & Builtjes,P. (2010). SMOKE for Europe - adaptation, modification and evaluation of a comprehensive emission model for Europe. *Geoscientific Model Development Discussions.*, 3, 949-1007.

Binkowski, F. S., & Roselle, S. J. (2003). Models-3 Community Multiscale Air Quality (CMAQ) model aerosol component 1. Model description. *Journal of Geophysical Research - Atmospheres*, 108(D6), 4183.

Bogo, H., Negri, R. M., & Roman, E. S. (1999). Continuous measurement of gaseous pollutants in Buenos Aires city. *Atmospheric Environment*, 33(16), 2587-2598.

Borge, R., Lumbreras, J., & Rodriguez, E., 2008, Development of a high-resolution emission inventory for Spain using the SMOKE modelling system: A case study for the years 2000 and 2010. *Environmental Modelling and Software,* 23, 1026–1044.

Brunekreef, B., & Holgate, S. T. (2002). Air pollution and health. *Lancet*, 360(9341), 1233-42.

Carter, W. P. L. (2000).Documentation of the SAPRC-99 Chemical Mechanism for VOC Reactivity Assessment. *Report to the California Air Resources Board,* Available at http://cert.ucr.edu/~carter/absts.htm#saprc99

Cimorelli, A. J., Venkatram, A., Weil, J. C., Paine, R., Wilson, R., Lee, R., et al. (2003). AERMOD : Description of model formulation. Environmental Protection Agency.

Davis, N., Lents, J., Osses, M., Nikkila, N., & Barth, M. (2005). Development and Application of an International Vehicle Emissions Model. *Transportation Research Record,* 51-59.

Demuzere, M., Trigo, R. M., Arellano, J. V.-guerau D., & Lipzig, N. P. M. V. (2009). The impact of weather and atmospheric circulation on O3 and PM10 levels at a rural mid-latitude site. Atmospheric Chemistry and Physics, 9, 2695-2714.

Dodge, M. C. (2000). Chemical oxidant mechanisms for air quality modeling: critical review. *Atmospheric Environment*, 34(12-14), 2103-2130.

D'Angiola, A., Dawidowski, L. E., Gomez, D. R., & Osses, M. (2010). On-road traffic emissions in a megacity. *Atmospheric Environment*, 44(4), 483-493.

EMEP/EEA. (2009). EMEP/EEA Air Pollutant Emission Inventory Guidebook — 2009. Technical guidance to prepare national emission inventories. Copenhagen. Available from: www.eea.europa.eu/help/infocentre/enquiries

Emmons, L., Walters, S., Hess, P. G., Lamarque, J., Pfister, G. G., Fillmore, D., et al. (2009). Description and evaluation of the Model for Ozone and Related chemical Tracers, version 4 (MOZART-4). *Geosci. Model Dev. Discuss.*, 2, 1157-1213. Max-Planck-Institute for Meteorology, Hamburg, Germany.

Endlicher, W., Zahnen, B., Schultz, E., Alessandro, M., Mikkan, R., & Polimeni, M. (1998). Clima Urbano y Contaminación Atmosférica en Mendoza/Argentina. In : *Clima y Ambiente Urbano en Ciudades Ibéricas e Iberoamericanas* (pp. 115-134). Madrid.

Fenger, J. (1999). Urban air quality. *Atmospheric Environment*, 33(29), 4877-4900.

Fernandez, R., Allende, D., Castro, F., Cremades, P., Puliafito, E. (2010). Modelado regional de la calidad de aire utilizando el modelo WRF/Chem : implementación de datos globales y locales para Mendoza. *Avances en Energías Renovables y Medio Ambiente,* 14, 01.43-01.50.

Ferretti, M., Andrei, S., Caldini, G., Grechi, D., Mazzali, C., Galanti, E., et al. (2008). Integrating monitoring networks to obtain estimates of ground-level ozone concentrations -- A proof of concept in Tuscany (central Italy). *Science of the Total Environment*, 396(2-3), 180-192.

Fowler, D., Pilegaard, K., Sutton, M. A., Ambus, P., Raivonen, M., Duyzer, J., et al. (2009). Atmospheric composition change: Ecosystems-Atmosphere interactions. *Atmospheric Environment*, 43(33), 5193-5267.

Freitas, S.R., Longo, K.M., Alonso,M.F., Pirre, M., Marecal, V., Grell, G., Stockler,R., Mello, R.F., & Sánchez Gácita,M.. (2010). A pre-processor of trace gases and aerosols emission fields for regional and global atmospheric chemistry models. *Geoscientific Model Development Discussions*, 3, 855-888.

GEIA/ACCENT. (2005). GEIA-ACCENT database, an international cooperative activity of AIMES/IGBP. ACCENT EU Network of Excellence. http://www.geiacenter.org

Gery, M.W., Whitten, G.Z., Killus, J.P. & Dodge, M.C. (1989). A photochemical kinetics mechanism for urban and regional scale computer modeling. *Journal of Geophysical Research*, 94, 1292-12956.

Gilliland, A. B., Hogrefe, C., Pinder, R. W., Godowitch, J. M., Foley, K. L., & Rao, S. T. (2008). Dynamic evaluation of regional air quality models: Assessing changes in O_3 stemming from changes in emissions and meteorology. *Atmospheric Environment*, 42(20), 5110-5123.

Grell, G. A, Peckham, S. E., Schmitz, Rainer, McKeen, S. a, Frost, G., Skamarock, W. C., et al. (2005). Fully coupled "online" chemistry within the WRF model. *Atmospheric Environment*, 39(37), 6957-6975.

Hanna, S. R., Lu, Z., Frey, H. C., Wheeler, N., Vukovich, J., Arunachalam, S., et al. (2001). Uncertainties in predicted ozone concentrations due to input uncertainties for the UAM-V photochemical grid model applied to the July 1995 OTAG domain. *Atmospheric Environment*, 35(5), 891-903.

IPCC. (2006). 2006 IPCC Guidelines for National Greenhouse Gas Inventories. Hayama, Kanagawa, Japan. Available from http://ipcc-nggip.iges.or.jp

Jiménez-Guerrero, P., Jorba, O., Baldasano, J. M., & Gassó, S. (2008). The use of a modelling system as a tool for air quality management: annual high-resolution simulations and evaluation. *The Science of the Total Environment*, 390(2-3), 323-40.

Leeuw, F. A. A. M. de, Moussiopoulos, N., Sahm, P., & Bartonova, A. (2001). Urban air quality in larger conurbations in the European Union. *Environmental Modelling & Software*, 16(4), 399-414.

Martin, R. V. (2008). Satellite remote sensing of surface air quality. *Atmospheric Environment*, 42(34), 7823-7843.

Mazzeo, N. A., Venegas, L. E., & Choren, H. (2005). Analysis of NO, NO_2, O_3 and NO_x concentrations measured at a green area of Buenos Aires City during wintertime. *Atmospheric Environment*, 39(17), 3055-3068.

Michalakes, J., Dudhia, J., Gill, D., Henderson, T., Klemp, J., Skamarock, W., et al. (2005). the Weather Research and Forecast Model: Software Architecture and Performance. *Use of High Performance Computing in Meteorology - Proceedings of the Eleventh ECMWF Workshop*, (June), 156-168. Singapore: World Scientific Publishing Co.

Muller, N. Z., & Mendelsohn, R. (2007). Measuring the damages of air pollution in the United States. *Journal of Environmental Economics and Management*, 54(1), 1-14.

Ntziachristos, L., Samaras, Z., & Kouridis, C. (2000). COPERT III Computer programme to calculate emissions from road transport. Copenhagen.

OAQPS. (1977). Guideline for development of control strategies in areas with fugitive dust problems. EPA-405/2- 77-029.

Olivier, J. G. J., & Berdowski, J. J. M. (2001). Global emissions sources and sinks. *Berdowski, J., Guicherit, R. and B.J. Heij (eds.) "The Climate System"* (pp. 33-78). Lisse, The Netherlands. A.A. Balkema Publishers/Swets & Zeitlinger Publishers.

Oxley, T., Valiantis, M., Elshkaki, a, & ApSimon, H. M. (2009). Background, Road and Urban Transport modelling of Air quality Limit values (The BRUTAL model). *Environmental Modelling & Software*, 24(9), 1036-1050.

Pearce, J. L., Beringer, J., Nicholls, N., Hyndman, R. J., Uotila, P., & Tapper, N. J. (2011). Investigating the influence of synoptic-scale meteorology on air quality using self-organizing maps and generalized additive modelling. *Atmospheric Environment*, 45(1), 128-136.

Peckham, S., Grell, G., McKeen, S., Fast, J., Gustafson, W., Ghan, S., et al. (2010). WRF/Chem Version 3.2 User's Guide. Available from http://ruc.noaa.gov/wrf/WG11/Users_guide.pdf

Perez Gunella, F., Puliafito, S.E., & Pirani, K. (2009). Calculo de las emisiones del transporte para la Ciudad de Buenos Aires usando un Sistema de Informacion Geografico. *Avances en Energías Renovables y Medio Ambiente*, 13, 57-64.

Puliafito, E, Guevara, M., & Puliafito, C. (2003). Characterization of urban air quality using GIS as a management system. *Environmental pollution*, 122(1), 105-17.

Puliafito, E., & Allende, D. G. (2007). Emission patterns of urban air pollution. *Revista Facultad de Ingeniería Universidad de Antioquia*, 42, 38-56.

Puliafito, E., Castro, F., & Allende, D. G. (2011). Air quality impact of PM_{10} emission in urban centers. *International Journal of Environment and Pollution*, (in press).

Puliafito, E., Perez Gunella, F., & Allende, D. G. (2010). Modelo de Calidad del Aire para la Ciudad de Buenos Aires. *World Congress & Exhibition ENGINEERING 2010-ARGENTINA*. Buenos Aires, Argentina.

Reich, S., Magallanes, J., Dawidowski, L., Gómez, D., Grošelj, N., & Zupan, J. (2006). An Analysis of Secondary Pollutants in Buenos Aires City. *Environmental Monitoring and Assessment*, 119(1), 441-457. Springer Netherlands.

Russell, A., & Dennis, R. (2000). NARSTO critical review of photochemical models and modeling. *Atmospheric Environment*, 34(12-14), 2283-2324.

Schlink, U., Herbarth, O., Richter, M., Rehwagen, M., Puliafito, J. L., Puliafito, S. E., et al. (1999). Ozone-monitoring in Mendoza, Argentina : Initial results. *Journal of the Air & Waste Management Association*, 49(1), 82-87.

Schürmann, G. J., Algieri, A., Hedgecock, I. M., Manna, G., Pirrone, N., & Sprovieri, F. (2009). Modelling local and synoptic scale influences on ozone concentrations in a topographically complex region of Southern Italy. *Atmospheric Environment*, 43(29), 4424-4434.

Schultz, M. G., Backman, L., Balkanski, Y., Bjoerndalsaeter, S., Brand, R., Burrows, J. P., et al. (2008). REanalysis of the TROpospheric chemical composition over the past 40 years. A long-term global modeling study of tropospheric chemistry funded under the 5th EU framework programme. *Final Report*. (Vol. 30, pp. 281-8). Hamburg.

Scire, J. S., Strimaitis, D. G., & Yamartino, R. J. (2000a). A User's Guide for the CALPUFF Dispersion Model. Concord, MA.

Scire, J. S., Robe, F. R., Fernau, M. E., & Yamartino, R. J. (2000b). A User's Guide for the CALMET Meteorological Model. Concord, MA.

Seaman, N. L. (2003). Future directions of meteorology related to air-quality research. *Environment International*, 29(2-3), 245-252.

Seigneur, C., Pun, B., Pai, P., Louis, J., Solomon, P. A., Emery, C., et al. (2000). Guidance for the Performance Evaluation of Three-Dimensional Air Quality Modeling Systems for Particulate Matter and Visibility. *Journal of the Air & Waste Management Association*, 50(4), 588-599.

Skamarock, W. C., Klemp, J. B., Gill, D. O., Barker, D. M., Wang, W., & Powers, J. G. (2008). A Description of the Advanced Research WRF Version 3. Mesoscale and Microscale Meteorology Division, National Center for Atmospheric Research.

Stockwell, W. R., Middleton, P., & Chang, J. S. (1990). The second generation regional acid deposition model chemical mechanism for regional air quality modeling. *Journal of Geophysical Research*, *95*, 16343-16367.

Stockwell, William R, Kirchner, F., Kuhn, M., & Seefeld, S. (1997). A new mechanism for regional atmospheric chemistry modeling. *Journal of Geophysical Research.*, *102*(D22), 25847-25879.

Tuia D., Ossés de Eicker M., Zah R., Osses M., Zarate E., Clappier A. (200). Evaluation of a simplified top-down model for the spatial assessment of hot traffic emissions in mid-sized cities. *Atmospheric Environment*, 41, 3658–3671.

U.S. EPA. (1995). User's Guide for the Industrial Source Complex (ISC3) Dispersion Model. Description of model algorithms. Research Triangle Park, NC 27711.

U.S. EPA. (2000). Meteorological Monitoring Guidance for Regulatory Modeling Applications (p. 171). Research Triangle Park, NC 27711.

U.S. EPA. (2005). 2005 National Emissions Inventory Data & Documentation. Retrieved January 1, 2010, from http://www.epa.gov/ttn/chief/net/2005inventory.html

U.S. EPA. (2008). Technical Issues Related to use of the CALPUFF Modeling System for Near-field Applications. Research Triangle Park, NC 27711.

U.S. EPA. (2009a). Motor Vehicle Emission Simulator (MOVES) 2010 User Guide (p. 150).

U.S. EPA. (2009a). SPECIATE 4.2 Speciation Database Development Documentation. EPA600-R-09/038. Available at: http:/www.epa.gov/ttn/chief/software/speciate

U.S. EPA. (2010). AP 42, Fifth Edition Compilation of Air Pollutant Emission Factors. Retrieved January 1, 2010, from http://www.epa.gov/ttnchie1/ap42/

U.S. Geological Survey (USGS). (2010). The Global Land Cover Characterization (GLCC) Database. Available from: http://www.src.com/datasets/GLCC_Info_Page.html

Van der Werf, G. R., Randerson, J. T., Giglio, L., Collatz, G. J., Mu, M., Kasibhatla, P. S., Morton, D. C., DeFries, R. S., Jin, Y., & van Leeuwen, T. T. (2010). Global fire emissions and the contribution of deforestation, savanna, forest, agricultural, and peat fires (1997–2009), *Atmospheric Chemistry and Physics Discussions.*, 10, 11707-11735.

Wang, Xuemei, Wu, Z., & Liang, G. (2009). WRF/CHEM modeling of impacts of weather conditions modified by urban expansion on secondary organic aerosol formation over Pearl River Delta. *Particuology*, *7*(5), 384-391.

Wang, Xueyuan, Liang, X.-Z., Jiang, W., Tao, Z., Wang, J. X. L., Liu, H., et al. (2010). WRF-Chem simulation of East Asian air quality: Sensitivity to temporal and vertical emissions distributions. *Atmospheric Environment*, 44(5), 660-669.

WHO. (2002). *The World Health Report 2002: Reducing Risks, Promoting Healthy Life*. World Health (p. 13). Geneva.

Willmott, C. (1982). Some Comments on the Evaluation of Model Performance. *Bulletin of the American Meteorological Society*, 63, 1309-1369.

Zannetti, P. (1990). *Air Pollution Modelling*. Van Nostrand Reinhold. New York.

7

Investigation on the Carbon Monoxide Pollution Over Peninsular Malaysia Caused by Indonesia Forest Fires from AIRS Daily Measurement

Jasim M. Rajab, K. C. Tan, H. S. Lim and M. Z. MatJafri
School of Physics, Universiti Sains Malaysia, Penang, Malaysia

1. Introduction

Carbon monoxide (CO) is an important pervasive atmospheric trace gas affecting climate and more than 50% of air pollution nationwide and worldwide, which also plays as a significant indirect greenhouse gases due to its influences on the budgets of hydroxyl radicals (OH) and ozone (O_3). We present a study on Atmosphere Infrared Sounder (AIRS), onboard NASA's Aqua Satellite, detection of CO emission from large forest fire in the year 2005 in the Sumatra, Indonesia. AIRS daily coverage of 70% of the planet symbolizes an important evolutionary advance in satellite trace gas remote sensing. AIRS is one of several instruments onboard the Earth Observing System (EOS) launched on May 4, 2002, with its two companions microwave instruments the Advanced Microwave Sounding Unit (AMSU) and the Humidity Sounder for Brazil (HSB) form the integrated atmospheric sounding system. AIRS providing new insights into weather and climate for the 21st century, as well as AIRS' channels include spectral features of the key carbon trace gases CO_2, methane(CH_4), and CO. AIRS is an infrared spectrometer/radiometer that covers the 3.7–15.4 m spectral range with 2378 spectral channels. Troposphere CO abundances are retrieved from AIRS 4.55 mm spectral region, and measure CO total column by 52 channels with the uncertainty, which is estimated approximately 15-20 % at 500mb. Results from the analysis of the retrieved CO daily Level 3 standard (AIR×3STD) and Monthly product (AIRX3STM) were utilized in order to study the impact of Indonesia forest fire on CO distribution, and the monthly CO distributions in Peninsular Malaysia. AIRS daily CO maps from 12 – 25 August 2005 for study area show large-scale, long-range transport of CO from anthropogenic and natural sources, most notably from forest fire biomass burning. The sequence of daily maps shows the CO advection from central Sumatra to Malaysia. AIRS can also capture the temporal variation in CO emission from forest fires through 6-day composites so it may offer a chance to enhance our knowledge of temporal fire emission over large areas. The result was compared with daily CO emission (13-24) August 2007. The daily measurements of CO concentration on August 2005 are higher than August 2007. The northern region (uppers the latitude 4) was more affected by forest fires than the rest area. Substantial seasonal variations demonstrate season-to-season changes in rainfall and drought patterns in different seasons. We see such seasonal variations in the biomass burning emissions in the late dry season, while industrial contributions are evident at smaller magnitudes on monthly distributions. The study shows that AIRS can reliably detect CO plumes from forest fires in 1°*1° spatial resolutions. The CO maps were generated using Kriging Interpolation technique.

CO is an important atmospheric constituent affecting climate and air quality, which also acts as an important indirect greenhouse gases as it significantly impacts the OH budget, and thus indirectly affects the CH_4 and O_3 concentrations. This is due to the characteristic of CO by the indirect radiative forcing effect and increases the concentrations of troposphere O_3 and CH_4 through chemical reactions with other atmospheric constituents. It is approximately 75 % of OH sinks, which is capable to influence the concentration of greenhouse gases, the oxidizing agent in troposphere and finally contributes to climate change (Daniel and Solomon, 1998). CO is not a significant greenhouse gas because of it weak absorption of infrared radiation from the Earth. In addition, CO is useful in many modern technology applications, such as in bulk chemicals manufacturing.

It is very important to observe and document changes in the forcing terms such as gases, in order to understand and assess their influences of climate change, and to achieve more dependable longer range projections. Over the past three decades, the abundances of the atmosphere parameters (gases) were obtained from a lot of sources such as balloons, airplane and sparsely distributed measurement sites. While the observations were mostly limited to the surface and give important insights on flux variability the estimates have large uncertainties because there is not much atmospheric concentration data and more sensitive to sources and sinks. In situ measurements normally have the best accuracy from ground and aircraft, but the major shortfall is not being able to make daily global variation evaluations as well as cost a lot of money and strenuous efforts. There is a lack of data both in the lower - particularly over land - and upper troposphere (Tiwari et al., 2005).

A potentially favourable measurement method for closing some of these data gaps are the retrieval of gases from space. The satellite remote sensing has very good global coverage increase our capability to access the influence of human activities on the chemical composition of the atmosphere and on the climate changes (Clerbaux et al., 2003). Furthermore, can provide the quantitatively data with high spatial or temporal resolution (Dousset & Gourmelon, 2003). In addition, the free download satellite data provided by infrared measurements from the Atmospheric InfraRed Sounder (AIRS) on the NASA AQUA satellite to establish and produce the high quality spectrally resolved radiance climatologic for observing and detecting climate change, to understand the distribution of trace gases, sources and sinks, and to validate weather and climate models.

From 705 km above the Earth's surface, the AIRS measure the integrated impact of numerous atmospheric molecules emitting and absorbing radiation at various temperatures throughout the atmospheric path from the surface to the instrument. The AIRS, included on the EOS Aqua satellite launched on May 4, 2002, is the first of the new generation of meteorological advanced sounders for operational and research use (Aumann et al., 2003).

In Malaysia, one of Southeast Asia country, industrialization, urbanization and rapid traffic growth have contributed significantly to economic growth. Pockets of heavy pollution are being created by emissions from major industrial zones, a dramatic increase in the number of residences, office buildings, manufacturing facilities, increases in the number of motor vehicles and trans-boundary pollution. Besides that, Malaysia is situated in a humid tropical zone with heavy rainfall and high temperatures (Mahmud & Kumar, 2008; Tangang et al.,

2007); the cloudy conditions cover the study area becomes the obstacle to acquire a high quality and resolution for satellite data.

The AIRS ability to provide simultaneous observations of the Earth's atmospheric temperature, water vapour, land surface temperature and the distribution of greenhouse gases (GHG's) with cloud clearing system, makes AIRS/AMSU a very functional space instrument to observe and study the atmosphere reaction to increased several gases such as CO. AIRS with its two companion microwave instruments, (AMSU) and (HSB), form the integrated atmospheric sounding system (Lambrigtsen, 2003). AIRS is an infrared spectrometer/radiometer that covers the 3.7–15.4 m spectral range with 2378 spectral channels and used to measure the upwelling radiance from Earth (Strow et al., 2003).

In addition, AIRS is the first hyperspectral infrared radiometer designed, in order to provide a valuable data for National Ocean and Atmospheric Administration's National centers for Environmental Prediction (NCEP) and other weather forecasting centers, for the operational requirements in terms of weather forecasting in medium-range (Aumann, 2003). AMSU is a 15-channel microwave radiometer operating between 23 and 89 GHz. HSB is a four-channel microwave radiometer that makes measurements between 150 and 190 GHz (Lambrigtsen, 2003). It also provides information for several greenhouse gases, CO_2, CH_4, CO and O_3 as well as to improve weather prediction and study the water and energy cycle. AIRS measure CO total column by 52 channels with the uncertainty, which is estimated approximately 15-20 % at 500mb.

Between 1 August 2005 and 15 August 2005, the central, northern and eastern parts of Peninsular Malaysia experienced severe haze (DOE, 2005). Immense plumes of the gas emitted from forest and grassland burning in Indonesia forest fires in 2005 caused serious air pollution in Malaysia, and the air pollution reached extremely hazardous levels and forced schools and an airport to close. The fires affected 10,000 hectares of peat forest in Sumatra between Riau and north Sumatra. NOAA recorded 5420 hotspots from satellite images over the area between mid-July and mid-August (Jasim et al., 2009). The land and forest fires in the Riau Province of central Sumatra, Indonesia were the primary cause of transboundary pollution which was aggravated by the stable atmosphere conditions during the period. The hazy conditions reached its peak on 11-12 August 2005 when a haze emergency was declared in two areas in the Klang Valley and Kuala Selangor (DOE, 2005).

This study is based on the results from the analysis of CO total column retrievals from AIRS, Standard Level-3 Daily gridded product (AIRX3STD) and Monthly product (AIRX3STM) $1\times1^{\circ}$ Spatial resolution, Version 5 ascending data, using AIRS IR and AMSU, without-HSB, to investigate the daily distribution map of CO and the impact of Indonesia's forest fire, and the monthly CO distributions in the Peninsular Malaysia.

The presented AIRS daily Peninsular Malaysia CO maps from 12 - 25 August 2005 for study area show large-scale, long-range transport of CO from Indonesia forest fires. The sequence of daily maps shows the CO advection from central Sumatra to Malaysia. The daily measurements of CO concentration on August 2005 are higher than August 2007. The northern region (uppers the latitude 4) was more affected by forest fires than the rest area. Substantial seasonal variations demonstrate season-to-season changes in rainfall and drought patterns in different seasons. We see such seasonal variations in the biomass burning emissions in the late dry season, while industrial contributions are evident at smaller magnitudes on monthly distributions. The CO maps were generated using Kriging Interpolation technique. In this study, the CO concentration was accurately and precisely mapped from AIRS data over Peninsular Malaysia.

2. Southeast Asia, climatology, and air pollution

Southern Asia is one of the most heavily populated regions of the world with a vibrant mixture of cultures, which comprises over a quarter of the world's population. One activity which is prevalent to all people amidst this massive diversity is energy consumption, from biomass burning in cook stoves to fossil fuel usage in trucks and rickshaws (Lawrence, 2004). Southeast Asia is experiencing a similar rapid economic growth to that in Northeast Asia. Furthermore, a large source of several air pollutants may make an important contribution to regional and global pollution because of increasing anthropogenic emissions associated with biogenic emissions from large tropical forests. The greater oxidizing capacity in the tropical regions is due to a higher UV intensity, humidity, rapid development and industrialization (Streets et al., 2001).

In addition, Southeast Asia has many social, economic, and environmental impacts caused by forest and land fires. Fires considered one of the largest anthropogenic influences on terrestrial ecosystems after agricultural activities and urban, and its indeed critical elements in the Earth system, vegetation, linking climate, and land use (Lavorel et al., 2007). Tropical haze from peatland has serious negative impacts on the human health and regional economy, and peatland fires affect global carbon dynamics (De Groote et al., 2007).

The strong monsoon and the associated movement of the inter-tropical convergence zone (ITCZ) were dominated the climatology of continental Southeast Asia which mostly located in the equatorial; seasons are not as discrete as in more temperate zones. When the (ITCZ) relocates southern with the sun across Southeast Asia into the Southern Hemisphere, the wintertime winds over considerable of southern Asia usually come from the NE to SW. This region experiences a dry season for about six months (November - April) before the (ITCZ) moves back to the Northern Hemisphere and long-range transport of continental air masses from the Indian Ocean in the summer monsoon prevails during the subsequent wet season (May - October). In the Southeast Asia the dry and wet seasons occur at opposite times of the year due to the difference circulations in the southern and northern hemisphere (Pochanart, 2004).

In the onset of the northeast monsoon, cold surges originating from Siberia and northeast Asia brings significant amount of pollution to southeast Asia while crossing through the heavily-polluted regions of east Asia (Pochanart et al., 2003) and also caused heavy rains which lead to grave flooding in December of some years along east coast and Johor (Tangang et al., 2007). At the same time with strong subsidence in the early dry season, there are important increases in air pollutant levels in continental Southeast Asia (Pochanart, 2004). Maximum air pollutants levels occurs over continental Southeast Asia in the late months of dry season due to strong regional biomass burning and long-range transport of air masses from the western Asia and Middle East. The tropical biomass burning in Southeast Asia is a major source of atmospheric pollutants and strongly influenced by anthropogenic post-agricultural waste burning (Pochanart et al., 2003). In some years the high pollutant levels can be observed in late dry season, were results of La Niña influences combined with the convection in the ITCZ, was considerably weaker than normal, the resulted in incompetent ventilation of the pollutants out of the continental outflow and accumulation of aerosol levels and trace gas in the lower troposphere region (Lawrence, 2004).

3. Carbon monoxide pollution

Carbon monoxide is a colourless, odourless and at high concentration, a poisonous gas. The resulting from fossil fuel combustion has become a major issue, especially in air pollution. CO is emitted mainly from motor vehicle exhaust, industrial processes and open burning activities (Buchwitz et al., 2007). The situation become worsens for the countries with the instinctive fuel consumption and the increasing energy demand. Carbon monoxide is an important atmospheric constituent affecting climate and a major troposphere air pollutant (Buchwitz et al., 2006). Therefore, it is very important to acquire information regarding the distribution of carbon monoxide. The investigation of carbon monoxide globally has gained attention from researchers recently. It also affects the concentration of greenhouse gases such as methane and ozone. Besides that, carbon monoxide also being a secondary pollutant regarding the respiratory problem and affect the crop yields (Buchwitz et al., 2006).

World-wide, the anthropogenic sources produce about 50% of CO emissions with the remainder coming from biomass burning and oxidation of naturally occurring volatile hydrocarbons. CO is a product of incomplete produced by combustion of fossil fuel and biomass, and having an average lifetime of 2-4 months in the atmosphere (McMillan et al., 2005).

It is generally agreed that biomass burning accounts for about one quarter of CO emission to the atmosphere and its concentrations in the northern Hemisphere are much higher, where human population and industry are much greater than in the southern Hemisphere. CO emissions are generally 5-15% of CO_2 emissions from burning, depending on the intensity of the burn (Liu et al., 2005). Andreae & Merlet (2001) estimated the annual emission of CO from vegetation fires in tropical forests and savannas' is 342 Mt (1 Mt=109 kg) CO per year, while the total CO emission for all non-tropical forest fires is 68 Mt CO per year. A concentration of as little as 400ppm (0.04%) CO in the air can be fatal. The levels of normal carboxyhemoglobin in an average person are less than 5%, whereas cigarette smokers (two packs/day) may have to levels up to 9% (Delaney et al., 2001).

CO can direct observe from space due to its strong absorption properties in the thermal infrared (4.7 μm) and in the solar shortwave infrared (2.3 μm). Since past decade, there are a number of satellite instruments available, in order to measure troposheric CO globally, including Measurements Of Pollution In The Troposphere (MOPITT) [Emmons et al., 2007, 2009], SCanning Imaging Absorption spectroMeter for Atmospheric CartograpHY (SCIAMACHY) [Bovensmann et al., 1999; Burrows et al., 1995], AIRS (McMillan et al., 2005, 2008), Atmospheric Chemistry Experiment-Fourier Transform Spectrometer (ACE-FTS) [Clerbaux et al., 2005, 2008], Technology Experiment Satellite (TES) [Lopez et al., 2008; Luo et al., 2007], and Infrared Atmospheric Sounding Interferometer (IASI) [Fortems-Cheiney et al., 2009; Turquety et al., 2009]. These satellite data overcome the problem faced by in-situ measurements and commercial aircraft (Nedelec et al., 2003).

The AIRS CO retrievals shown here were produced as operational products by the NASA Goddard Earth Sciences (GES) Data Information and Service Center (DISC) have been employed in many studies. The first observation of tropospheric CO was presented by McMillan et al. (2005) during track the impact of major south American fire 22-29 September 2002, and the daily global maps showed the advection of a large CO plume with forward trajectories conforming to long-rang transport, from biomass burning, as far east as the southern Indian Ocean. The comparisons to in-situ aircraft profiles indicate AIRS CO retrievals at 500 mbar are accurate to at least 10% in the northern hemisphere and approaching the 15% accuracy target set by per-launch simulations. McMillan et al. (2007) investigated the wealth of AIRS CO retrievals information contained more than five years, and improved decisively both the characterization of the vertical distribution of CO information in the AIRS retrievals and the sensitivity to the lower tropospheric CO. The analysis showed the capabilities of AIRS see Megacities and large-scale interannual variations in emissions linked to El Nino.

Warner et al. (2007), Kopacz et al. (2010) compared AIRS retrievals against MOPITT for AIRS validation of global CO mixing ratios at 500 mbar. Furthermore, presented are the comparisons of the satellite profiles with in-situ profiles. Because of the simultaneous measurements of troposphere CO from instruments, the comparison and a cross reference are necessary to understand these two data sets and their effects to the scientific conclusions developed from them. Yurganov et al. (2010) compared troposphere CO profiles from AIRS against Atmospheric Emitted Radiance Interferometer (AERI) over North America, the Northern Hemisphere (NH) mid-latitudes, and over the tropics for the period 2002-2009, and there is a good agreement between AIRS and AERI in the surface and total column measurements. They showed that the three months temporal of CO burden minima between the NH mid-latitudes and tropics could effect from transport of lower CO burdens from south to north. In addition, tropics contain about half of the global air mass and CO burden anomaly are higher there than in the NH.

4. Study area

The study area is Peninsular Malaysia, which is located between 1° to 7° latitudes north and 99° to 105° longitudes east, south of Thailand, north of Singapore and east of the Indonesian island of Sumatra. An area (Fig. 1), covering 3.575×105 km^2, with a center at Pahang (102° E and 4° N) was selected for this study. The central dimensions of the study domain are 550 km E-W and 650 km N-S. The Titiwangsa Mountain is a range from the Malaysia–Thai border in the north running approximately south-southeast over a distance of 480km forms the backbone of the Peninsular and separating the western part from the western part. Surrounding the central high regions are the coastal lowlands (Suhaila & Jemain, 2007).

As a characteristic for a humid tropical climate, in Peninsular Malaysia the weather is warm and humid throughout the year with temperatures ranging from 21° C to 32° C. In Malaysia never experienced the excessive day temperatures which are found in continental tropical areas, very rarely been recorded air temperature of 38°C, the days are oftentimes hot and the nights are reasonably cool everywhere (Dasimah, 2009). The highest average temperatures are at April-May and July-August in most places, and the lowest average monthly temperatures are at November-January. Although the variations of the spatial and seasonal temperature are relatively small, they are, nevertheless, equitably definite in some respects and are deserving mention (Shahruddin & Mohamed, 2005). There is a definite variation of monthly mean temperature coincided with the monsoons, the annual fluctuations of roughly 1.5-2° C. The monsoons significantly affect the climate of Peninsular Malaysia. It experiences two rainy seasons throughout the year associated with the Northeast Monsoon (NEM) from November to February and the Southwest Monsoon (SWM) from May to August (Wong at el., 2009).

During the inter-monsoon months (usually occur in April and October), the wind light and variable, and thunderstorms develop causing Substantial rainfall in each of the two transition periods, especially in the west coast states. Monsoon changes as well as the effects of topography are the main factors that affect the rainfall distributions. The monsoon rainfalls form 81% of annual rain that falls in the entire Peninsular Malaysia, which was estimated approximately 2300mm (Suhaila & Jemain, 2009). Maximum rainfall is occurred near the end of the year during the early NEM in most of the areas, while second maximum rainfall during the intermonsoon months (April or May). The high intensity rainfall is absolutely absent during the SWM period except in the west coastal stations. The lowest monthly rainfall occurs in February, and the highest monthly rainfall in December (Varikoden et al., 2010).

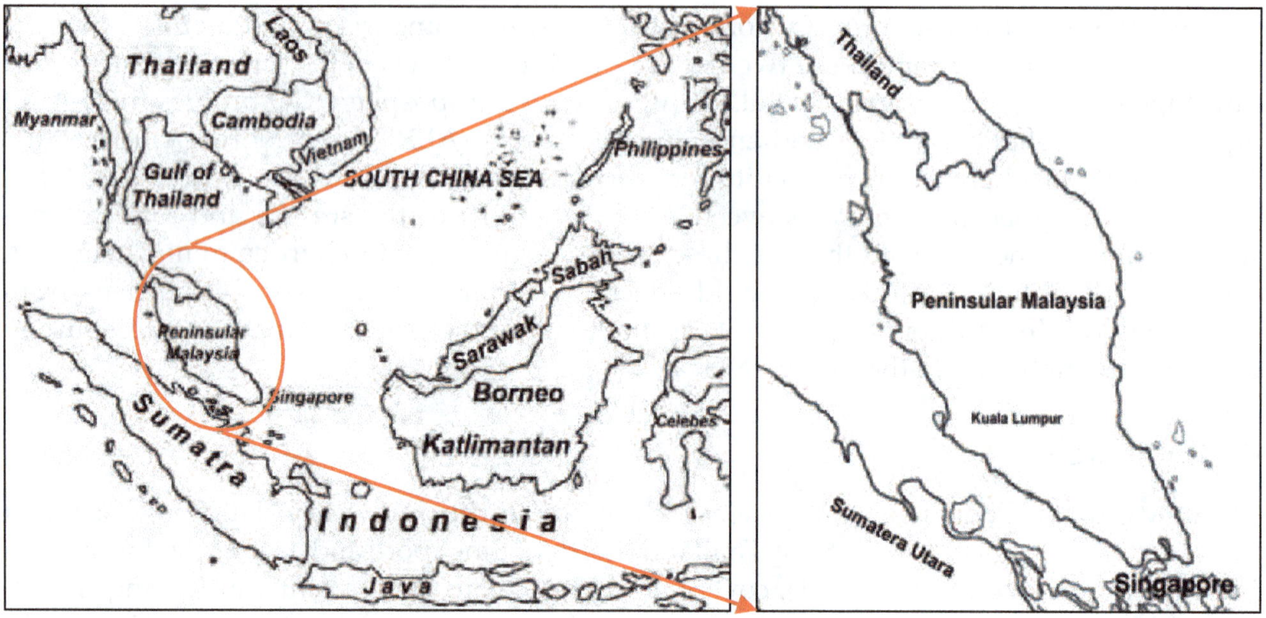

Fig. 1. The geographical feature of study area.

5. Acquisition and specification

The first new generation of meteorological advanced sounders for operational and research use was AIRS, one of several instruments onboard the EOS Aqua spacecraft launched May 4, 2002. AIRS instrument and its two companions microwave instruments, the Advanced Microwave Sounding Unit (AMSU) and the Humidity Sounder for Brazil (HSB), form the integrated atmospheric sounding system are characterizing and observing the entire atmospheric column from the surface to the top of the atmosphere in terms of temperature and surface emissivity, cloud amount and height, atmospheric temperature and humidity profiles and the spectral outgoing infrared radiation (Fishbein et al., 2007). The Aqua spacecraft coverage is pole-to-pole, and covers the globe two times a day, at orbit is polar sun-synchronous with a nominal altitude of 705 km (438 miles) and an orbital period of 98.8 minutes. The platform equatorial crossing local times are 1:30 in the afternoon (ascending) and 1:30 in the morning (descending). The respect cycle period is 233 orbits (16 days) with a ground track repeatability of +/- 20 km (Aumann et al., 2003).

The AIRS is a "facility" instrument developed by NASA as an experimental demonstration of the benefits of high resolution infrared spectra to science investigations and advanced technology for remote sensing (Pagano et al., 2006). The capability of AIRS/AMSU/HSB to supply simultaneous observations of the Earth's atmospheric temperature, land surface temperature, and ocean surface temperature, water vapor, cloud amount and cloud height, albedo, as well as new research products of greenhouse gas and aerosols, makes AIRS the most important EOS instrument for investigating several interdisciplinary issues to be addressed in Earth science (Pagano et al., 2003). The instruments suite was designed to measure the Earth atmospheric water vapor to an accuracy of 10% in 2 km layers in the lower troposphere and an accuracy of 50% in the upper troposphere. It also provides atmosphere temperature profiles with 1K/km accuracy in the troposphere and 1K/4 km layer in the stratosphere up to an altitude of 40 km. Also, provides integrated column burden for several trace gases (Chahine et al., 2006). Figure 2.1 shows the scanning geometry of AIRS and AMSU-A.

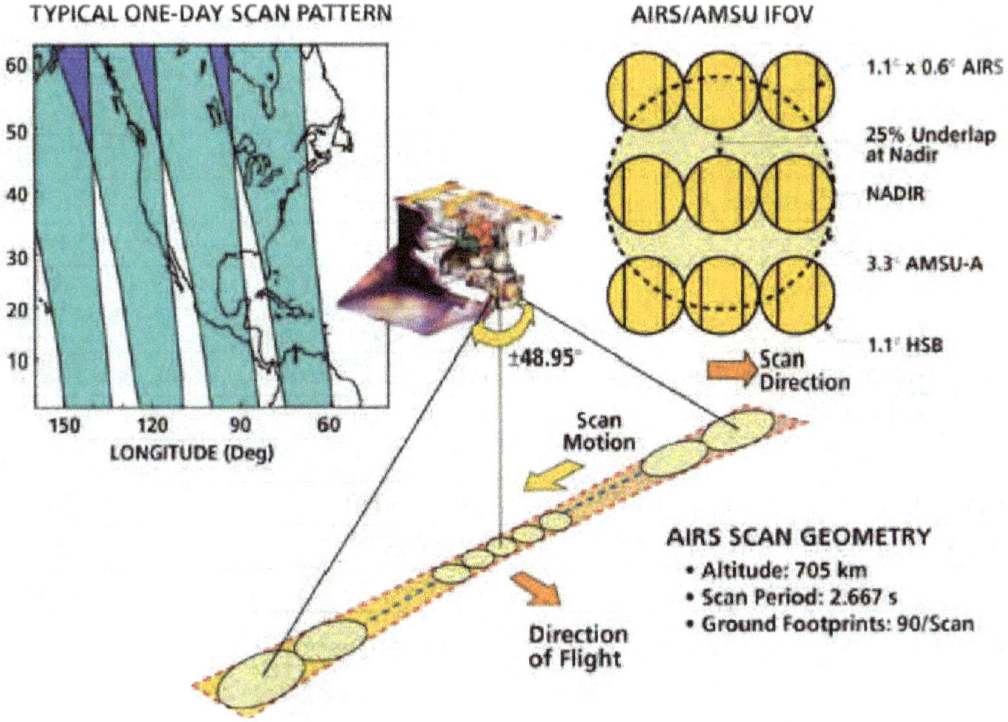

Fig. 2. Schematic showing the AIRS and AMSU scan geometrics [Jason, 2008].

AIRS is a continuously operating cross-track scanning sounder, consisting of a telescope that feeds a scale spectrometer. The AIRS instrument views the atmospheric infrared spectrum in 2378 channels with a nominal spectral resolving power $\lambda/\Delta\lambda$ ranging from 1086 to 1570 covering more than 95% of the earth surface and returning about three million spectra daily, in the 3.74-4.61 μm, 6.20-8.22 μm and 8.8-15.4 μm infrared wavebands at a nominal spectral resolution, also includes four visible/near-IR (Vis/NIR) channels between 0.40 and 0.94 μm, with a 2.3-km FOV (Strow et al., 2003).

The AIRS Science Processing System (SPS) is a set of programs, or Product Generation Executives (PGEs), utilize to process AIRS Science Data. These PGEs process raw, AIRS Visible (VIS), low level AIRS Infrared (AIRS), AMSU, and HSB instrument data to acquire temperature and humidity profiles. There are four distinct processing phases for processing the AIRS PGEs: Level 1A, Level 1B, Level 2 and Level 3 (Fishbein et al., 2007). Level 1A, Level 1B and Level 2 produce 240 granules of different products every day. Each product granule contains six minutes of data (Ye et al., 2007).

The L3 data are created from the L2 data product by binning them in 1°x1° grids. Level 3 products are statistical summaries of geophysical parameters that have been temporally aggregated and spatially re-sampled from lower level data products (e.g., Level 2 data) (Pagano et al., 2006). There are three AIRS Level 3 data products separately derived from Microwave-Only (MW-Only) retrievals and combined Infrared/Microwave (IR/MW) retrievals: daily, weekly and monthly as summarized in Table 1 Each product provides separate ascending (daytime) and descending (nighttime) binned data sets. When there is no coverage for that day, the daily Level 3 products will have gores between the satellite paths. The weekly Level 3 products may have missing data because of data dropouts. Monthly Level 3 products will probably contain complete global coverage with little missing data and without gores (Aumann et al., 2001).

Data Set	Short Name	Granule Size
L1B AMSU-A radiances	AIRABRAD	0.5 MB
L1B HSB radiances	AIRHBRAD	1.7 MB
L1B AIRS radiances	AIRIBRAD	56 MB
L1B VIS radiances	AIRVBRAD	21 MB
L1B AIRS QA	AIRIBQAP	5.6 MB
L1B VIS QA	AIRVBQAP	1.1 MB
L2 Cloud-cleared radiances	AIRI2CCF	10 MB
L2 Standard Product	AIRX2RET	5.4 MB
L2 Support Product	AIRX2SUP	20 MB
L3 standard daily product	AIRX3STD	~70 MB
L3 8-day standard product	AIRX3ST8	~103 MB
L3 monthly standard product	AIRX3STM	~105 MB

Table 1. Level 1, Level 2 and Level 3 data set products. (Jason, 2008).

6. Materials and methods

The study has been carried out for daily [six-day (12-23) August 2005 and six-day (13-24) August 2007], and monthly 2009 data. In order to evaluate and analysis the impact of Indonesia forest fire on CO distribution, and the monthly CO distributions in the study area. Five dispersed stations were selected across Peninsular Malaysia: Subang, Penang, Kuantan, Johor, and Kota Bharu to compare the daily data from August 2005 and August 2007. Results from the analysis of the retrieved for the CO obtained from AIRS ascending Level 3 daily standard (AIR×3STD) and Monthly standard product (AIRX3STM) data. Generally, twelve daily and twelve monthly L3 ascending granules were downloading to obtain the desired output. Extract the (AIR×3STD) and (AIRX3STM) product's files from the Satellite using the AIRS website, and saves in HDF-EOS4 files; this is a convenient file extension that can be easily extracted data from it and arrange in table using MS Excel.

Data including the corresponding location and time along the satellite track in a HDF (Hierarchical Data Format) format on monthly basis. Map of the study area was conducted by using Photoshop CS and SigmaPlot 11.0 software to analyze CO data distribution along the study period. To better assess the impacts and distribution of CO above Peninsular

Malaysia, the maps of CO were generated by using Kriging interpolation technique for the daily (August 2005 and 2007), and monthly 2009. The CO data were obtained from 1°*1° degree (latitude × longitude) spatial resolution ascending orbits.

7. Data analysis and results

7.1 Daily peninsular Malaysia CO maps and comparisons

The skies over Peninsular Malaysia were noticeably hazier than normal on mid-August 2005. In this section, we examine the impact of one smoke transport event on CO levels in Peninsular Malaysia. The elevated levels of CO found in smoke plumes from incomplete combustion of forest fires can lead to important further CO production downwind of the fires. To better assess the influences of forest fires and the transport smoke filled air mass on the local air quality in Peninsular Malaysia. We examined the data from AIRS for two different periods daily (12-23) August 2005 and (13-24) August 2007.

The six maps in Fig. 3 (12-23) August 2005 illustrate the extent of AIRS Daily coverage of Total column CO, the nominal peak of AIRS vertical sensitivity and the magnitude of the variations in atmospheric CO over Peninsular Malaysia. It observed an elevation in the CO values higher than normal rates during the period from mid-July to mid-August 2005 in Malaysia (Rajab et al., 2009).

The AIRS data in Fig. 3 indicate numerous fires in the regions of enhanced CO over Malaysia. In Indonesia, the fire counts peaked on 13-15 August in Sumatra (DOE, 2009). From Fig. 3 (12, 14 and 16 August), can see the advection of the CO plume from Sumatra, Indonesia, to Peninsular Malaysia. The CO Pollution for this event was characterized the increase in CO by 60% in the northern region (uppers the latitude 4, e.g. Penang and Kota Bahru), 35% in the central and east coast district (e.g. Kuala Lumpur and Kuantan), and 20% in the southern area (e.g. Johor). Plainly evident the CO total column values are high in 12 August, increase to be highest in 14 August, and a gradual decrease of CO values after 16 with a reduction impact of the forest fire, compared to previous days (DOE, 2005). The highest value was when the hazy conditions reached its peak on 12 August over Selangor (2.68×10^{18} molecules/cm^2).

Fig. 3 (18, 21 and 23), while the CO value is still high in the Sumatra, shows the gradual decrease of the concentration of CO in most areas in coinciding with the declaration of the end of the hazy conditions. The lowest value of CO was on 23 August over Terengganu (1.75×10^{18} molecules/cm^2).

From Fig. 4, illustrate the CO values from 13 to 24 August 2007, were normal circumstances in the absence of any event can observe the highest CO total column over central areas (e.g. Selangor and Negeri Sembilan), Penang and Johor. These regions represent biomass burning sources, long-range transport and industrial/domestic fuel sources. Enhanced CO values in Selangor on 13 August (1.95×10^{18} molecules/cm^2). The greater draw down of CO total column occurs over pristine marine environment in the north east coast of Terengganu and in the Inland region of north Kelantan (1.56×10^{18} molecules/cm^2).

To illustrate the biases between the two data sets, which are caused by biomass burning, the daily August 2005 and 2007 zonal averaged CO total column is showed in Fig. 5. In Fig. 5, a daily series graph of CO August 2005 (solid line) compared to daily CO August 2007 (dots line) from observed AIRS for five stations; kota Bahru, Penang, Kuantan, Subang, and Johor, where the day is on the x-axis and CO total column (molecules/cm^2) on the y-axis. It can be seen clearly, in all the stations the daily CO concentration on August 2005 is higher than August 2007. There is a clear difference between two lines, especially in 14 August when the impacts of forest fire up to a maximum. The discrepancies become less after 17 August with a lower impact of fires. In general, south to north the CO values differences are larger than West to East between two lines over study areas. The largest differences are in the north regions (uppers the latitude 4).

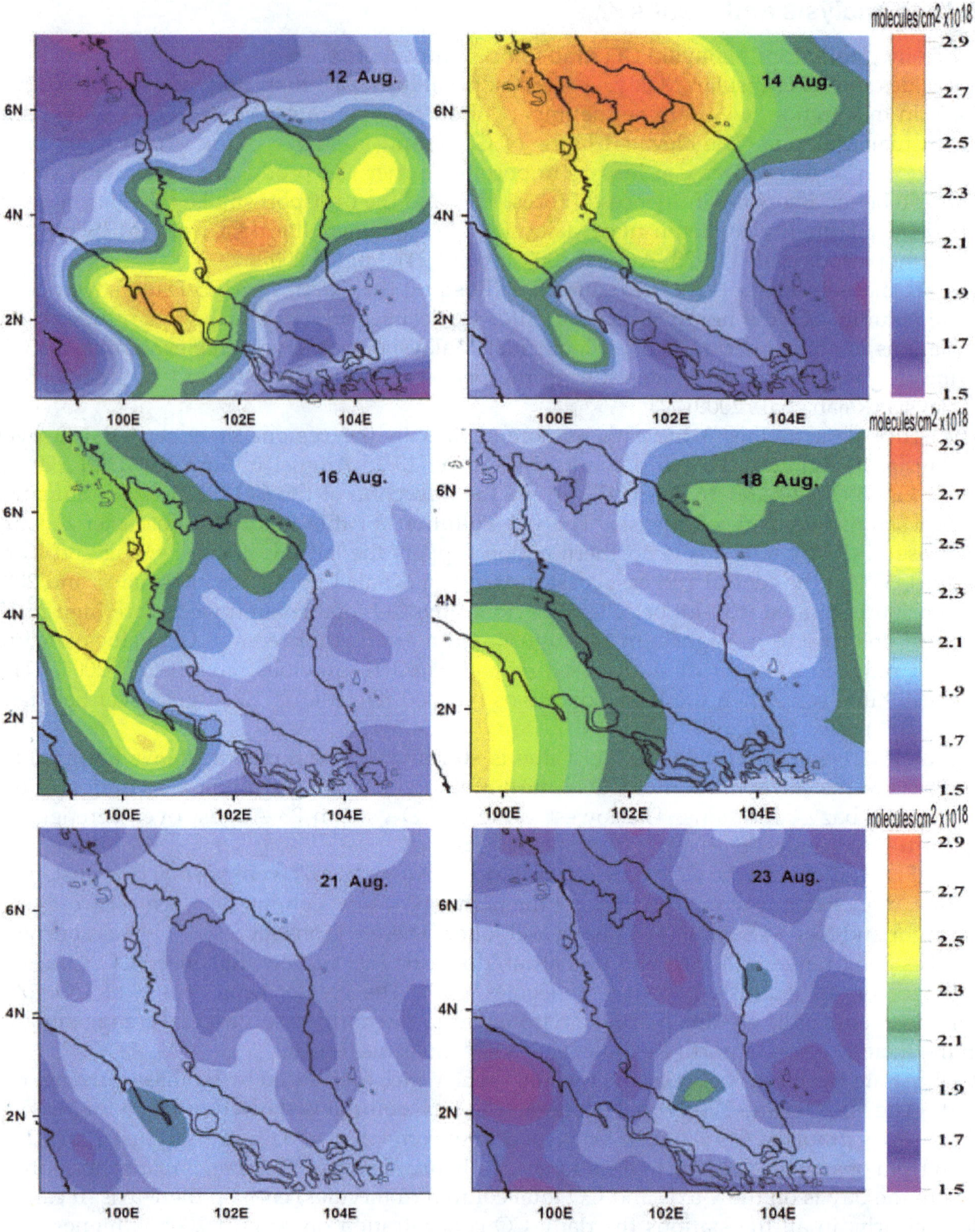

Fig. 3. CO total column (molecules/cm^2) for daily (12-23) August 2005 in Peninsular Malaysia.

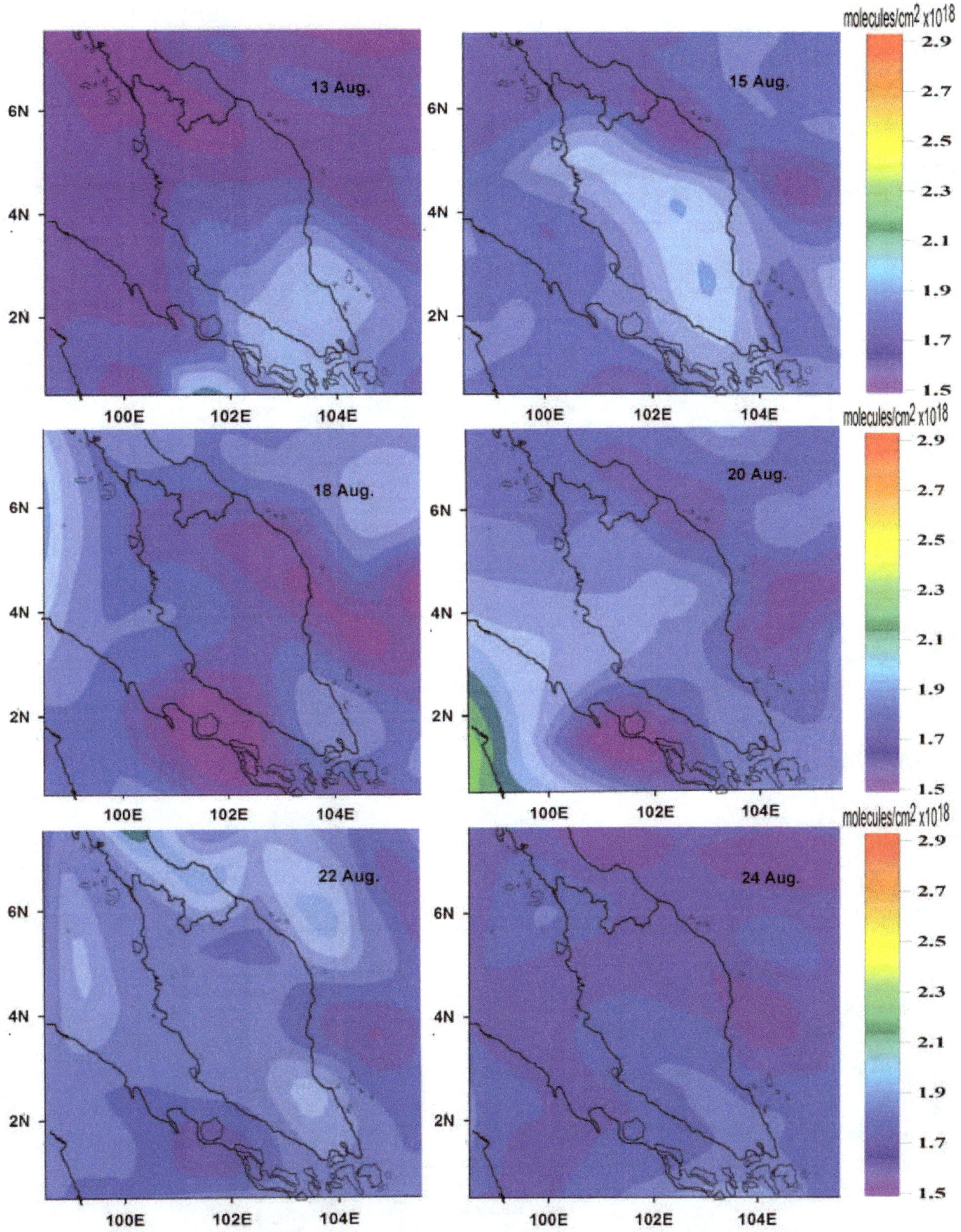

Fig. 4. CO total column (molecules/cm^2) for daily (13-24) August 2007 in Peninsular Malaysia.

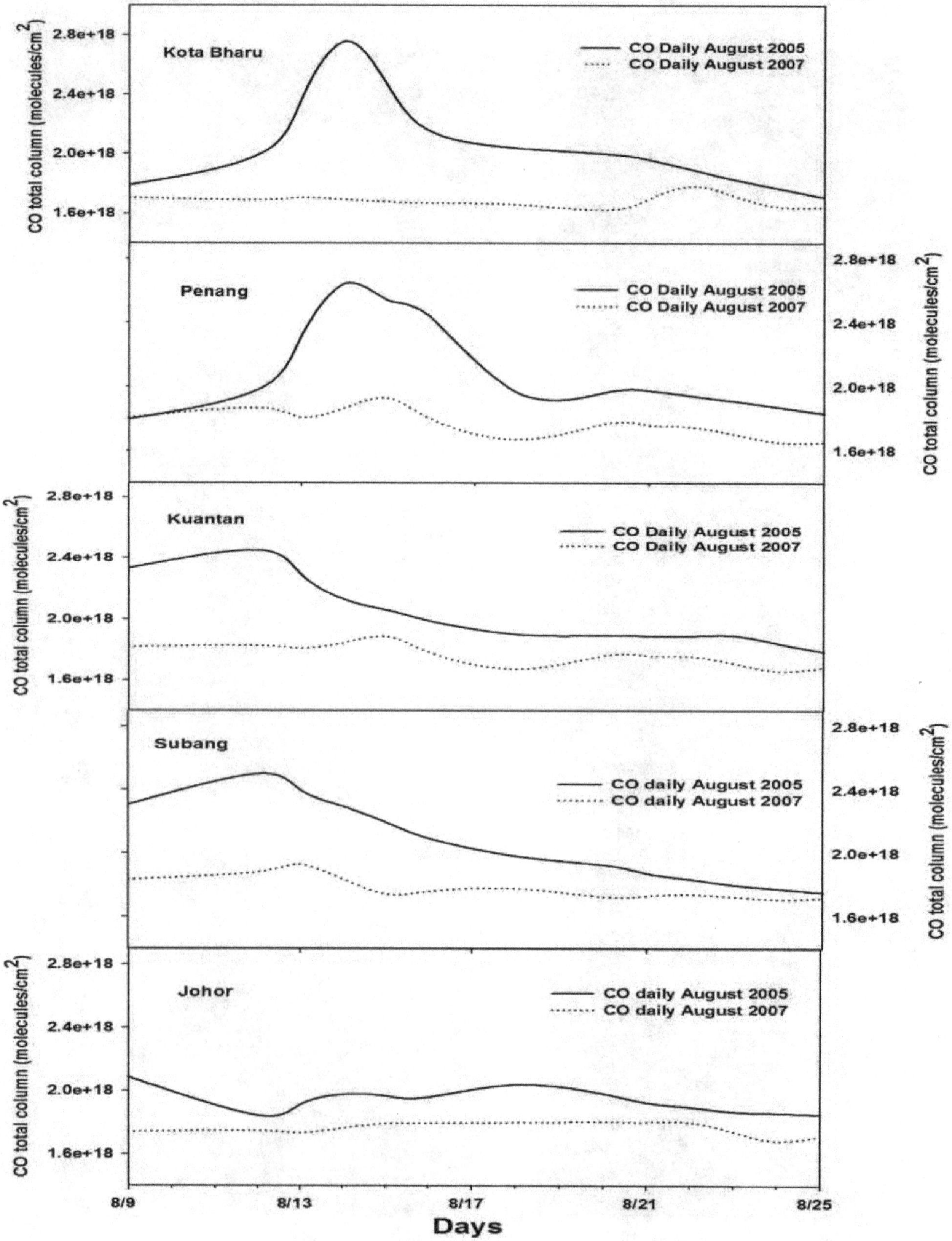

Fig. 5. The CO total column August 2005 (solid line) compared to daily CO total column August 2007 (dots line) from observed AIRS for five stations; Kota Bahru, Penang, Kuantan, Subang, and Johor.

7.2 Monthly peninsular Malaysia CO maps for 2009

The retrieved CO total column amount (CO_total_column_A) Level-3 monthly (AIRX3STM) 1°×1° spatial resolution used for mapping CO for 2009. Based on the combination of rich local sources of CO along with the transport of additional significant amount of pollution from Siberia and northeast Asia brings in the beginning of the northeast monsoon led to increase the values of CO at early dry season (Pochanart et al., 2004), that are shown in Fig. 6 for dry season.

Plainly evident the impact of peatland fires in several areas in Sumatra, Indonesia, on the values of CO in the north and central of Peninsular Malaysia during January, with continuing its impact on the central and southern regions during February. Unusually, there was a decrease in the values of CO at the end of the dry season during March-April due to unexpected reasonable rainfall during this period throughout the Peninsular Malaysia (Malaysian Meteorological Department, 2009). The highest value of CO in dry season was (2.19×10^{18} molecules/cm^2) over Selangor on January {at 101.5°×3.5°, red pixels}, while the less value was (1.75×10^{18} molecules/cm^2) over Perak on April {at 101.5°×5.5°, violet pixels}. The strong effects of regional biomass burning during the dry season, a regular annual occurrence in January to April, makes it extremely difficult to detect any evidence of the long-range transport of air pollution to Southeast Asia from other regions (Pochanart et al., 2004).

Normally, in the late dry season, large regional biomasses burning occur and the impact of air masses pollution transport has significant from the western Asia, Middle East and East Asia (Pochanart et al., 2003). The El-Niño episodes can lead to immense displacement of rainfall regions in the tropics, bringing torrential rain to otherwise arid regions and drought to vast areas. Uniform warming prevail all the regions in Malaysia during an El-Niño event, especially during the November to March (Cheang, 1993). When El-Niño has a drought effect, there will be large biomass burning and increase CO emission (McMillan et al., 2007).

Fig. 7 shows that the CO have low value in the north regions, upper the latitude 5, and low to moderate in the rest of regions during the early wet season (May-July), due to the high levels of moisture and incident solar irradiation causes the production of the OH, the primary oxidizer for CO. When OH abundance is near a maximum; CO is near minimum (McMillan et al., 2007). Considerable values of OH are sufficient to provide significant spoilage rates of trace gases in the haze outflow, outputting for example in a CO lifetime of only ~15 days near the surface. The subtle differences peak of biomass burning indicate differences in transport patterns as well as differences in rainfall patterns across the region, the enhanced CO emission correlates with occasions of less rainfall (McMillan et al., 2007). Additionally, the lack of rain results in long carbonaceous lifetimes, this lengthens the time during which oxidation reactions can take place (Rasch et al., 2001).

In the late wet season (August-October), observed moderate value in the north regions, upper the latitude 5, while slightly high to high values in the rest of regions due to the impact of peatland fires from Sumatra and transboundary pollution which was aggravated by hot weather conditions (DOE, 2009). The highest value was (2.153×10^{18} molecules/cm^2) over Johor on October {at 103.5°×1.5°, red pixels}, while the less value was (1.702×10^{18} molecules/cm^2) over Perak on June {at 101.5°×5.5°, violet pixels} during the wet season.

During the summer monsoon (wet season) May - October, the marine air masses from the middle and low latitudes of the Southern Hemispheric Indian Ocean are dominated continental Southeast Asia. These marine air masses carry small to moderate amounts of air pollution to southwest Asia. In addition, less regional biomass burning occurred and the impact of air masses is less significant from the Pacific and Indian Oceans (Pochanart et al., 2003).

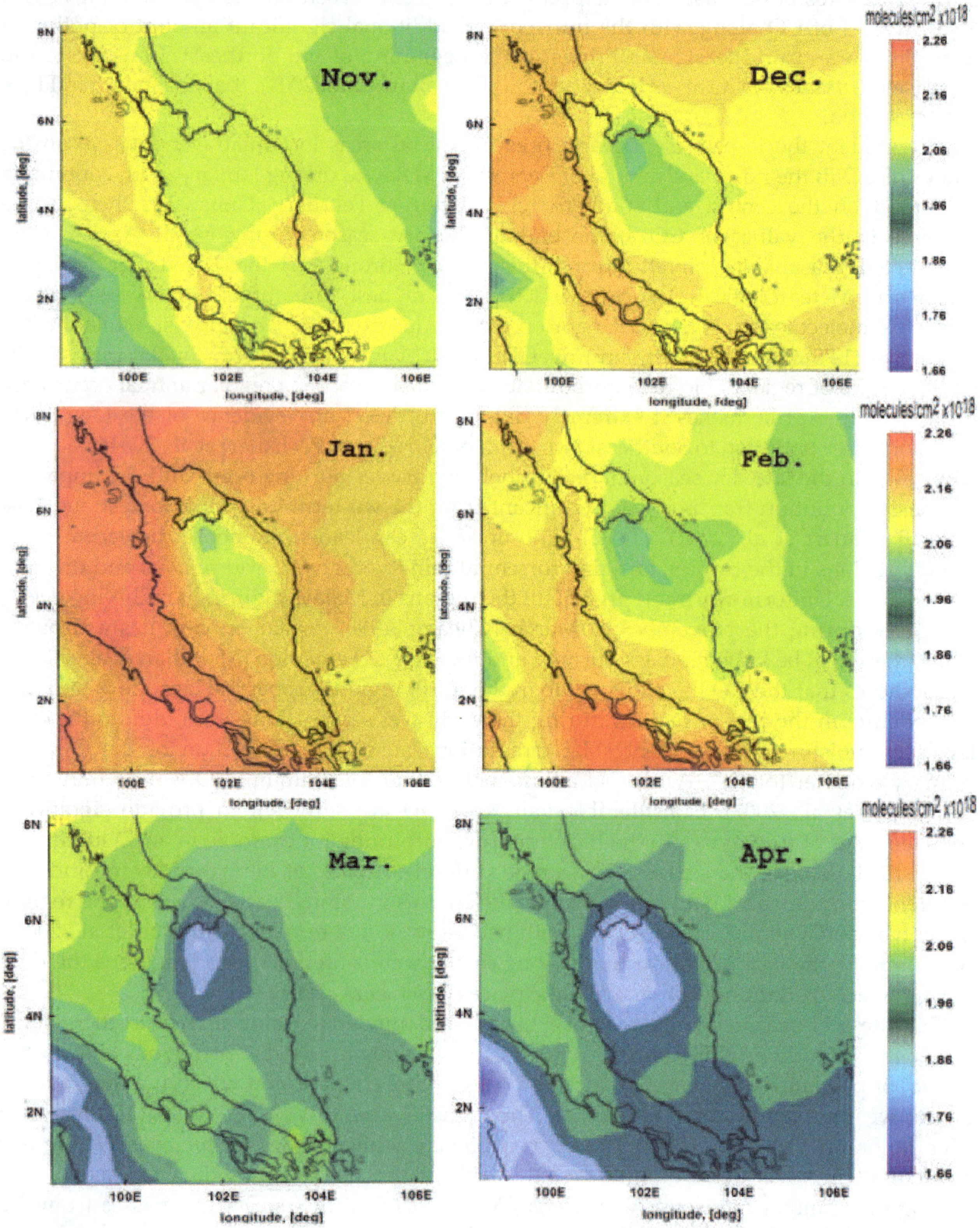

Fig. 6. AIRS monthly coverage retrieved total column carbon monoxide (CO) for dry season [November to April] 2009.

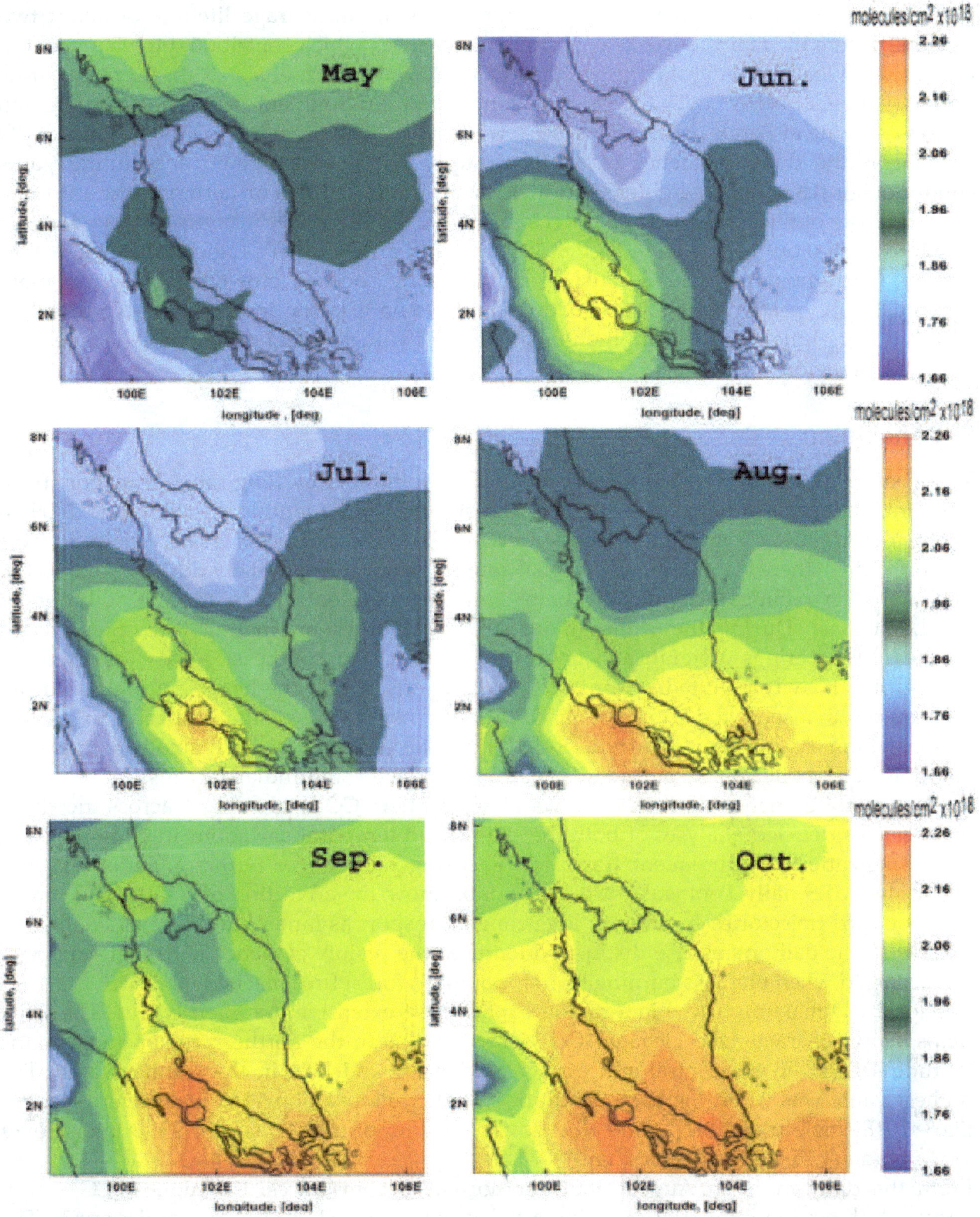

Fig. 7. AIRS monthly coverage retrieved total column carbon monoxide (CO) for wet season [May to October] 2009.

In short, observed the highest values of CO occurred when biomass burning during dry season, and also over the industrial and congested urban zones (it is the most abundant pollutant in urban atmosphere and very stable, having an average lifetime of about two months) where the main source of emission was vehicles which contributed to 95 percent of CO emission load in Malaysia during 2009 (DOE, 2009; Kopacz, et al., 2010). A greater draw down of the CO occurs over the pristine continental environment in the northeast region regions on June at Perak (101.5°×5°) during wet season. This was due to lack of sources CO as well as the direct influence of south westerly wind, which remove polluting gases continuously (CO slightly lighter than air) (Jasim et al., 2010). Furthermore, the rain is a great cleanser of the atmosphere so the (Hoskins, 2001).

In addition, the CO values were higher in the central and southern regions than the rest of regions throughout the year because there are many sources, crowded cities, and more affected by the peatland fires in Sumatra. Can be seen also the impact of El-Nino on CO values during June-August from the unusually high values of CO in this period, especially in southern region.

8. Conclusion

This chapter has reviewed the impact of the CO pollutants by long-range transport from Sumatra, Indonesia, forest fires during August 2005, and monthly distributions 2009 in Peninsular Malaysia. While an immense amount about these issues has been analyzed over the forest fires, there are still many regional sources making it extremely difficult to detect it. The monsoon regimes provide the main climatological controls on the air pollution in Southeast Asia. During the winter monsoon, cold surges originating from Siberia and northeast Asia brings a significant amount of pollution to Southeast Asia while crossing through the heavily-polluted regions of East Asia and caused the heavy rain falls. During the summer monsoon, the impact of air pollution transport is less significant due to the less regional biomass burning occurred, and the marine air masses carry small to moderate amounts of air pollution to southwest Asia from the Pacific and Indian Oceans.

As demonstrated here, AIRS' daily views of atmosphere CO total column across the study area enables detailed analyses of both the spatial and temporal variations in emissions and the visualization of subsequent transport. Focusing on a major Sumatra, Indonesia, fire event, the AIRS daily Peninsular Malaysia maps show the advection of a large CO plume with forward trajectories confirming long-range transport as far as north east as Peninsular Malaysia. The daily maps also distinctly identified the plainly evidence with high values of CO occurred when biomass burning from Indonesia's forest fires reached its peak.

The local CO maximum was in a region experienced extensive the intense fires. The daily maps shows characterized elevated CO values by 60% in the northern region (uppers the latitude 4), 35% in the central and east coast district, and 20% in the southern area. The highest value was when the hazy conditions reached its peak on 12 August over Selangor (2.68×1018 molecules/cm2). And the lowest value of CO was on 23 August over Terengganu (1.75×10^{18} molecules/cm^2).

From the daily maps in August 2007 for normal circumstances, the highest CO values occurred above the industrial and congested urban zones (1.95×10^{18} molecules/cm^2). The greater draw down of CO total column occurs over pristine marine environment in the north east coast of Terengganu and in the Inland region of north Kelantan (1.56×10^{18} molecules/cm^2).

It can be seen clearly from the comparison between the daily August 2005 and 2007; in all the stations, the daily CO concentration on August 2005 is higher than August 2007. There is a clear large difference between two lines, especially in 14 August when the impacts of forest fire up to a maximum. The discrepancies become less after 17 August coincides with a reduction effect of the forest fires.

The AIRS/Aqua Level 3 Monthly gridded product (AIRX3STD) 1×1° spatial resolution, Version 5 data using AIRS IR and AMSU, without-HSB, employed to investigate a monthly distribution map of CO total column over study area for 2009. The high CO values observed during the dry season, due to the strong influences of regional biomass burning, a regular annual occurrence in January to April, and coupled with the significant impacts of air mass pollution transport from the East Asia to by winter monsoon. The highest value was (2.19×10^{18} molecules/cm^2) over Selangor, while the less value was (1.75×10^{18} molecules/cm^2) over Perak.

The low CO values prevail on the wet season due to the less biomass burning occurred, and the marine air masses from the middle and low latitudes of the Southern Hemispheric Indian Ocean carried small to moderate amounts of air pollution. Furthermore, the high levels of moisture and incident solar irradiation lead to the production of the OH, the primary oxidizer for CO. The OH abundance is near a maximum; CO is near minimum. The highest value was (2.153×10^{18} molecules/cm^2) over Johor.

In short, the highest CO values was during biomass burning occurred in dry season, and over industrial and congested urban zones. The cleaner areas and the less CO value areas throughout the year in Peninsular Malaysia was Perak at (long. 101.5°× lat. 3.5°), where the green lands, vast forest and lack sources of pollution.

This study has provided evidence for the impact of remote CO total column emissions and forest fire on CO pollution levels above study area using satellite data. Furthermore, it enhanced our knowledge on AIRS detection of CO emission from forest fire, the utility and accuracy of remotely sensed atmospheric CO total column and abundances from AIRS. Satellite measurements are able to measure the increase of atmospheric CO values over different areas. CO maps from AIRS will lead to a more understanding of the CO budgets.

9. Acknowledgement

The authors gratefully acknowledge the financial support from the Relationship between Heavy Rain, Flash Floods and Central Pressure in Malaysia Grant, account number: 1001/PFIZIK/811152 and USM-RU-PRGS Grant, account number:1001/PFIZIK/841029. We would like to thank the technical staff who participated in this study. Thanks are also extended to USM for support and encouragement.

10. References

Andreae, M. O. & Merlet, P. (2001). Emission of Trace Gases and Aerosols From Biomass Burning. *Global Biogeochemical Cycles*, Vol.15, pp.955–966.

Aumann, H. H., Goldberg, M., Mcmillin, L., Rosenkranz, P., Staelin, D., Strow, L. & Susskind, J. (2001). AIRS-Team Retrieval For Core Products and Geophysical Parameters. In AIRS Team - NASA (Ed., Jet Propulsion Laboratory.

Aumann, H. H., Chahine, M. T., Gautier, C., Goldberg, M. D., Kalnay, E., McMillin, L. M., Revercomb, H., Rosenkranz, P. W., Smith, W. L., Staelin, D. H., Strow, L. L. & Susskind, J. (2003). AIRS/AMSU/HSB on the Aqua mission: Design, Science Objectives, Data Products, and Processing Systems. *IEEE Transaction On Geoscience And Remote Sensing*, Vol.41, No.2, pp.253-264.

Bovensmann, H., Burrows, J. P., Buchwitz, M., Frerick, J., Noël, S., Rozanov, V. V., Chance, K. V. & Goede, A. P. H. (1999). SCIAMACHY - Mission Objectives and Measurement Modes. *Atmospheric Chemistry and Physics*, Vol.56, pp.127-150.

Buchwitz, M., Khlystova, I., Bovensmann, H. & Burrows, J. P. (2007). Three Years of Global Carbon Monoxide From SCIAMACHY: Comparison With MOPITT and First Results Related To The Detection of Enhanced CO Over Cities. *Atmospheric Chemistry and Physics*, Vol.7, pp.2399-2411.

Buchwitz, M., de Beek, R., Noël, S., Burrows, J. P., Bovensmann, H., Schneising, O., Khlystova, I., Bruns, M., Bremer, H., Bergamaschi, P., Körner, S. & Heimann, M. (2006). Atmospheric Carbon Gases Retrieved From SCIAMACHY By WFM-DOAS: Version 0.5 CO and CH_4 and Impact of Calibration Improvements on CO_2 Retrieval. *Atmospheric Chemistry and Physics*, Vol.6, pp.2727-2751.

Burrows, J. P., Hölze, E., Goede, A. P. H., Visser, H. & Fricke, W. (1995). SCIAMACHY - Scanning Imaging Absorption Spectrometer for Atmospheric Chartography. *Acta Astronaut*, Vol.35, No.7, pp.445-451.

Chahine, M. T., Thomas, S. P., Hartmut, H. A., Robert, A., Christopher, B., John, B., Luke, C., Murty, D., Eric, J. F., Mitch, G., Catherine, G., Stephanie, G., Scott, H., Fredrick, W. I., Ramesh, K., Eugenia, K., Bjorn, H. L., Sung-Yung, L., John, L. M., McMillan, W. W., Larry, M., Edward, O. T., Henry, R., Philip, R., William, S. L.,David, S., Strow, L. L., Joel, S., David, T., Walter, W. & Lihang, Z. (2006). AIRS Improving Weather Forecasting and Providing New Data on Greenhouse Gases. *American Meteorological Society*, Vol.87, pp.911 -926.

Cheang, B. K. (1993). Interannual Variability of Monsoons In Malaysia and Its Relationship With ENSO. *Proc. Indian Acad. Sci. (Earth Planet. Sci.),* Vol 102, pp. 219-239.

Clerbaux, C., Hadji-Lazaro, J., Turquety, S., Mégie, G. & Coheur, P. –F. (2003). Trace Gas Measurements From Infrared Satellite For Chemistry and Climate Applications. *Atmospheric Chemistry and Physics*, Vol.3, pp.1495-1508.

Clerbaux, C., Coheur, P. F., Hurtmans, D., Barret, B., Carleer, M., Colin, R., Semeniuk, K., McConnell, J. C., Boone, C. & Bernath, P. (2005). Carbon Monoxide Distribution From The ACE-FTS Solar Occulation Measurements. *Geophysical Research Letters*, Vol.32, pp.1-4, DOI: 10.1029/2005GL022394.

Clerbaux, C., George, M., Turquety, S., Walker, K. A., Barret, B., Bernath, P., Boone, C., Borsdorff, T., Cammas, J. P., Catoire, V., Coffey, M., Coheur, P. -F., Deeter, M., De Mazière, M., Drummond, J., Duchatelet, P., Dupuy, E., de Zafra, R., Eddounia, F., Edwards, D. P., Emmons, L., Funke, B., Gille, J., Griffith, D. W. T., Hannigan, J., Hase, F., Höpfner, M., Jones, N., Kagawa, A., Kasai, Y., Kramer, I., Le Flochmoën, E., Livesey, N. J., López-Puertas, M., Luo, M., Mahieu, E., Murtagh, D., Nédélec, P., Pazmino, A., Pumphrey, H., Ricaud, P., Rinsland, C. P., Robert, C., Schneider, M., Senten, C., Stiller, G., Strandberg, A., Strong, K., Sussmann, R., Thouret, V., Urban, J. & Wiacek, A. (2008). CO Measurements from The ACE-FTS Satellite Instrument: Data Analysis and Validation Using Ground-Based, Airborne and Spaceborne Observations. *Atmospheric Chemistry and Physics*, Vol.8, pp.2569–2594.

Daniel, J. S. & Solomon, S. (1998). On The Climate Forcing of Carbon Monoxide. *Journal of Geophysical Research*, Vol.103, pp.13249-13260.

Dasimah, O. (2009). Urban Form and Sustainability of a Hot Humid City of Kuala Lumpur. European Journal of Social Sciences vol.8, pp.353-359.

De Groot, W. J., Field, R. D., Brady, M. A., Roswintiarti, O. & Mohamad, M. (2007). Development of The Indonesian and Malaysian Fire Danger Rating Systems. *Mitig Adapt Strat Glob Change*, Vol.12, pp.165–180.

Delaney, K., Ling, L., & Erickson, T. (2001). In Ford Md, Clinical Toxicology, WB Saunders Company, ISBN 0-7216-5485-1.

Department of Environment (DOE), M. (2005). Malaysia Environmental Quality Report. Petaling Jaya.

Department Of Environment (DOE), M. (2009). Malaysia Environmental Quality Report. Petaling Jaya.

Dousset, B., & Gourmelon, F. (2003). Satellite Multi-sensor Data Analysis of Urban Surface Temperatures and Landcover. *ISPRS Journal of Photogrammetry and Remote Sensing*, Vol 58, pp. 43-54.

Emmons, L. K., Pfister, G. G., Edwards, D. P., Gille, J. C., Sachse, G., Blake, D., Wofsy, S., Gerbig, C., Matross, D. & Nedelec, P. (2007). Measurements of Pollution in the Troposphere (MOPITT) Validation Exercises During Summer 2004 Field Campaigns Over North America. *Journal of Geophysical Research*, Vol.112, D12S02, DOI: 10.1029/2006JD007833, 2007.

Emmons, L. K., Edwards, D. P., Deeter, M. N., Gille, J. C., Campos, T., Nédélec, P., Novelli, P. & Sachse, G. (2009). Measurements of Pollution In The Troposhere (MOPITT) validation through 2006. *Atmospheric Chemistry and Physics*, Vol.9, pp.1795-1803.

Fishbein, E., Granger, S., Lee, S. Y., Manning, E., Weiler, M., Blaisdell, J., & Susskind, J. (2007). AIRS/AMSU/HSB Version 5 Data Release User Guide. In Atmospheric Infrared Sounder, A., EOS, Ed., California Institute of Technology.

Fortems-Cheiney, A., Chevallier, F., Pison, I., Bousquet, P., Carouge, C., Clerbauz, C., Coheur, P. –F, George, M., Hurtmans, D. & Szopa, S. (2009). On The Capability of IASI Measurements to Inform about CO Surface Emissions. *Atmospheric Chemistry and Physics*, Vol.9, pp.8735-8743.

Hoskins, A. J. (2001). Ozone Matters. *Indoor and Built Environment*, Vol.10, pp.1-2.

Jasim, M. R., MatJafri, M. Z., Lim, H. S. & Abdullah, K. (2009). Indonesia Forest Fires Exacerbate Carbon Monoxide Pollution Over Peninsular Malaysia During July to September 2005. *Proceedings of Sixth International Conference on Computer Graphics, Imaging and Visualization*, DOI: 10.1109/CGIV.2009.96, Tianjin University, Tianjin, China, August, 2009.

Jasim, M. R., Lim, H. S., MatJafri, M. Z. & Abdullah, K. (2010). Daily Carbon Monoxide (CO) Abundance from AIRS Over Peninsular Malaysia. *Journal of Materials Science and Engineering*, Vol.4, pp.93-99.

Jason, L. (2008). README Document for AIRS Level-2 Version 005 Standard Products. In Goddard Earth Sciences Data And Information Services Center (Ed., National Aeronautics and Space Administration (NASA).

Kopacz, M., Jacob, D. J., Fisher, J. A., Logan, J. A., Zhang, L., Megretskaia, I. A., Yantosca, R. M., Singh, K., Henze, D. K., Burrows, J. P., Buchwitz, M., Khlystova, I., McMillan, W. W., Gille, J. C., Edwards, D. P., Eldering, A., Thouret, V. & Nedelec, P. (2010). Global Estimates of CO Sources with High Resolution by Adjoint Inversion of Multiple Satellite Datasets (MOPITT, AIRS, SCIAMACHY, TES). *Atmospheric Chemistry and Physics*, Vol.10, pp.855-876.

Lambrigtsen, B. H. (2003). Calibration of The AIRS Microwave Instruments. *IEEE Transaction On Geoscience And Remote Sensing*, Vol.41, No.2, pp.369-378.

Lavorel, S., Flannigan, M. D., Lambin, E. F. & Scholes, M. C. (2007). Vulnerability of Land Systems to Fire: Interactions Among Humans, Climate, the Atmosphere, and Ecosystems. Mitig Adapt Strat Glob Change, Vol. 12, pp. 33–53.

Lawrence, M. G. (2004). Export of Air Pollution From Southern Asia and Its Large-Scale Effects. IN STOHL, A. (Ed.) The Handbook of Environmental Chemistry. Berlin, Springer.

Liu, J., Drummond, J. R., Li, Q., Gille, J. C. & Ziskin, D. C. (2005). Satellite Mapping of CO Emission from Forest Fires in Northwest America Using MOPITT Measurements. *Remote Sensing of Environment*, Vol.95, pp. 502–516.

Lopez, J. P., Luo, M., Christensen, L. E., Loewenstein, M., Jost, H., Webster, C. R. & Osterman, G. (2008). TES Carbon Monoxide Validation during Two AVE Campaigns Using the Argus and ALIAS Instruments on NASA's WB-57F. *Journal of Geophysical Research Atmospheres*, Vol.113, D16S47, DOI: 10.1029/2007JD008811.

Luo, M., Rinsland, C. P., Fisher, B. M., Sachse, G., Diskin, G., Logan, J. A., Worden, H. M., Kulawik, S. S., Osterman, G., Eldering, A., Herman, R. & Shephard, M. W. (2007). TES Carbon Monoxide Validation with DACOM Aircraft Measurement During INTEX-B 2006. *Journal of Geophysical Research*, Vol.112, D24S48, DOI: 10.1029/2007JD008803.

Mahmud, M. & Kumar, T. S. V. V. (2008). Forecasting Severe Rainfall in the Equatorial Southeast Asia. *GEOFIZIKA*, Vol. 25, pp. 109-127.

Malaysian Meteorological Department (2009). General Climate of Malaysia. Selangor, Ministry of Science, Technology and Innovation (MOSTI).

McMillan, W. W., Barnet, C., Strow, L., Chahine, M. T., McCourt, M. L., Warner, J. X., Novelli, P. C., Korontzi, S., Maddy, E. S. & Datta, S. (2005). Daily Global Maps of Carbon Monoxide From NASA's Atmospheric Infrared Sounder. *Geophysical Research Letters*, Vol.32, L11801, DOI: 10.1029/2004GL021821.

McMillan, W. W., Yurganov, L., Evans, K., & Barnet, C. (2007). Global Climatology of Tropospheric CO from the Atmospheric InfraRed Sounder (AIRS). *20th Conference on Climate Varibility and Change,* Vol. 5B.3, pp. 217 - 228.

McMillan, W. W., Warner, J. X., McCourt Comer, M., Maddy, E., Chu, A., Sparling, L., Eloranta, E., Hoff, R., Sachse, G., Barnet, C., Razenkov, I. & Wolf, W. (2008). AIRS Views Transport from 12 to 22 July 2004 Alaskan/Canadian fires: Correlation of AIRS CO and MODIS AOD with Forward Trajectories and Comparison of AIRS CO Retrievals with DC-8 in Situ Measurements During INTEX-A/ICARTT. *Journal of Geophysical Research*, Vol.113, D20301, DOI: 10.1029/2007JD009711.

Nedelec, P., Cammas, J. –P, Thouret, V., Athier, G., Cousins, J. –M., Legrand, C., Abonnel, C., Lecoeur, F., Cayez, G. & Marizy, C. (2003). An Improved Infrared Carbon Monoxide Analyzer for Routine Measurements aboard Commercial Airbus Aircraft: Technical Validation and First Scientific Results of The MOZAIC III Programme. *Atmospheric Chemistry and Physics*, Vol.3, pp.1551-1564.

Pagano, T. S., Aumann, H. H., Hagan, D. E. & Overoye, K. (2003). Prelaunch and In-Flight Radiometric Calibration of The Atmospheric Infrared Sounder (AIRS). *IEEE Transactions on Geoscience and Remote Sensing*, Vol.41, pp.265 - 273.

Pagano, T. S., Chahine, M. T., Aumann, H. H., Tian, B., Lee, S. Y., Olsen, E., Lambrigtsen, B., Fetzer, E., Irion, F. W., McMillan, W., Strow, L., Fu, X., Barnet, C., Goldberg, M., Susskind, J. & Blaisdell, J. (2006). Remote Sensing of Atmospheric Climate Parameters from the Atmospheric Infrared Sounder. IEEE International Geosience and Remote Sensing Symposium (IGARSS), pp.2386-2389, ISBN: 0780395107.

Pochanart, P., Akimoto, H., Kajii, Y. & Sukasem, P. (2003). Carbon Monoxide, Regional-Scale Transport, and Biomass Burning in Tropical Continental Southeast Asia: Observations in Rural Thailand. *Journal of Geophysical Research*, Vol.108, pp. 4552, 15.

Pochanart, P., Wild, O. & Akimoto, H. (2004). Air Pollution Import To and Export From East Asia. IN STOHL, A. (Ed.) The Handbook of Environmental Chemistry. Berlin, Springer.

Rasch, P. J., Collins, W. D. & Eaton, B. E. (2001). Understanding the Indian Ocean Experiment (INDOEX) Aerosol Distributions with an Aerosol Assimilation. *Journal of Geophysical Research,* Vol.106, pp.7337–7355.

Shaharuddin, A. & Mohamed, E. Y. (2005). Urban climate research in Malaysia. IAUC Newsletter, 5-10.

Streets, D., Tsai, N., Akimoto, H. & Oka, K. (2001). Trends in Emissions of Acidifying Species in Asia, 1987–1997. *Water, Air, and Soil Pollution*, Vol.130, pp. 187–192.

Strow, L. L, Hannon, S. E., Souza-Machado, S. D., Motteler, H. E. & Tobin, D. (2003). An Overview of the AIRS Radiative Transfer Model. *IEEE Transaction on Geoscience and Remote Sensing*, Vol.41, No.2, pp.303-313.

Suhaila, J. & Jemain, A. A. (2007). Fitting Daily Rainfall Amount in Malaysia using the Normal Transform Distribution. *Journal of Applied Sciences*, Vol.7, pp.1880 - 1886.

Suhaila, J. & Jemain, A. A. (2009). Investigating the Impacts of Adjoining Wet Days on the Distribution of Daily Rainfall Amounts in Peninsular Malaysia. *Journal of Hydrology*, Vol.368, pp.17–25.

Tangang, F. T., Juneng, L. & Ahmad, S. (2007). Trend and Interannual Variability of Temperature in Malaysia: 1961–2002. *Theoretical and Applied Climatology*, Vol.89, pp.127-141.

Tiwari, Y. K., Gloor, M., Engelen, R., Rodenbeck, C. & Heimann, M. (2005). Comparing Model Predicted Atmospheric CO_2 with Satellite Retrievals and In-situ Observations - Implications for the Use of Upcoming Satellite Data in Atmospheric Inversions. *Geophysical Research Abstracts*, Vol. 7, 09823.

Turquety, S., Hurtmans, D., Hadji-Lazaro, J., Coheur, P. –F, Clerbauz, C., Josset, D. & Tsamalis, C. (2009). Tracking the Emission and Transport of Pollution from Wildfires Using the IASI CO Retrievals: Analysis of the Summer 2007 Greek fires. *Atmospheric Chemistry and Physics*, Vol.9, pp.4897-4913.

Varikoden, H., Samah, A. A. & Babu, C. A. (2010). Spatial and Temporal Characteristics of Rain Intensity in The Peninsular Malaysia Using TRMM Rain Rate. *Journal of Hydrology*, Vol.387, pp.312–319.

Warner, J., Comer, M. M., Barnet, C. D., McMillan, W. W., Wolf, W., Maddy, E. & Sachse, G. (2007). A Comparison of Satellite Tropospheric Carbon Monoxide Measurements from AIRS and MOPITT during INTEX-A. *Journal of Geophysical Research*, Vol.112, D12S17.

Wong, C. L., Venneker, R., Uhlenbrook, S., Jamil, A. B. M. & Zhou, Y. (2009). Variability of Rainfall in Peninsular Malaysia. *Journal Hydrology and Earth System Sciences Discussions*, Vol. 6, pp.5471–5503.

Ye, H., Fetzer, E. J., Bromwich, D. H., Fishbein, E. F., Olsen, E. T., Granger, S. L., Lee, S. Y., Chen, L. & Lambrigtsen, B. H. (2007). Atmospheric Total Precipitable Water from AIRS and ECMWF during Antarctic Summer. *Geophysical Research Letters*, Vol.34, L19701.

Yurganov, L., McMillan, W., Wilson, C., Fischer, M. & Biraud, S. (2010). Carbon Monoxide Mixing Ratios over Oklahoma Between 2002 and 2009 Retrieved from Atmospheric Emitted Radiance Interferometer Spectra. *Atmospheric Measurement Techniques*, Vol. 3, pp. 1263–1301.

8

Method for Validation of Lagrangian Particle Air Pollution Dispersion Model Based on Experimental Field Data Set from Complex Terrain

Boštjan Grašič, Primož Mlakar and Marija Zlata Božnar
MEIS environmental consulting d.o.o.
Slovenia

1. Introduction

Validation of air pollution dispersion model is very important process. It determines performances and efficiency of model in well defined conditions. Conditions consist of type of terrain orography (flat or complex), size of domain (local, regional, continental, global), number of grid cells in domain, meteorological conditions (strong or weak winds, etc.) and emission types (stacks, traffic, domestic heating). Results of validation give good guidelines how, where and when model can be successfully applied.

Validation is especially important when model is used for regulatory purposes. FAIRMODE European guidelines for air pollution modelling explicitly require that modeling tool must be successfully validated in similar environment (FAIRMODE, 2010). Slovenian legislation (Ur.l. RS, št. 31/2007, 2007) that is following European Council Directive of 28th June 1984 on combating air pollution from industrial plants (EUR-Lex 84/360/EEC, 1984) requires that the modeling tool for reconstructions of air pollution around stationary industry sources meet the requirements of complex terrain because most of Slovenian industry is located in the bottom of basins, river canyons and valleys. Complex terrain defines a set of specific atmospheric conditions: low wind speeds, temperature inversions, flow over topography, presence of terrain obstacles or discontinuities (land-sea, urban-rural environment), etc. Lagrangian particle dispersion model is the only air pollution model at the moment that is successfully achieving these requirements (Wilson and Sawford, 1996, Schwere et al., 2002). It has significantly evolved in last years and moved from research usage to usage for operational regulatory purposes (Tinarelli et al., 2000, Graff, 2002).

Validations over complex terrain are still very rare. They are very important for research community and governmental environment agencies. Research community use the results for further developments and improvements of modeling techniques and environment agencies for setting up and implementation of regulatory policies.

A study has been made to improve traditional air pollution model validation methodology. It is upgraded to estimate inaccuracy in position and time of the Lagrangian particle air pollution dispersion model. New validation methodology has been demonstrated on a field from a very complex terrain from Šaleška region (Slovenia). For validation Lagrangian particle air pollution dispersion model *SPRAY* produced by ARIANET Srl from Milano, Italy is selected. It has been chosen for validation because it follows Slovenian legislation about air pollution modeling over complex terrain. Validation is performed on one very complex terrain air pollution situation that is very difficult for reconstruction and includes phenomenon of air pollution accumulation. Traditional statistical indexes are determined at four locations in different directions from the point of view of air pollution source. To estimate model's inaccuracy in position and time new enhanced validation methodology is demonstrated and described in details. Results of this validation will serve for future improvements of selected air pollution dispersion model.

2. Methodology

2.1 Traditional validation methodology

Traditional validation methodology for air pollution modeling is based on statistical comparison between measured and reconstructed data about air pollution concentrations in environment. It is well described in model validation framework named "Model evaluation toolkit" that has been established and maintained by Olesen (1996).

Measured data are collected from automatic environmental measuring stations located on the area of interest (domain) usually around sources of air pollution. Reconstructed concentrations are obtained from the air pollution modeling simulation.

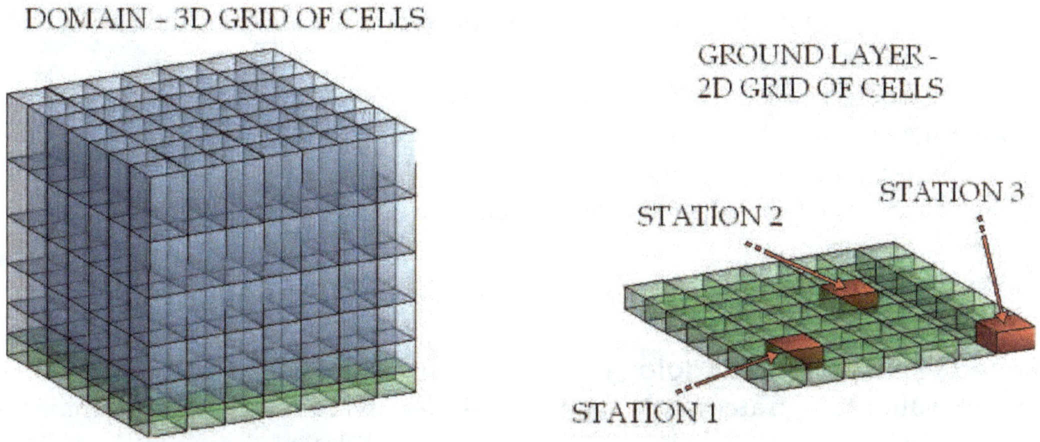

Fig. 1. The domain split in 3D grid of cells is presented on the left side where ground layer is colored in green; on the right side only ground layer is presented where the cells where stations are located are highlighted in red color.

In the air pollution model usually area of interest consists of a grid of cells where each cell describes average air pollution situation in certain part of the domain (i.e. in presented study case in next chapter domain is split into 100 x 100 cells in horizontal and in 20 layers in vertical which give 200 000 cells for the domain). For the comparison reconstructed average concentration from the ground cell where measuring station is located is taken. An example is presented on Figure 1.

Statistical analysis of data is performed for selected time interval where measured and reconstructed data are available. For this time interval a set of data patterns must be prepared. Each data pattern from this set consists of a pair of measured and reconstructed concentraion obtained at time step *t* as presented in equation (1).

$$\{C_{meas}(t), C_{recon}(t)\} \tag{1}$$

Using traditional validation methodology most often three statistical indexes are determined:

- the correlation coeficient (CR):

$$CR = \frac{\frac{1}{T}\sum_{t=0}^{T}\left(C_{meas}(t) - \hat{C}_{meas}\right)\left(C_{recon}(t) - \hat{C}_{recon}\right)}{\sigma_{C_{meas}}\sigma_{C_{recon}}} \tag{2}$$

- the normalized mean square error (NMSE):

$$NMSE = \frac{\frac{1}{T}\sum_{t=0}^{T}\left(C_{meas}(t) - C_{recon}(t)\right)^2}{\hat{C}_{meas} \cdot \hat{C}_{recon}} \tag{3}$$

- the fractional bias (FB):

$$FB = 2\frac{\hat{C}_{meas} - \hat{C}_{recon}}{\hat{C}_{meas} + \hat{C}_{recon}} \tag{4}$$

Definitions of variables and functions for determination of statistical indexes:

$C_{meas}(t)$ ...measured concentration at time step t
$C_{recon}(t)$ ...reconstructed concentration at time step t
$\hat{C}_{meas}$ ...average measured concentration
$\hat{C}_{recon}$ ...average reconstructed concentration
σ_C ...standard deviation of (measured or reconstructed) concentrations
t ...time step index
T ...length of full time interval (number of measured concentrations)

2.2 Enhanced validation methodology

In the model validation framework named "Model evaluation toolkit" maintained by Olesen (1996) difficulties that can arise in model validation are outlined. Differences between measured and reconstructed concentrations are caused by measuring errors, inherent uncertainty, input uncertainty and model formulation error. In the paper by Grašič et al. (2007) it has been determined that inaccuracy in position and time exists in the model. To estimate these inaccuracies enhanced validation methodology is presented. It is based on methodology where additionally reconstructed ground level concentrations in neighboring cells of the cell where station is located are also used in validation. Each measured value is during enhanced validation compared with one reconstructed concentration selected from a set of reconstructed concentrations. Set of this reconstructed concentrations NC as described in equation (5) consists of average concentration in the cell where station is located and neighboring cells. Neighborhood is defined in position (space) (i.e. for neighborhood of 1 cells in position we create a set of 9 cells as presented on Figure 2 and equation (6)) and in time scale (i.e. neighborhood of 1 time interval consist of 3 time intervals as presented on Figure 3 and equation (7)).

$$NC(t,m,n)=\left\{\begin{matrix} C_{recon}(t,m,n); \\ t-\Delta T<t<t+\Delta T; \\ m-\Delta H<m<m+\Delta H; \\ n-\Delta H<n<n+\Delta H \end{matrix}\right\} \tag{5}$$

Definitions of variables for determination of set of neighborhood concentrations *NC*:

NC	...set of reconstructed concentrations in the station's neighborhood
t	...time step index
ΔT	...length of neighborhood in time scale (number of time steps)
m	...index (number) of cell in east-west direction
n	...index (number) of cell in east-west direction
ΔH	... length of neighborhood in position (space) (number of cells)

$$NC_{position}(t,m,n)=\left\{\begin{matrix} C_{recon}(t,m,n); \\ m-\Delta H<m<m+\Delta H; \\ n-\Delta H<n<n+\Delta H \end{matrix}\right\} \tag{6}$$

$$NC_{time}(t,m,n)=\left\{\begin{matrix} C_{recon}(t,m,n); \\ t-\Delta T<t<t+\Delta T \end{matrix}\right\} \tag{7}$$

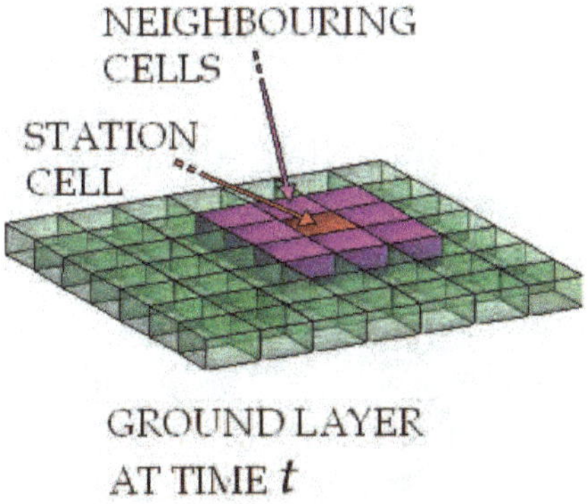

Fig. 2. Example of neighboring cells in position ($\Delta H=1$) where set of neighborhood concentrations *NC* consists of 9 cells

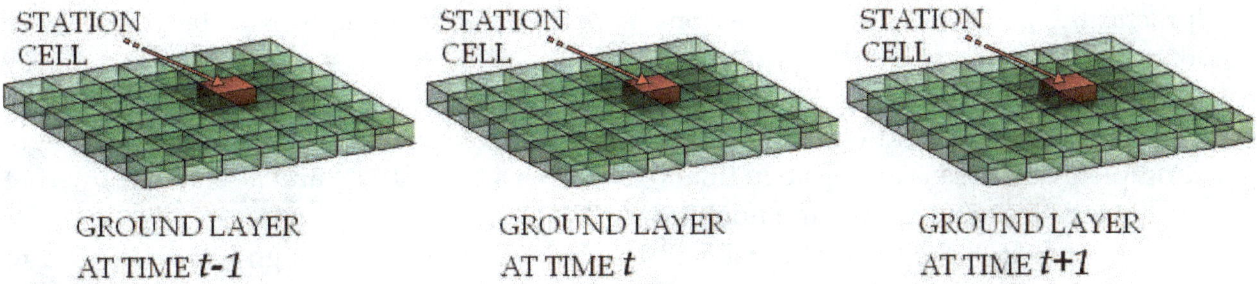

Fig. 3. Example of neighboring cells in time ($\Delta T=1$) where set of neighborhood concentrations *NC* consists of 3 cells

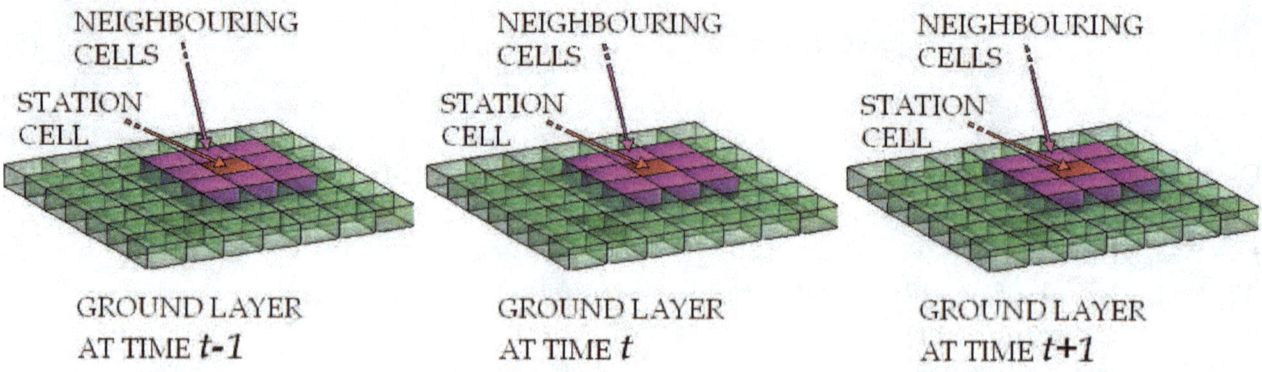

Fig. 4. Example of neighboring cells in position and time ($\Delta H=1$ and $\Delta T=1$) where set of neighborhood concentrations *NC* consists of 27 cells

Finally in enhanced validation methodology each measured value is compared with one reconstructed concentration selected from a set of neighborhood concentrations *NC*. From this set of reconstructed concentrations one concentration $C_{BMrecon}$ is selected using best matching function according to measured concentration as described in equation (8). Best matching function selects one element from *NC* set where difference of this element and measured concentration in lowest compared to other elements in *NC* set.

$$C_{BMrecon}(t,m,n) = BM(NC(t,m,n)) \tag{8}$$

3. Study case

Presented method is demonstrated on a field data set from a complex terrain. In the following sub-chapters field data set from Šaleška region (Slovenia) is described. Field data set from Šaleška region has been chosen for several reasons:

- The first reason is the complex terrain of the region where all typical complex terrain meteorological conditions occur (Grašič, 2007, Blumen et al, 1990).
- The second reason are high emissions from thermal power plant which were about 100000 tons of sulphur dioxide SO_2 and 12400 tons of nitrogen oxides NO_x per year (Elisei, 1991) because no desulphurization plant has been installed at that time. These high emissions represented the main air-pollution source in the region where ambient SO_2 concentrations higher than 1 mg/m^3 were measured at surrounding automatic environmental measuring stations. All other local source of air pollution can be practically neglected for this reason. Experimental campaign had been therefore organized as a tracer experiment.
- The third reason is the availability of all measured data from environmental automatic measuring stations and emission station for the whole period of measuring campaign. Complete database is available in final report (Elisei et al., 1991) and also on the internet web page (Šoštanj91 Campaign home page, 2007).
- And the fourth reason is that database obtained during the campaign had been used to validate several available air pollution models (Grašič, 2007).

Study case continues with description of air pollution modelling and comparison of validation results using standard and new presented method.

3.1 "Šoštanj91" field data set description

An experimental measuring campaign named had been performed in spring of year 1991 from 15th of March till 5th of April 1991 in surrounding of Thermal power plant Šoštanj (TPPŠ). Main purpose of the campaign was determination of environmental impact of the air pollution from the three stacks of thermal power plant. The emphasis has been on the meteorological conditions that cause severe air pollution episodes.

TPPŠ is located in the centre Šaleška valley as presented on Figure 5. In the central part of Šaleška valley there is a plain located north of Paka river. Average altitude of the valley is three hundred meters above sea level. Valley is surrounded by hills on the south side and by high mountains Karavanke Alps on the west, north and east side. There are two towns and several small villages in the valley and its surrounding where approximately 36000 people lived in the time when campaign had been performed (Elisei et al., 1991). Map on Figure 6 shows the location of Šaleška valley in the north-eastern part of Slovenia.

The experimental campaign had been performed by researchers from three research institutions: ENEL-CRAM and CISE, Milano, Italy and Jozef Stefan Institute, Ljubljana, Slovenia. Data obtained during the campaign had been used to validate several available air pollution models: standard and advanced Gaussian models, Gaussian puff model and Lagrangian particle dispersion model (Brusasca et al., 1992, Božnar et al., 1993, Božnar et al., 1994). Final results of this studies proved that the Lagrangian particle dispersion model is the most effective tool for air pollution modelling in very complex terrain. Campaign was described in details in a final report (Elisei et al., 1991) where also all measured data is available. Database consists of measurements from different measuring systems: automatic measuring stations of Environmental Information System (EIS) maintained by TPPŠ, automatic mobile laboratory, one mobile Doppler SODAR and DIAL. Pictures of some of equipment are presented on Figure 7.

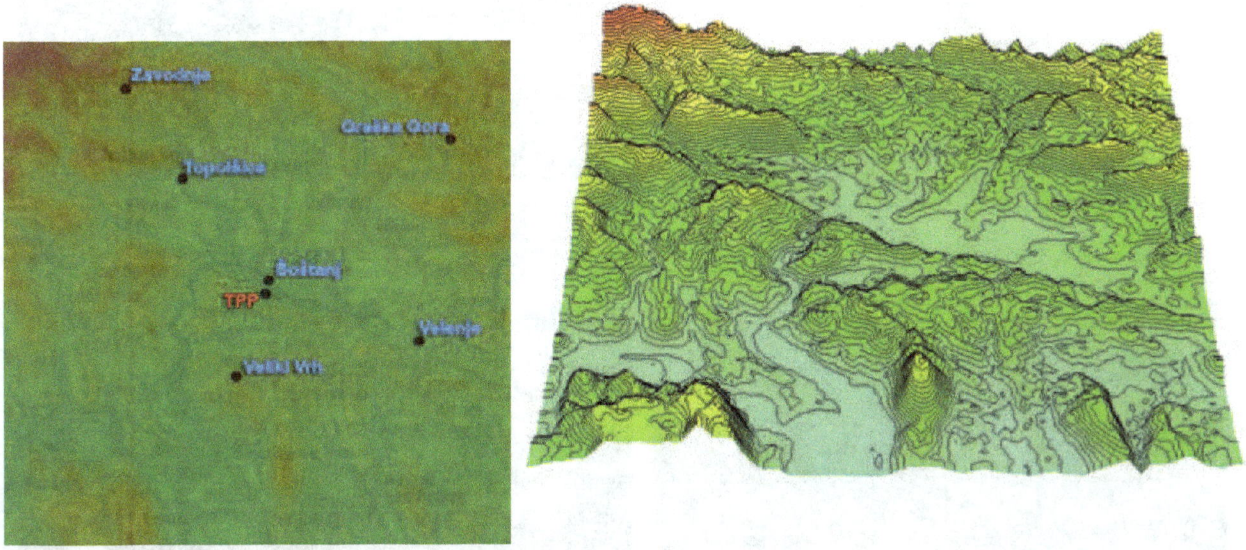

Fig. 5. Map of Šaleška region with locations of automatic environmental stations and location of the Thermal power plant Šoštanj in the centre (left picture) and the topography of the region (right picture)

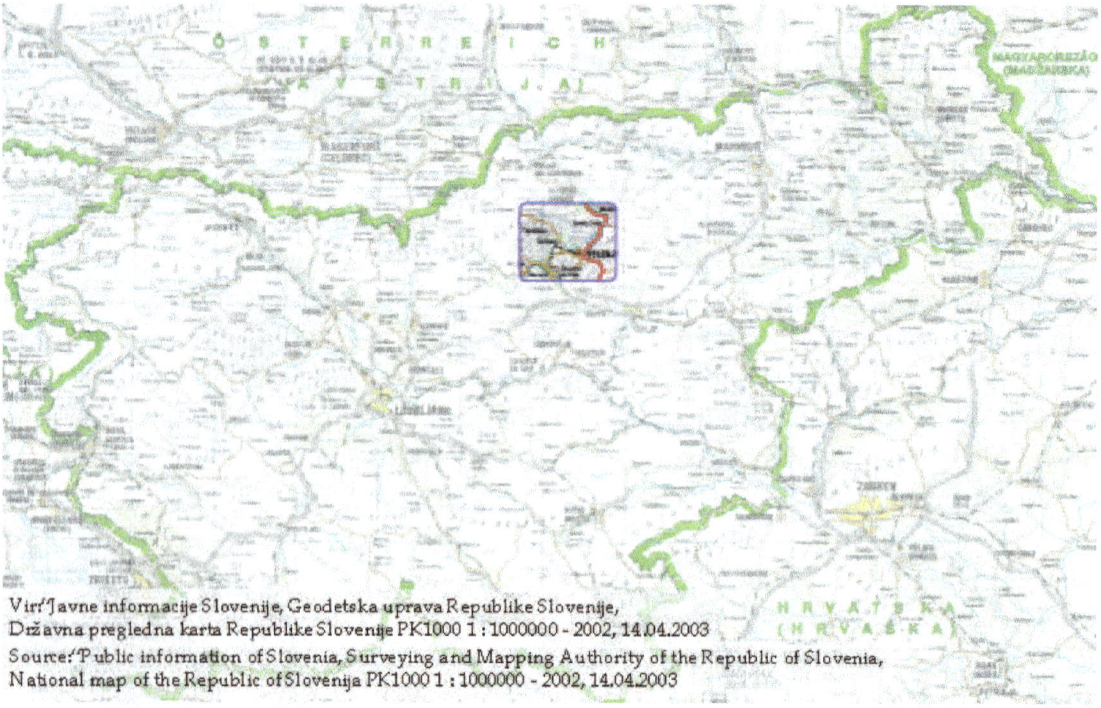

Fig. 6. Location of Saleška region in the north-eastern part of Slovenia

Fig. 7. Pictures of some of equipment used in campaign in spring 1991: environmental automatic measuring station (left), mobile SODAR (right - upper) and DIAL (right -lower)

Environmental Information System of TPPŠ consisted of six stationary automatic measuring stations and one mobile station. Locations of the stations are presented on Figure 5. Environmental parameters measured on stations are presented in Table 1.

Station name / Parameter name	Zavodnje	Graška Gora	Topolšica	Veliki Vrh	Šoštanj	Velenje
Air temperature	x	x	x	x	x	x
Relative humidity	x	x	x	x	x	x
Global solar radiation						x
Precipitation					x	
Air pressure					x	
Wind	x	x	x	x	x	x
SO2	x	x	x	x	x	x
NO	x					
NOx	x					
O3	x					

Table 1. List of parameters measured at automatic environmental stations (x denotes that parameter is measured at certain station)

TPPŠ had during the campaign three operating stacks of different heights: 100m, 150 m and 230 m. Neither of the stacks had installed desulphurization plant during the experimental campaign. Measured emissions are presented in Table 2 where static and dynamic parameters are given. Emissions from generators Block 1, Block 2 and Block 3 are emitted from one stack named Block 1,2,3. Picture of TPP Šoštanj is presented on Figure 7.

Stack name	Height	Diameter	Location
Block 1,2,3	100 m	6.50 m	46.373N 15.052E
Block 4	150 m	6.34 m	46.372N 15.053E
Block 5	230 m	6.20 m	46.371N 15.055E

Stack name	Emission rate	Exit temperature	Exit velocity
Block 1,2,3	0.01 ÷ 0.24 kg/s	155 ÷ 171°C	0.7 ÷ 2.9 m/s
Block 4	0.87 ÷ 2.05 kg/s	155 ÷ 183°C	8.8 ÷ 12.3 m/s
Block 5	0.53 ÷ 2.46 kg/s	172 ÷ 202°C	8.6 ÷ 12.7 m/s

Table 2. List of emission parameters for all TPP Šoštanj stacks that were operating during experimental campaign in spring 1991, upper table presents static and lower dynamic parameters.

Fig. 8. Thermal power plant Šoštanj

3.2 Air pollution modelling

In this study case the Lagrangian particle air pollution dispersion model has been chosen for validation. The name of the model is SPRAY and its detailed description is given in papers by its authors (Brusasca et al, 1992, Tinarelli et al., 2000).

Model has been chosen for validation due to several reasons:

- First version of the model has already been validated on the "Soštanj91" field data set (Brusasca et al., 1992, Božnar et al., 1993, Božnar et al., 1994). Model has significantly evolved in last years. It has moved from research usage to usage for operational regulatory purposes (Tinarelli et al., 2000).
- Model follows new Slovenian legislation where for complex terrain it is required to use Lagrangian particle dispersion model. Model is coupled with corresponding meteorological pre-processor module which is able to reconstruct three dimensional diagnostic non-divergent wind fields.
- Most of Slovenian industry is located in the complex terrain at the bottom of valleys, river canyons or basins. The results of validation can therefore be applied also on the other similar cases in Slovenia or anywhere else where complex terrain is present.
- Latest version has also been recently validated on "Soštanj91" field data set (Grašič, 2007). The validation results in this study are extended and focused on the validation method rather than on the validation of the model.

Detailed description of model parameters and settings for this study are described in paper about latest validation of the model (Grašič, 2007). For input into meteorological pre-processor measured data mobile Doppler SODAR and from automatic environmental measuring stations as described in Table 1 has been used. All measured data are available in half-hour intervals. Mobile Doppler SODAR has been located in the centre of the domain.

Meteorological fields have been reconstructed at 150 m horizontal resolution. The same resolution has also been used to describe the complex topography (i.e. orography, Corine land use, etc.). Given all this topography and local meteorological data three dimensional mass consistent wind fields have been generated and used in Lagrangian particle dispersion model for air pollution reconstruction.

Lagrangian particle dispersion model has been generating half-hour average ground concentration fields at the same resolution 150 m as meteorological pre-processor. It has been using Thomson's 1987 scheme with Gaussian random forcing (Thomson, 1987). The number of emitted virtual particles has been set in order to assure minimum resolution for ground level concentrations less than 1 $\mu g/m^3$. Anfossi's formulation (Anfossi, 1993) has been used for plume rise of hot stack plumes where horizontal and vertical variations of both mean wind and atmospheric stability had been taken into consideration.

Air pollution reconstruction has been made for the full duration of the experimental campaign: from 15th of March till 5th of April 1991. Results from simulation are available in half-hour intervals. Each half-hour result represents average air pollution situation over complete domain for one half-hour interval. This result is a three-dimensional (3D) concentration field describing concentrations for each cell of the domain. Domain consists of 100x100 grid cells in horizontal and of 20 layers in vertical that is 200000 grid cells in one 3D concentration field. For validation of the model only two-dimensional (2D) concentration field is relevant representing ground level concentrations. This ground-level concentration field consists of 100x100 cells from ground layer where each cell size is 150 m x 150 m in horizontal and 10 m in vertical.

For demonstration of new validation method only one very complex air pollution situation has been selected. It is a typical complex terrain situation, very difficult for reconstruction and still represents greatest challenge to all available air pollution dispersion models. The situation is described in details in paper by Grašič et al. (2007). It lasted from 1st of April 1991 at 20:00 until 2nd of April 1991 at 20:00.

Spreading of reconstructed plume in three-dimensional domain is presented on Figure 9 where it is shown that plume has been spreading in all directions over domain during a relatively short period of time. This is also seen from the Doppler SODAR measurements presented on Figure 10. This graph represents measurements from SODAR for each half-hour time interval at different heights. Each arrow on the graph represents direction of horizontal wind component at certain height. The length of the arrow represents the magnitude of horizontal wind speed component.

Air pollution spreading in all directions is also proven by measurements of half-hour average SO_2 concentrations at four environmental stations at different directions from TPPŠ as presented on Figures 11, 12, 13, 14.

In the paper by Grašič et al. (2007) it is also reported that during this selected period the phenomenon of air pollution accumulation occurred. Very stable meteorological situation was main cause for very slow mixing of plume with air. Pollution plume was moving very slowly according to measured average wind speed and direction. At the beginning of this situation the air pollution from the point of view of a measuring station came from the direction of the source. But when the main wind changed its direction to opposite direction, also the air pollution cloud changed its direction. From now on from the point of view of measuring station it appeared that the air pollution cloud is coming from the virtual emission source located on the other side. In our case selected domain was not wide enough to capture this phenomenon by Lagrangian particle dispersion model. Part of the air pollution cloud has been lost out of domain which should be taken into account when model is being validated. Lagrangian particle model could reproduce this phenomenon correctly if the domain would be widened but in this case we would have to decrease the final resolution of the domain due to computational limits of the model. Decrease of the resolution (i.e. from 150 m grid cell to 500 m grid cell) would result in more coarse results and also some local complex terrain effects could be lost.

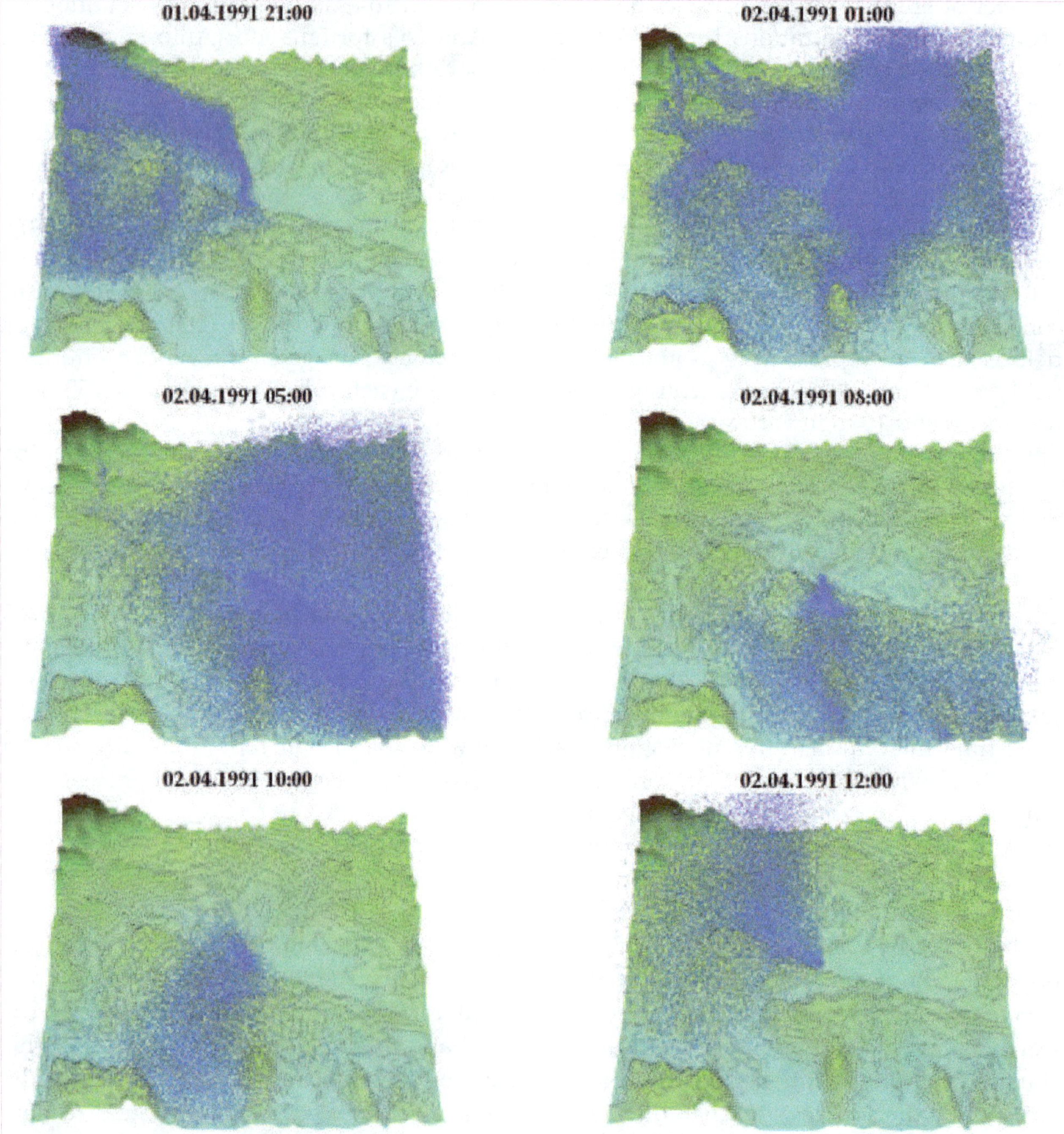

Fig. 9. 3D reconstruction of plume spreading in all directions during selected air pollution situation

Fig. 10. SODAR measurements during selected air pollution situation from 1st of April 1991 at 20:00 until 2nd of April 1991 at 20:00 when air pollution accumulation occurred; direction of arrows presents horizontal wind direction at certain height; length and color of arrow presents horizontal wind speed at certain height

3.2 Validation results

Validation of modelling results is performed at four stations located in different directions from the point of view of thermal power plant Šoštanj. Four locations are selected according to positions of four environmental automatic measuring stations as presented on Figure 5: Graška Gora, Šoštanj, Veliki Vrh and Zavodnje. From all these stations measurements of half-hour average SO_2 concentrations are available for selected air pollution situation from 1st of April 1991 at 20:00 till 2nd of April 1991 at 20:00.

As presented on Figure 11 measured SO_2 concentration was increased due to wind change at the beginning of selected air pollution situation. Wind at approximate height 250 m changed its direction from north-west to south-east. Next wind change was toward the south which caused an increase of SO_2 concentrations at Šoštanj (Figure 12) and Veliki Vrh (Figure 13) stations. Figure 14 presents measured SO_2 concentrations at Zavodnje station which is the most distant station from the TPPŠ. This result is interesting especially because

of the measured SO_2 concentration peak at the ending of air pollution situation. This peak was caused by air pollution accumulation phenomenon as describe in previous sub-chapter about air pollution modelling. Because the station is located near the border of domain (Figure 5) it is expected that the model results will be underestimated in this case.
In the following sub-chapters a comparison between measured and reconstructed SO_2 concentrations at the locations of presented four stations is made using traditional and enhanced validation methods. Within validation using traditional validation methodology modelling problems will be described that cause under or over estimations of reconstructed concentrations. And the sub-chapter using enhanced validation methodology is presenting different comparison results which can be used as a good estimation of model's inaccuracy of position and time.

3.2.1 Traditional validation results

Figure 11 shows comparison between measured and reconstructed SO_2 concentrations at station Graška Gora. Reconstructed concentrations agree very well with measured ones. Also comparison using traditional statistical indexes for complete duration of experimental measuring campaign from 15th of March till 5th of April 1991presented in Table 3 shows good correlation where correlation reaches value higher than 0.3.
Same comparison of SO_2 concentrations at station Šoštanj is presented on Figure 12. Comparison on the graph shows underestimation of reconstructed concentration values. The first reconstructed peak at 11:30 hour is underestimated due to model's inaccuracy of position. In the paper by Grašič et al. (2007) it is shown that correct peak has been reconstructed just two cells away from the station. The second underestimated concentration peak is caused by short distance between station and stacks (approximately 500 m). There are two effects that are not well captured due to this short distance. First is the stack tip down-wash effect. And the second is the combination of low-wind speed in direction towards the station and convective turbulences (Grašič et al., 2007). Comparison using traditional statistical indexes presented in Table 3 shows almost no correlation and medium underestimation of reconstructed concentrations
The comparison of SO_2 concentrations at Veliki Vrh station are presented on Figure 13. During the air pollution situation two concentration peaks have been reconstructed (from 00:00 till 04:00 and from 06:00 till 12:00). Both peaks are not correctly reconstructed due to inaccuracy of the model in position. Such peaks can appear in real situation just few meters from the measuring station without being detected (Grašič et al., 2007). Comparison presented in Table 3 shows poor correlation between measured and reconstructed concentrations.
Even more obvious phenomenon of model's inaccuracy in position is presented on Figure 14 where comparison between measured and reconstructed SO_2 concentrations at the location of Zavodnje station is made. The phenomenon is more expressed because of the long distance between the station and thermal power plant. It generated first reconstructed peak in time interval from 00:00 till 04:00 hour. The second measured concentration peak has been underestimated due to air pollution accumulation that has been lost because the domain was not wide enough to capture the phenomenon. This event occurred at the end of air pollution situation when the wind changed direction from south back to north-west direction (Grašič et al., 2007). Comparison presented in Table 3 shows none correlation between measured and reconstructed concentrations and very high normalized mean square error.

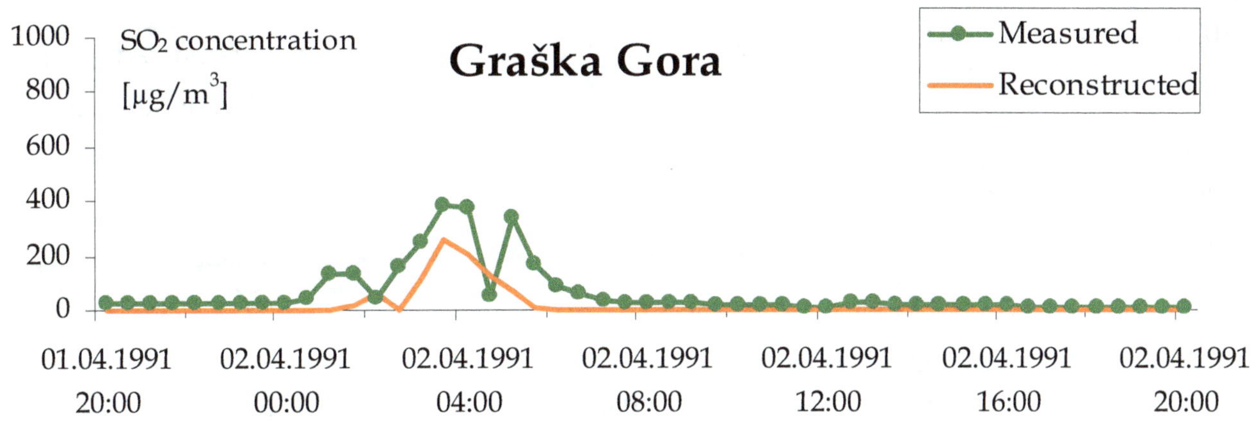

Fig. 11. Comparison of measured and reconstructed ambient SO_2 concentrations at Graška Gora station

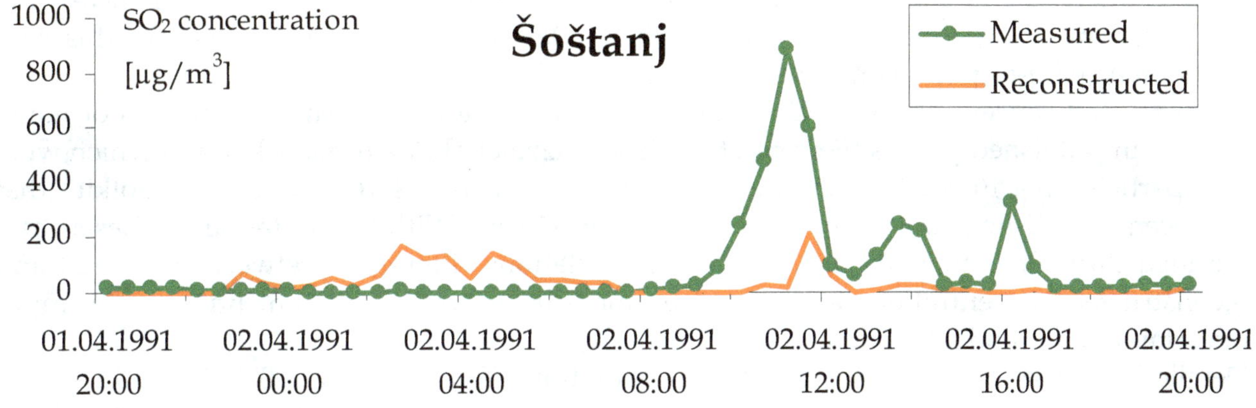

Fig. 12. Comparison of measured and reconstructed ambient SO_2 concentrations at Šoštanj station

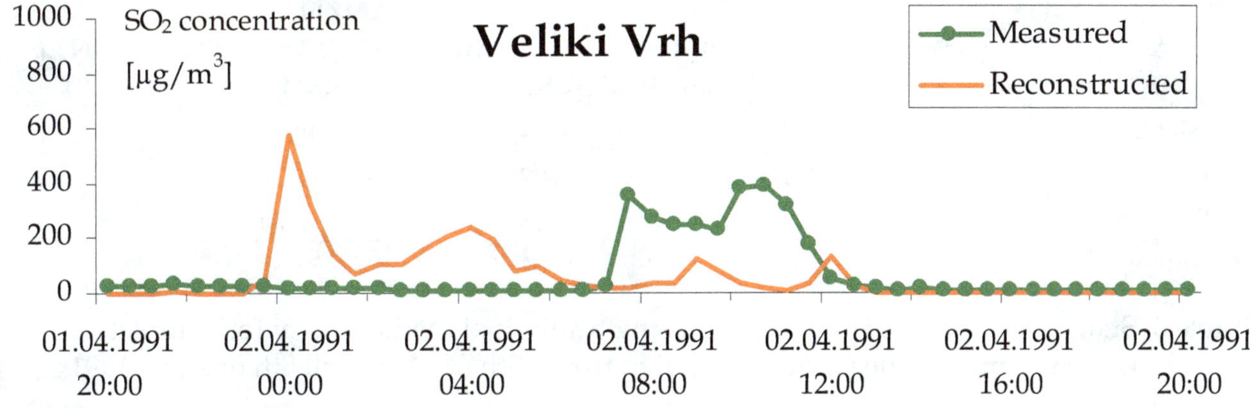

Fig. 13. Comparison of measured and reconstructed ambient SO_2 concentrations at Veliki Vrh station

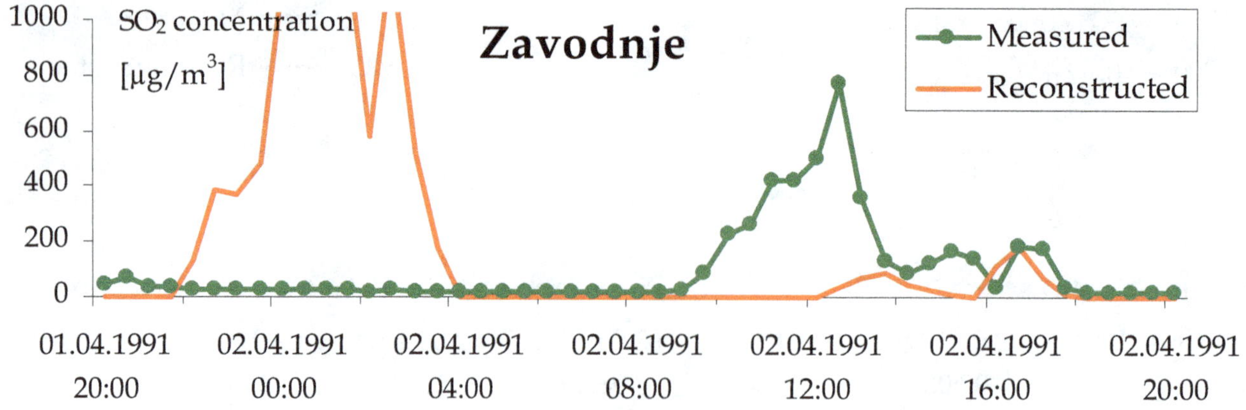

Fig. 14. Comparison of measured and reconstructed ambient SO_2 concentrations at Zavodnje station

Table 3 presents results of comparison between measured and reconstructed SO_2 concentrations using statistical indexes for complete duration of experimental measuring campaign in spring of year 1991 from 15th of March till 5th of April 1991. In this case traditional point-to-point comparison has been made. From the results seems that only the reconstructed concentrations at Graška Gora are satisfactory comparing to results of many authors in published papers (Ferrero et al., 1996, Rizza et al., 1996, Kaasik, 2005) which were also participating in model validation framework named "Model evaluation toolkit" that has been established and maintained by Olesen (1996). Within this research Olesen aslo outlined difficulties that can arise in model validation: differences between measured and reconstructed concentrations are caused by measuring errors, inherent uncertainty, input uncertainty and model formulation error.

Detailed analysis of selected air pollution situation (Grašič et al., 2007) determined that inaccuracy in position and time of reconstructed concentrations have been caused mostly by model's sensitivity to wind speeds and directions measured at different stations and by SODAR. Model's sensitivity strongly depends on the complexity of the terrain which is in our case highly complex.

STATION NAME	NUMBER OF PAIRS	CORRELATION COEFICIENT (CR)	NORMALIZED MEAN SQUARE ERROR (NMSE)	FRACTIONAL BIAS (FB)
Graška Gora	884	0,34	40,42	1,60
Šoštanj	881	0,02	17,32	0,37
Veliki Vrh	839	0,13	8,70	0,09
Zavodnje	858	-0,004	38,35	0,10

Table 3. Statistical indexes of comparison using traditional methodology for complete duration of experimental measuring campaign from 15th of March till 5th of April 1991

3.2.2 Enhanced validation results

Figure 15, 16, 17 and 18 shows comparison between measured and reconstructed SO_2 concentrations at stations Graška Gora, Šoštanj, Velikih Vrh and Zavodnje. There are three types of reconstructed concentrations with different size of neighborhood as described in section 3.2 *Enhanced validation methodology*:

- Recon. (ΔH=0,ΔT=0) - size of neighborhood is 0, only 1 cell where station is located is used for comparison, results are identical to traditional validation method
- Recon. (ΔH=1,ΔT=1) - size of neighborhood is 27 cells (9 cells in horizontal scale and 3 cells in time scale)
- Recon. (ΔH=2,ΔT=2) - size of neighborhood is 125 cells (9 cells in horizontal scale and 3 cells in time scale)

Agreement between measured and reconstructed concentrations is significantly improving when neighborhood is expanding. Similar result is obtained within comparison using traditional statistical indexes for complete duration of experimental measuring campaign from 15th of March till 5th of April 1991 presented in Tables 4 and 5. Comparison of results presented in Tables 3, 4 and 5 show significant improvement of all statistical indexes.

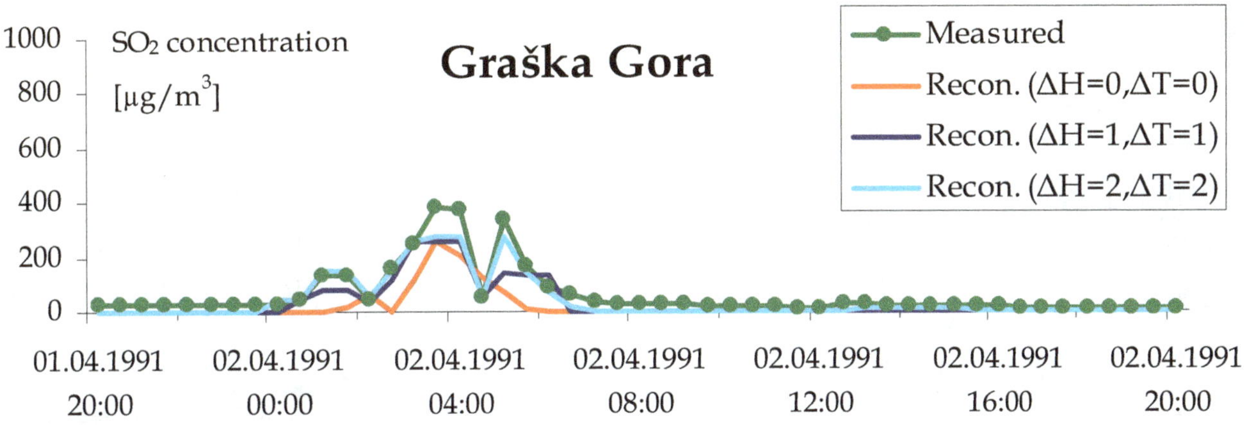

Fig. 15. Comparison of measured and reconstructed ambient SO_2 concentrations at Graška Gora station using different sizes of neighborhood

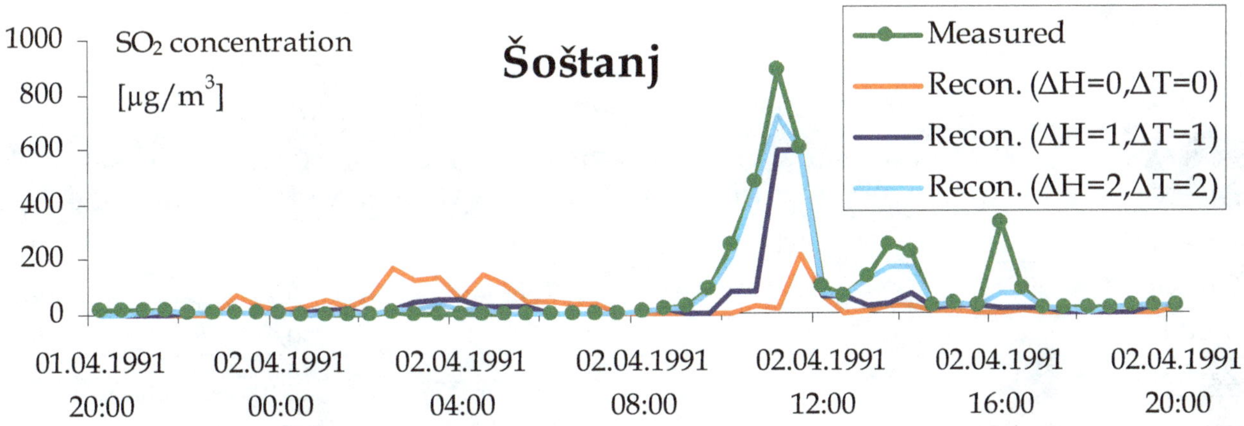

Fig. 16. Comparison of measured and reconstructed ambient SO_2 concentrations at Šoštjan station using different sizes of neighborhood

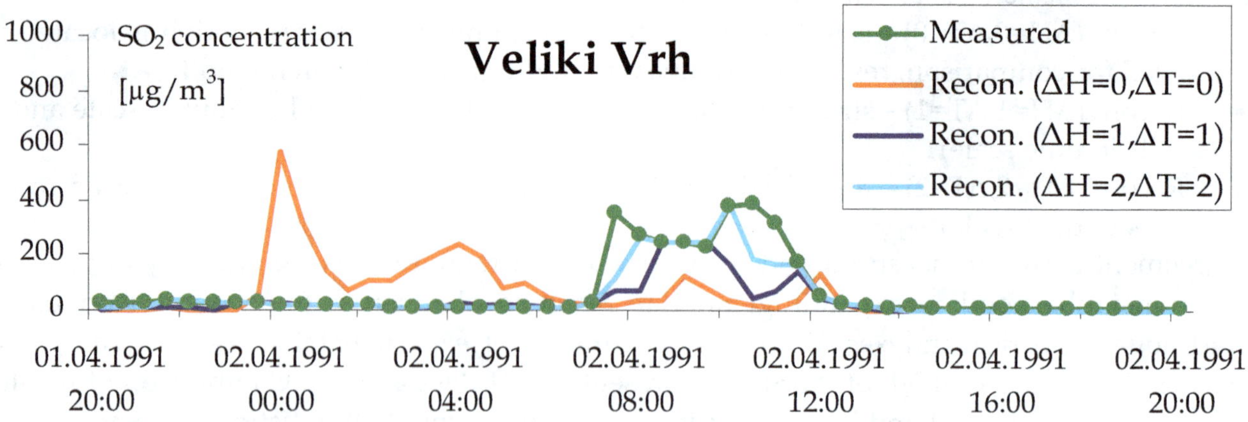

Fig. 17. Comparison of measured and reconstructed ambient SO_2 concentrations at Veliki Vrh station using different sizes of neighborhood

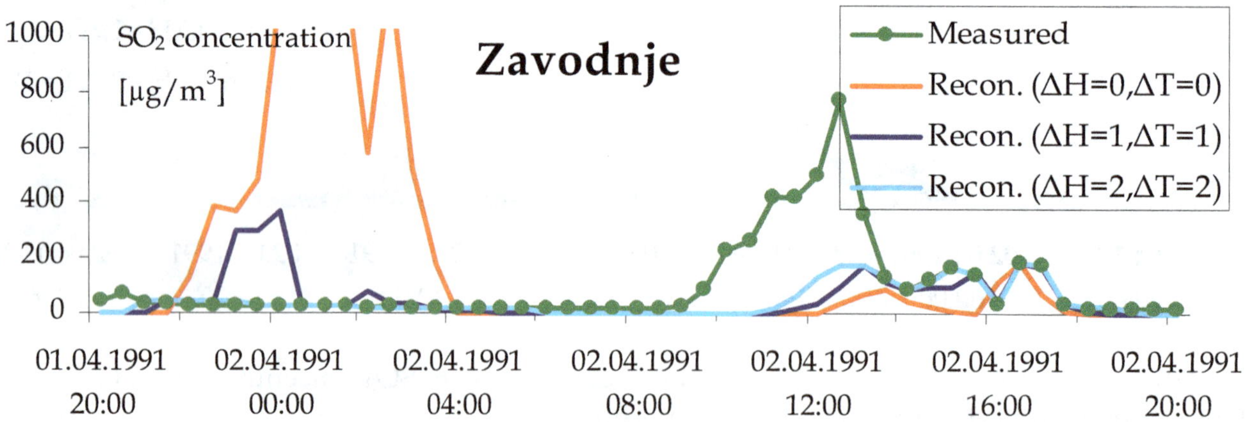

Fig. 18. Comparison of measured and reconstructed ambient SO_2 concentrations at Zavodnje station using different sizes of neighborhood

STATION NAME	NUMBER OF PAIRS	CORRELATION COEFICIENT (CR)	NORMALIZED MEAN SQUARE ERROR (NMSE)	FRACTIONAL BIAS (FB)
Graška Gora	881	0,70	1,08	9,07
Šoštanj	836	0,38	0,93	19,72
Veliki Vrh	878	0,75	0,61	3,10
Zavodnje	855	0,39	0,78	6,02

Table 4. Statistical indexes of comparison using enhanced validation methodology for complete duration of experimental measuring campaign from 15th of March till 5th of April 1991 where size of neighborhood consists of 27 cells (*ΔH=1* and *ΔT=1)*

STATION NAME	NUMBER OF PAIRS	CORRELATION COEFICIENT (CR)	NORMALIZED MEAN SQUARE ERROR (NMSE)	FRACTIONAL BIAS (FB)
Graška Gora	879	0,76	0,83	5,22
Šoštanj	876	0,47	0,79	14,92
Veliki Vrh	834	0,88	0,38	1,30
Zavodnje	853	0,67	0,65	3,38

Table 5. Statistical indexes of comparison using enhanced validation methodology for complete duration of experimental measuring campaign from 15th of March till 5th of April 1991 where size of neighborhood consists of 125 cells ($\Delta H=2$ and $\Delta T=2$)

4. Further improvements of enhanced validation methodology

In the paper by Grašič et al. (2007) validation has been performed using enhanced validation methodology as explained in chapter *3.2 Enhanced validation methodology*. For this validation neighborhood of 27 cells (one cell in each horizontal direction $\Delta H=1$ and one time step on time scale $\Delta T=1$) has been used. Figures 19, 20, 21 and 22 present comparisons of the results obtained in paper by Grašič et al. (2007) and results presented in previous sub-chapter *3.2.2 Enhanced validation results* where also neighborhood of 27 cells has been used.

This comparison shows slightly better results for recent study than for the previous study. This is also apparent from statistical indexes presented in Table 6. Main difference between previous and recent study is in removing of used reconstructed concentrations for further comparison in the old method. Main idea of enhanced validation methodology is to assign each measured concentration one reconstructed concentration from the neighborhood. Focusing on the time scale this means that one the same reconstructed concentration can be assigned to three measured concentrations when size of neighborhood is one time interval $\Delta T=1$. To avoid this in the previous study to each measured concentration only one unique reconstructed concentration has been assigned that appeared firstly in the set. The set was processed in time order from the oldest to youngest measured concentration. If processing had been performed in opposite order for the youngest to oldest the results would be slightly different. To find out the best order how to process this set very advanced optimizing algorithm performing in three dimensions would have to be developed which will be our main task in the future. Another task that will have to be solved in parallel will also be determination of appropriate criteria function to measure success of this optimization algorithm.

STATION NAME	NUMBER OF PAIRS	CORRELATION COEFICIENT (CR)	NORMALIZED MEAN SQUARE ERROR (NMSE)	FRACTIONAL BIAS (FB)
Graška Gora	881	0,69	1,14	10,38
Šoštanj	836	0,36	0,95	20,64
Veliki Vrh	878	0,74	0,61	3,26
Zavodnje	855	0,37	0,79	6,30

Table 6. Statistical indexes of comparison using enhanced validation methodology for complete duration of experimental measuring campaign from 15th of March till 5th of April 1991 where size of neighborhood consists of 27 cells ($\Delta H=1$ and $\Delta T=1$) and unique reconstructed concentrations are used

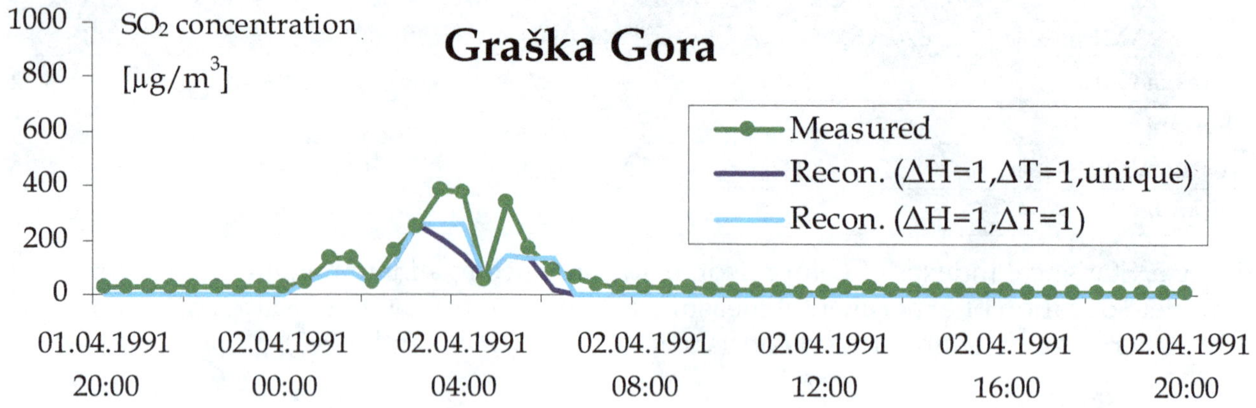

Fig. 19. Comparison of measured and reconstructed ambient SO_2 concentrations at Graška Gora station where for first comparison unique reconstructed concentrations are used

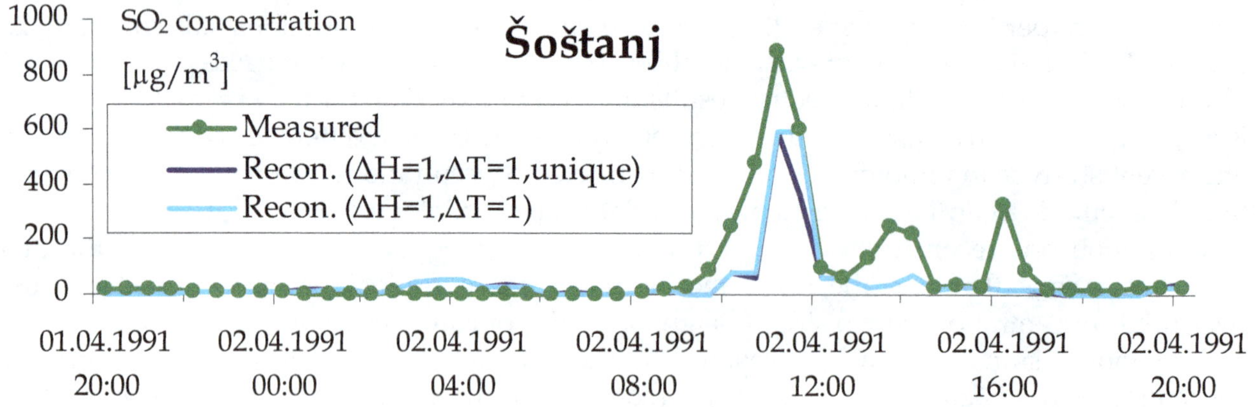

Fig. 20. Comparison of measured and reconstructed ambient SO_2 concentrations at Šoštanj station where for first comparison unique reconstructed concentrations are used

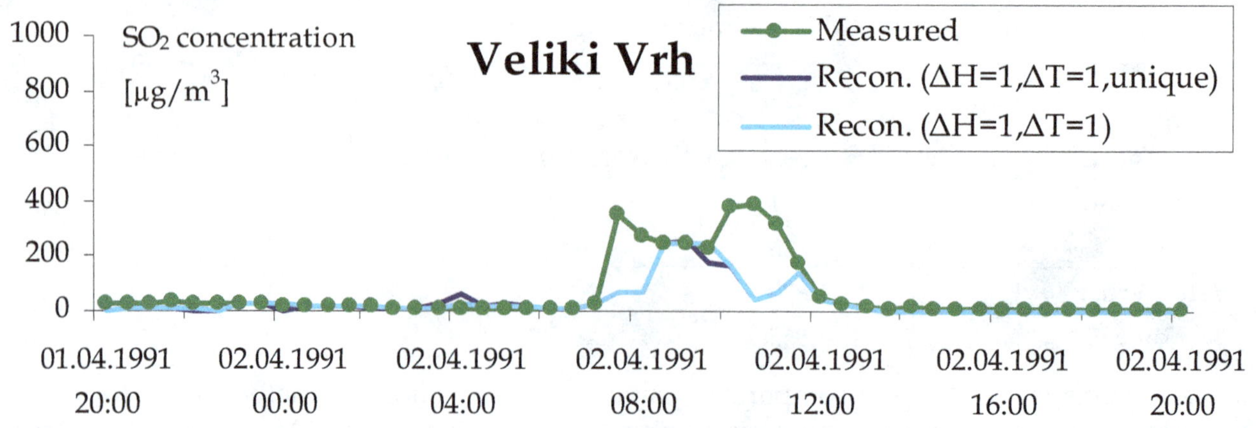

Fig. 21. Comparison of measured and reconstructed ambient SO_2 concentrations at Veliki Vrh station where for first comparison unique reconstructed concentrations are used

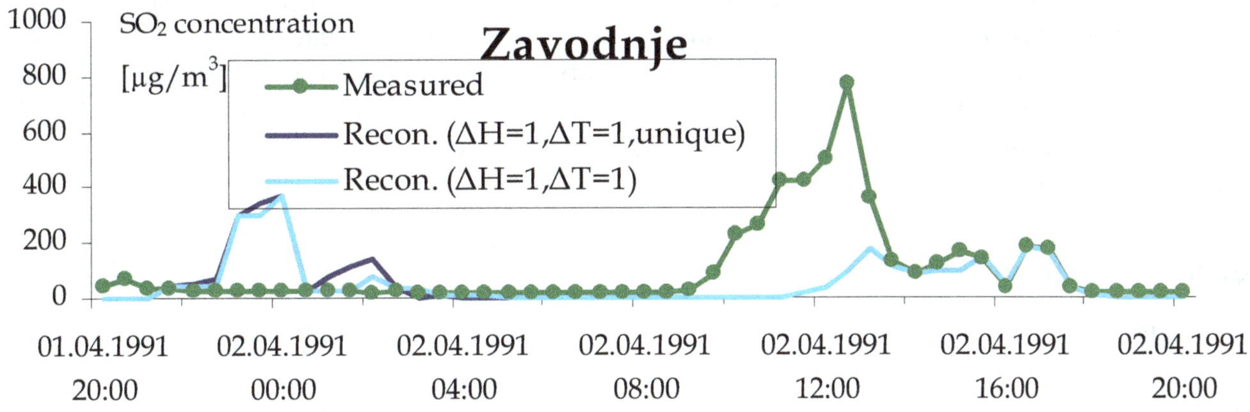

Fig. 22. Comparison of measured and reconstructed ambient SO_2 concentrations at Zavodnje station where for first comparison unique reconstructed concentrations are used

5. Conclusion

Traditional air pollution model validation methodology has been extended in this paper. It is based on statistical comparison between measured and reconstructed air pollution concentrations in the environment where different statistical indexes are determined. The method been upgraded to estimate inaccuracy in position and time of the Lagrangian particle air pollution dispersion model. To obtain these inaccuracies additional reconstructed air pollution concentrations from the neighborhood are used. Neighborhood is defined in spatial and time scale.

Enhanced validation methodology has been demonstrated on a field data set »Šoštanj91« from a very complex terrain from Šaleška region (Slovenia). Field data set is described in details and it has been selected mainly due to high emissions during experimental campaign where SO_2 air pollution situation behaved as tracer experiment.

Air pollution modeling has been made using Lagrangian particle air pollution dispersion model *SPRAY* produced by ARIANET Srl from Milano, Italy. This model has been chosen for validation because it follows Slovenian legislation about air pollution modeling over complex terrain where most of Slovenian industry is located.

For validation of the model only one very complex air pollution situation has been selected. It is a typical complex terrain situation, very difficult for reconstruction and includes phenomenon of air pollution accumulation and convective mixing afterwards. Validation using standard statistical indexes has been made at four locations in different directions from the point of view of air pollution source.

Validation begins using traditional validation methodology. Comparison between measured and reconstructed SO_2 concentrations gives relatively poor results. Only reconstructed concentrations at one station are satisfactory. It has been determined that these results are caused by model's sensitivity to measured wind speeds and directions.

To "measure" this model's inaccuracies in position and time enhanced validation methodology is demonstrated. It gives more satisfactory results at location of all stations and it also estimates inaccuracies. It has been estimated that model's inaccuracy in position is about +-300 m and in time +-1 hour which is indeed excellent result for such a complex terrain. These results give good information for future improvement of air pollution dispersion model.

On the other hand also inaccuracies of measurements should be taken into account during the validation process. It is very important to be aware that the measurements are made at certain location. In certain meteorological conditions these measurements are not representative for the nearest neighborhood. This effect is even more obvious for the complex terrain where the air pollution plume can be present in the nearest neighborhood of the station but it is not detected due to certain local phenomenon.

6. Acknowledgment

The study was partially financed by the Slovenian Research Agency, Project No. L1-2082.

7. References

Anfossi, D., Ferrero, E., Brusasca, G., Marzorati, A., & Tinarelli, G. (1993). A simple way of computing buoyant plume rise in Lagrangian stochastic dispersion models. *Atmospheric Environment* 27A, pp. 1443-1451.

Blumen, W., Banta, R. M., Berri, G., Carruthers, D. J., Dalu, G. A., Durran, D. R., Egger, J., Garratt, J. R., Hanna, S. R., Hunt, J. C. R., Meroney, R. N., Miller, W., Neff, W. D., Nicolini, M., Paegle, J., Pielke, R. A., Smith, R. B., Strimaitis, D. G., Vukicevic, T., & Whiteman, C. D. (1990). Atmospheric processes over complex terrain. *Meteorological monographs,* Volume 23, Number 45, American Meteorological Society, Boston, USA

Božnar, M., Brusasca, G., Cavicchioli, C., Faggian, P., Finardi, S., Minella, M., Mlakar, P., Morselli, M. G., & Sozzi, R. (1993). Model evaluation and application of advanced and traditional gaussian models on the experimental Šoštanj (Slovenia, 1991) campaign. Editor: Cuvelier, C., *Intercomparison of Advances Practical Short-Range Atmospheric Dispersion Models* : Proceedings of the Workshop : August 30 - September 3, 1993, Manno-Switzerland, (Joint Research centre, EUR 15603 EN). Brussels: ECSC-EEC-EAEC, 1994, pp. 112-121.

Božnar, M., Brusasca, G., Cavicchioli, C., Faggian, P., Finardi, S., Mlakar, P., Morselli, M. G., Sozzi, R., & Tinarelli, G. (1994). Application of advanced and traditional diffusion models to en experimental campaign in complex terrain. Editor: Baldasano, J. M., *Second International Conference on Air Pollution,* Barcelona, Spain, 1994. Air Pollution II. Volume 1, Computer simulation. Southampton; Boston: Computational Mechanics Publications, 1994, pp. 159-166.

Brusasca, G., Tinarelli, G., & Anfossi, D. (1992). Particle model simulation of diffusion in low windspeed stable conditions. *Atmospheric Environment* Vol. 26, pp. 707-723

Elisei, G., Bistacchi, S., Bocchiola, G., Brusasca, G., Marcacci, P., Marzorati, A., Morselli, M. G., Tinarelli, G., Catenacci, G., Corio, V., Daino, G., Era, A., Finardi, S., Foggi, G., Negri, A., Piazza, G., Villa, R., Lesjak, M., Božnar, M., Mlakar, P., & Slavic, F. (1991). *Experimental campaign for the environmental impact evaluation of Sostanj thermal power plant, Progress Report,* ENEL S.p.A, CRAM-Servizio Ambiente, Milano, Italy, C.I.S.E. Tecnologie Innovative S.p.A, Milano, Italy, Institute Jozef Stefan, Ljubljana, Slovenia

EUR-Lex (2007). Council Directive 84/360/EEC of 28 June 1984 on the combating of air pollution from industrial plants. Available from: http://eur-lex.europa.eu/LexUriServ/LexUriServ.do?uri=CELEX:31984L0360:SL:NOT, 08.10.2007

FAIRMODE (2010). Guidance on the use of models for the European Air Quality Directive, working document of the Forum for Air Quality Modelling in Europe FAIRMODE ETC/ACC report Version 6.2, Editor: Bruce Denby, In: *FAIRMODE*. Available from: http://fairmode.ew.eea.europa.eu/fol429189/forums-guidance/model_guidance_document_v6_2.pdf, 08.03.2011

Ferrero, E., Anfossi, D., Brusasca, G., Tinarelli, G., Alessandrini, S., & Trini Castelli, S. (1993). Simulation of atmospheric dispersion in convective boundary layer: comparison between different Lagrangian particle models. *4th Workshop on Harmonisation within Atmospheric Dispersion Modelling for Regulatory Purposes*, Oostende, 6-9 May 1996, 67-74. International Journal of Environment and Pollution, Vol 8, Nos. 3-6, pp. 315-323.

Government of Republic Slovenia (October 2007). Decree on the emission of substances into the atmosphere from stationary sources of pollution, In: *Slovenian legislation register Ur.l. RS, št. 31/2007*, Available from: http://zakonodaja.gov.si/rpsi/r06/predpis_URED4056.html, 08.10.2007

Graff, A. (2002). The new German regulatory model - a Lagrangian particle dispersion model. *8th International Conference on Harmonisation within Atmospheric Dispersion Modelling for Regulatory Purposes*, October 14-17, Sofia, Bulgaria, pp. 153-158

Grašič, B., Božnar, M. Z., & Mlakar, P. (2007). Re-evaluation of the Lagrangian particle modelling system on an experimental campaign in complex terrain. *Il Nuovo Cimento C*, Vol. 30, No. 6, pp. 19-

Kaasik, M. (2005). Validation of the AEROPOL model against the Kincaid data set. *10th International Conference on Harmonisation within Atmospheric Dispersion Modelling for Regulatory Purposes*, October 17-20 2005, Sissi, Crete, pp. 327-331

Olesen, H.R. (1996). Toward the establishment of a common framework for model evaluation. *Air Pollution Modeling and Its Application XI*, Edited by S-E. Gryning and F. Schiermeier, Plenum Press, New York, pp. 519-528

Rizza, U., Mangia, C., & Tirabassi, T. (1996). Validation of an operational advanced Gaussian model with Copenhagen and Kincaid datasets. *4th Workshop on Harmonisation within Atmospheric Dispersion Modelling for Regulatory Purposes*, Oostende, 6-9 May 1996, 67-74. International Journal of Environment and Pollution, Vol 8, Nos. 3-6, pp. 41-48.

Schwere S., Stohl A., Rotach M. W. (2002). Practical considerations to speed up Lagrangian stochastic particle models. *Computers & Geosciences* 28, pp. 143-154

Šoštanj91 Campaign home page (2007). Experimental Campaign for the Environmental Impact Evaluation of Šoštanj Thermal Power Plant. In: *Šoštanj91 home page*. Available from: http://193.77.212.133/tes-campaign91/indexe.html

Thomson, D.J. (1987). Criteria for the selection of stochastic models of particle trajectories in turbulent flows. *Journal of Fluid Mechanics*, Vol. 180, pp. 529-556

Tinarelli, G., Anfossi, D., Bider, M., Ferrero, E., & Trini Casteli, S. (2000). A new high performance version of Lagrangian particle dispersion model SPRAY, some case studies. *Air pollution modelling and its Applications XIII*, S. E. Gryning and E. Batchvarova eds., Kluwer Academic / Plenum Press, New York, pp. 499-507

Wilson, J. D., & Sawford, B. L. (1996). Review of Lagrangian stochastic models for trajectories in the turbulent atmosphere. *Boundary-Layer Meteorology* 78, pp. 191-210

9

Modeling the Dynamics of Air Pollutants: Trans-Boundary Impacts in the Mexicali-Imperial Valley Border Region

Alberto Mendoza[1], Santosh Chandru[2], Yongtao Hu[2],
Ana Y. Vanoye[1] and Armistead G. Russell[2]
[1]*Tecnológico de Monterrey, Campus Monterrey,*
[2]*Georgia Institute of Technology,*
[1]*Mexico*
[2]*United States of America*

1. Introduction

Air pollution continues to be an increasing problem in the largest metropolitan areas and regional industrial and commercial corridors in the world. This is also the case in Mexico. Current air quality trends in Mexico indicate that major urban centers continue to exceed the Mexican Ambient Air Quality Standards (MAAQS) for ozone (O_3) and particulate matter with less than 10 microns of aerodynamic diameter (PM_{10}), while other cities are starting to show warning signs of future air quality problems (Zuk et al., 2007). $PM_{2.5}$ monitors are just starting to be deployed around the country, thus no extensive historical data is available on this pollutant.

Some of the urban centers of concern share a common airshed with twin cities across the international border with the United States of America (USA), bringing additional complexity to the study of air pollution dynamics in the region. In this sense, trans-boundary air pollution across USA and Mexico has become a rising problem due to increased commercial and industrial activities in the border region. Trans-boundary air pollution has been studied at different levels in different areas of the border region (Mukerjee, 2001). Two main areas can be identified as the ones that have drawn most of the attention. The first one is the Lower California Area: Tijuana/San Diego, Mexicali/Calexico-Imperial Valley (Figure 1). Here, most of the attention has been on primary PM (e.g., Osornio-Vargas et al., 1991; Chow et al., 2000; Sheya et al., 2000; Kelly et al., 2010), with some studies addressing secondary pollutants (e.g., Zielinska et al., 2001). The second area is the airshed formed by Ciudad Juarez-El Paso-Sunland Park. Perhaps, this area is the one that has received most of the attention regarding trans-boundary air pollution and in a more comprehensive fashion (Currey et al., 2005).

Two of the key steps to improving air quality in a region are identifying, quantitatively, the emissions from sources that affect the area, and assessing how those emissions evolve in the atmosphere to impact pollutant concentrations. Both are difficult, and both can be subject to uncertainties. Air quality modeling is key to both steps because it provides a means to do

Fig. 1. Location of the twin cities of Mexicali-Calexico and Tijuana-San Diego in the Mexico-US border region.

both in a consistent, supportable fashion (Russell & Dennis, 2000). Armed with such information, policy makers can then identify environmentally and economically effective strategies to improve air quality (McMurry et al., 2004).

As indicated, trans-boundary air pollution has been studied at different levels in different areas of the US-Mexico border region. However, limited modeling studies exist where comprehensive chemistry-transport air quality models (CTMs) have been applied at a regional level to understand trans-boundary air pollution in the Mexicali-Imperial Valley border region. In the present study we used a CTM to describe pollutant formation and transport around the Mexicali-Imperial Valley border area, as well as to estimate source contributions to O_3 and $PM_{2.5}$. Even though the principal attention in this study was on the Mexicali-Imperial Valley area, we also expanded our attention outside this area to track down pollutant transport from major urban centers and point sources outside it, but close enough to affect the air quality of the valley (e.g., Tijuana in Mexico, and San Diego and Los Angeles [LA] in the USA).

2. Past air pollution studies in the Mexicali-Imperial Valley Border Area

Tijuana-San Diego has been a border economical belt for a long time. However, over the last 15 years, Mexicali has been one of the fastest-growing cities in Mexico in terms of industrial development, job creation, and energy demand (Quintero-Núñez et al., 2006). This has resulted in that Mexicali on the Mexican side of the border is non-compliant with respect to CO, O_3 and PM_{10} MAAQS, as Calexico is in non-attainment for PM_{10}, $PM_{2.5}$, and O_3 USA standards.

Harmful contaminants in the Mexicali-Imperial Valley border region originate from a number of sources (Sweedler et al., 2003; Quintero-Núñez et al., 2006), including motor vehicles, farms, power plants (natural gas fired and geothermal), and factories. Light manufacturing operations, waste disposal sites, mining, and aggregate handling are also located near the border. In particular, poorly maintained vehicles contribute heavily to the levels of CO, NOx ($NO+NO_2$), and hydrocarbons (HCs) in the air; driving on unpaved roads contributes heavily to PM emissions. Burning of trash, tires, and other materials are also sources of PM, SO_2, and CO.

Several studies have been conducted to understand the composition, spatial variability, and sources of air pollution in the Mexicali-Imperial Valley region. Cerro Prieto, the largest geothermal plant in Latin America (720 MW) is located ~30 km to the south of downtown Mexicali. Since it started operations in the 1970's, H_2S emissions and transport from this facility to Mexicali and Imperial Valley has been a concern (Gudiksen et al., 1980; Deane, 1984). However, atmospheric conversion of H_2S to SO_2 was estimated as not significant. A major effort to understand PM_{10} pollution in Mexicali and Imperial Valley was undertaken in the early 1990's (Chow et al., 2000; Chow & Watson, 2001; Watson & Chow, 2001). This study demonstrated that PM_{10} in the region is mainly composed of crustal material (50% to 62% of the mass) and organic matter (over 25% of the mass). Receptor modeling gave evidence that pollution transport from LA to Mexicali and Calexico could be a concern. In addition, preliminary pollutant flux estimates indicated that the total PM_{10} flux from Mexico to the USA was about 1.5 times the total flux from the USA to Mexico. PM_{10} levels in Mexicali are consistently higher than in Imperial Valley, however wind patterns tend to be in a higher percentage from the north. Other studies have also given evidence of the potential transport of emissions originating in Mexicali and Imperial Valley to areas to the north like the Grand Canyon National Park (Eatough et al., 2001). However, these results have relied on the use of receptor models rather than comprehensive CTMs.

More recently, the fact that the Mexicali Valley and Imperial Valley continue to experience high air pollutants levels made it relevant to conduct an integrated study of the air quality problem in the region. Partial results of this integrated study have been published elsewhere, particularly on levels and chemical composition of fine PM (Mendoza et al., 2010), chemical speciation and source apportionment of VOCs (Mendoza et al., 2009), mobile source emissions characterization using a mobile laboratory (Zavala et al., 2009), and numerical experiments to address the meteorological patterns that foster air pollution episodes (Vanoye & Mendoza, 2009). Here we present our findings on the application of a regional three-dimensional comprehensive CTM to the Mexicali-Calexico border region to follow the dynamics of gas-phase and particulate-phase air pollutants. Particular emphasis is placed on the relevance of understanding trans-boundary air pollutants transport and the implications on emission control strategies on both sides of the border.

3. Description of the modeling system and its application

3.1 Modeling platform

Three-dimensional CTMs continue to be the most scientifically sound tool to assess how emissions from multiple sources impact air quality (Russell & Dennis, 2000). The modeling effort in this study consisted in the application of an advanced air quality and emissions modeling system to the border region to assess how particular sources impact O_3 and $PM_{2.5}$ levels. Specifically, we applied an extended version of the Models-3 suite, including the

Community Multiscale Air Quality model -CMAQ- (Byun & Ching, 1999) for air quality modeling, the PSU/NCAR 5th generation Mesoscale Meteorological model -MM5- (Grell et al., 1995) for meteorological modeling, and the Sparse Matrix Operator Kernel for Emissions model -SMOKE- (Houyoux & Vukovich, 1999) was used to process emissions.

3.2 Modeling domains

The modeling system was applied using nested grids (Figure 2). At the coarser scale, horizontally, 36 km × 36 km grid cells were used. This mother domain is the same as the one defined by the Regional Planning Organizations (RPOs) that oversee other major modeling efforts in North America (i.e., VISTAS, WRAP, CENRAP, LADCO, MANE-VU). At the horizontal mid-resolution level, we included a 12 km × 12 km grid (Figure 3). The coarse grid system allows relatively rapid simulation to set appropriate boundary conditions for the finer grid, serves to stabilize the meteorological and air quality model solutions, and to consider long-range transport from very particular sources. A 4 km × 4 km grid was specified in the Mexicali area for simulations that suggested that fine scale features existed and could not be accurately represented using the coarser grids (i.e., capture the dynamics at the urban scale in Mexicali-Calexico). In this work, we only present the results obtained with the 12 km x 12 km grid, which are ones than provide details on the mid-range pollutants transport in the border region of interest. Details on the extent of each modeling domain are presented in Table 1. The horizontal resolution (including grid nesting) was kept consistent between MM5, SMOKE and CMAQ.

3.3 Episodes selection

The adequate selection of modeling episodes constitutes a fundamental part of the modeling process. If representative episodes are not selected adequately, the modeling results might not characterize effectively the meteorological features that foster high pollution level episodes. Here we used Classification and Regression Tree (CART) Analysis (Breiman et al., 1998) as the formal statistical tool to select the modeling episodes of interest. In essence, CART is a recursive binary partition technique. It divides a set of observations in subgroups taking as reference the value of a particular variable defined by the user (e.g., maximum daily ozone concentration). Each partition in the decision tree is conducted to minimize the classification error of the decision variable. This technique has demonstrated its capacity to help in the selection of days with similar meteorological conditions that give rise to similar pollution levels, using a formal procedure and eliminating the effects of meteorological variability (Kenski, 2004).

CART was applied to obtain decision trees to classify daily maximum O_3, CO and PM_{10} concentrations (separately). The database used was composed of observations (chemical and meteorological) taken by air quality stations in the border region for the years 2001 and 2002. The purpose was to group days with similar O_3, CO, and PM_{10} levels and influenced by similar meteorological condition. The results obtained from CART application were compared against time series plots to corroborate that the episodes selected in fact represented a continuum of days with relatively high pollutant concentrations levels. One of the parameters that can be manipulated while applying the CART technique is the number of final bins that the decision tree will have. Typically, as the number of bin increases the error is reduced; however, if the number of bin increases the probability of having consecutive days in a bin decreases and thus it is harder to construct episodes for air quality

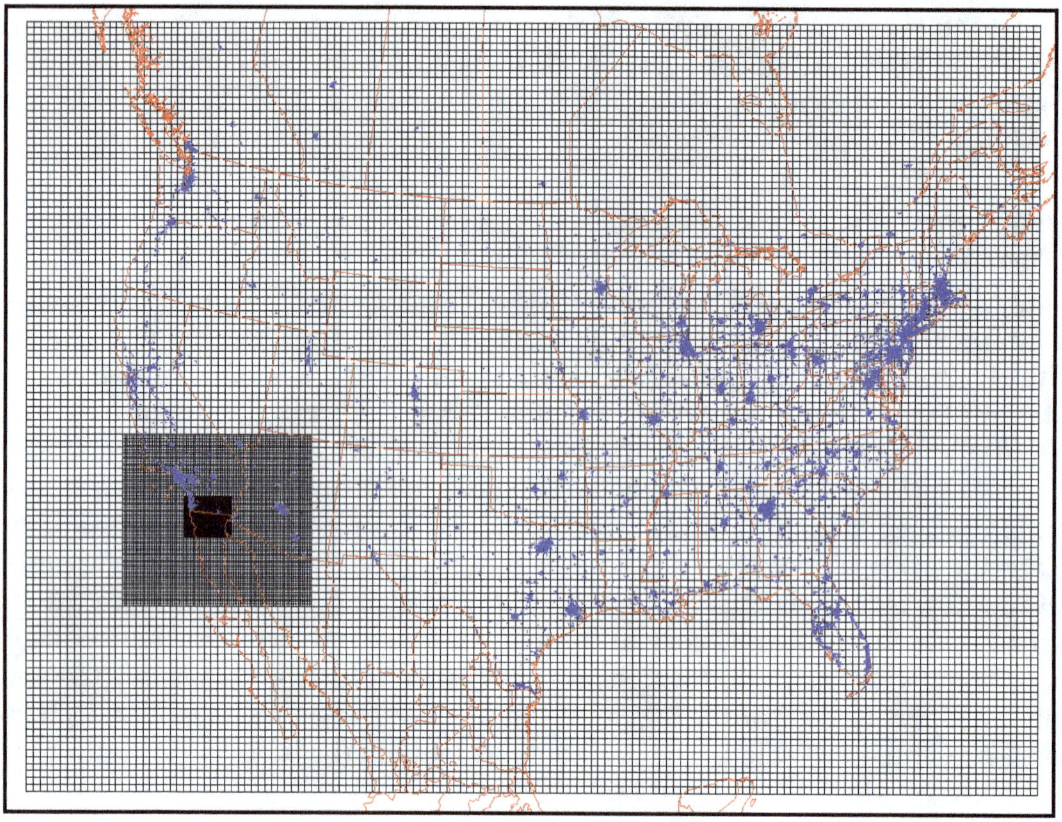

Fig. 2. Horizontal resolution of the nested modeling domain.

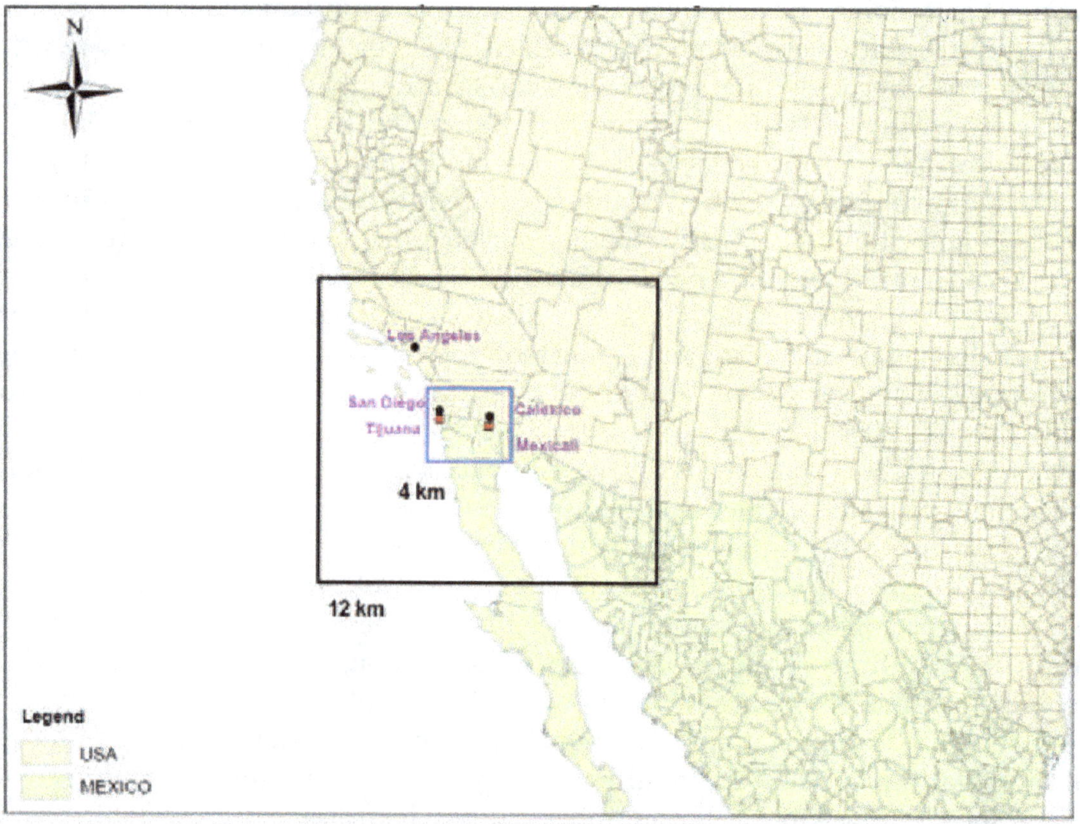

Fig. 3. Location of the 12 km and 4 km horizontal modeling domains.

Grid ID	Origin (*x*, *y*) in km	Horizontal domain (# columns, # rows)
USMEX36	(−2736.0, −2088.0)	(148,112)
USMEX12	(−2232.0, −1160.0)	(84,75)
USMEX4	(−1908.0, −756.0)	(63,54)

Table 1. Modeling domains specifications. Origin coordinates are based on a Lambert Conformal Conic projection with centre lat. and long. as 40 and −97 degrees, respectively

modeling purposes. A convenient number of consecutive days for a modeling episode is between 10 and 15, so with this in mind the number of bins was varied until decision trees with low classification errors and high number of consecutive days in the bins were obtained.

Based on the CART Analysis application, the following modeling episodes were defined: August 18-27, 2001 and July 17-25, 2001 for high O_3 events, and January 6-16, 2002 for high CO and PM events that are typical during autumn and winter times. Additional details on the episode selection process can be found elsewhere (Vanoye & Mendoza, 2009).

3.4 Emissions modeling

SMOKE is a computational engine used to generate the gridded emissions inventory, and its main purpose is to speciate and allocate spatially and temporally area and point emissions and to couple emission estimation tools for mobile and biogenic emissions to spatial and temporal allocation routines.

Base emissions inventory data for the USA side of the border were obtained from the 2001 US National Emissions Inventory (NEI) prepared for the Clean Air Interstate Rule (CAIR). Emissions for the Mexican side came from combining the 1999 BRAVO Mexican inventories (Pitchford et al., 2004) with the 1999 Six Border States Mexican inventory (MNEI) (ERG et al., 2004). Biogenic emissions for both sides of the border were prepared using BEIS3 (Vukovich & Pierce, 2002), and USA mobile emissions were prepared using MOBILE6 (US EPA, 2003). Mobile emissions for Mexico were directly obtained from BRAVO and MNEI.

The emissions inventory generated considers O_3 and PM precursors, as well as primary PM emissions and some toxics (particularly VOC species). The modeling episodes selected were not the same ones as the base years used to derive the raw emissions inventories used for the Mexican side of the border; thus, scaling was needed to update the emissions (e.i., MNEI base year is 1999 and modeling years for our applications were 2001 and 2002). This scaling was based primarily on population growth. VOCs speciation was conducted based on the chemical mechanism selected for the CTM application: SAPRC-99 chemistry (Carter, 2000). Spatial surrogate ratios used to allocate emissions on both sides of the border considered population, highways, total railroads, airport points, and marine ports.

As an example of the results obtained from the application of SMOKE, Figure 4 illustrates CO and biogenic isoprene emission inventories for the 12 km resolution domain. It can be seen, for example, that the CO emissions inventory contains the expected spatial structure (main roads are clearly shown and emissions follow general population patterns). Overall, mobile sources contribute to ~65% of the NOx and ~30% of the VOCs emitted in the LA area; area sources represent ~15% of the NOx and 25% of the VOCs emitted in LA. Values for San Diego are similar as the ones for LA. In contrast, 34% of the NOx and 13% of the VOCs emitted in Mexicali come from mobile sources; area sources (including non-road mobile sources) represent 37% of the NOx and 51% of the VOCs emitted in Mexicali. In Tijuana, mobile sources emit 61% of the NOx and 23 of the VOCs, whilst area sources emit 29% of the NOx and 60% of the VOCs. PM in the Mexicali-Calexico region comes from area sources that are dominated by wood-fuel combustion, agricultural burning, and paved and unpaved road dust.

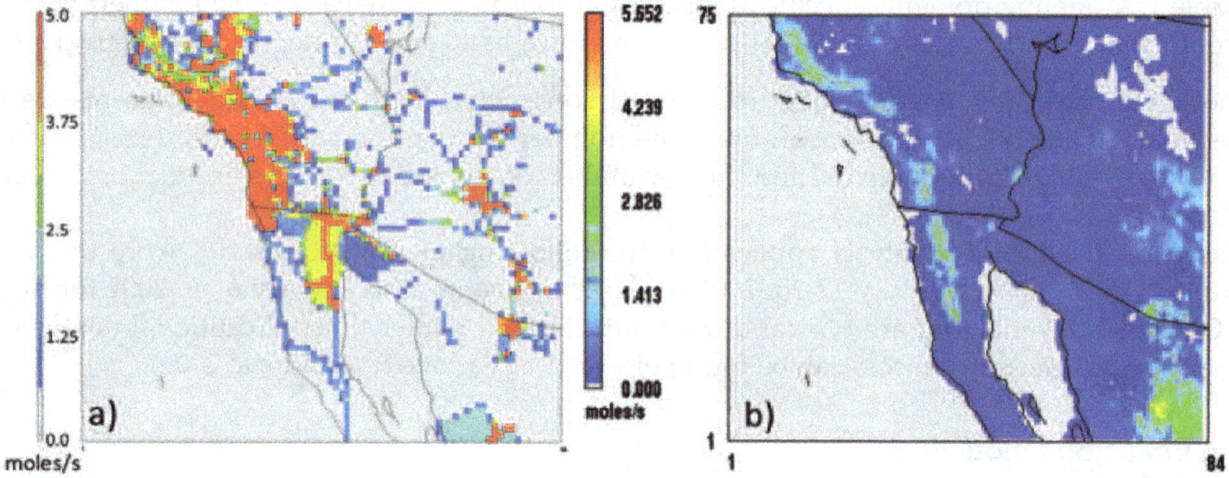

Fig. 4. Examples of (a) CO and (b) biogenic isoprene emissions allocated in the 12 km resolution modeling domain.

3.5 Meteorological modeling

MM5 (Grell et al., 1995) version 3.7 was the meteorological model used here to develop the fields needed to drive the CTM simulations and to provide meteorological information needed to estimate meteorological-variable emissions (e.g., biogenic emissions depend on solar radiation and temperature, while mobile emissions depend on temperature). MM5 is a non-hydrostatic mesoscale meteorological model with grid nesting and four-dimensional data assimilation capabilities. Here we briefly describe the model setup and the input data used to run the model. Additional details, including model performance statistics on the MM5 application, can be found elsewhere (Vanoye & Mendoza, 2009).

MM5 was run with 34 vertical layers with the top of the domain set at 70 mb; horizontal resolution was described earlier. Following a set of sensitivity tests, the MM5 parameterization configuration that gave the best statistical performance of the model for the July and August episodes is presented in Table 2. Of note, the Pleim-Xiu Land Surface model is the recommended scheme by the US Environmental Protection Agency (US EPA) and is the one that has demonstrated to give the best meteorological fields for CMAQ (Olerud & Sims, 2003; Morris et al, 2004). Another advantage of the Pleim-Xiu scheme is that it allows using CMAQ's dry deposition scheme which is technically superior to the conventional Wesley scheme.

MM5 was executed enabling its four-dimensional data assimilation capabilities for the 36 km and 12 km domains. One-way nesting was selected as the way MM5 transferred information from the outer grids to the inner grids. Finally, a relaxation scheme was chosen for the manipulation of the boundary conditions, i.e. the five outermost points are used to damp the information flowing from the boundaries to the inner domain.

Initial and boundary condition were prepared using the National Center for Atmospheric Research (NCAR)/National Centers for Environmental Prediction (NCEP) Eta analyses data. The data consist of regional meteorological analyses for North America based on the output of the Eta model, which generates data every 12 hours from observations of over 600 stations in the region. To complement this information and increase the effectiveness of the data assimilation step, additional observations with a temporal resolution of 6 hours were extracted from NCAR archives, through its Data Support Section of the Scientific Computing Division. This included observations from surface and marine stations, as well as from aerial soundings. Basic landuse, vegetation cover and topography was also obtained from NCAR. Landuse information was based on the USGS 24 categories, and topographic resolutions of 10 min, 5 min, 2 min, and 30 sec were used.

Parameter ID	Description	Selected option
IMPHYS	Explicity Moisture Scheme	Mix Phase
MPHYSTBL	Intrinsic Exponent for Calculating IMPHYS	Use Look-up table for moist physics
ICUPA	Cumulus Schemes	Grell
IBLTYP	Planetary Boundary Layer	Pleim-Xiu
FRAD	Radiation Cooling of Atmosphere	Rapid Radiative Transfer Model
ISOIL	Multilayer Soil Temperature Model	Pleim-Xiu Land Surface Model
ISHALLO	Shallow Convection Option	No Shallow Convection

Table 2. MM5 parameterization options that gave the best model performance for the simulation of meteorological conditions in the Mexicali-Imperial Valley border area.

3.6 CMAQ application

3.6.1 Base case simulations

CMAQ is an Eulerian photochemical model that simulates the emissions, transport, and chemical transformations of gases and PM in the troposphere (Byun & Ching, 1999). Similar to other photochemical models, CMAQ solves the species conservation equation:

$$\frac{\partial C_i}{\partial t} = -\nabla \cdot (\mathbf{u}C_i) + \nabla \cdot (\mathbf{K}\nabla C_i) + R_i + E_i \qquad (1)$$

where, C_i is the concentration of species i, **u** is the wind field, **K** is the eddy diffusivity tensor, R_i is the net rate of generation of specie i, and E_i is the emission rate of species i. Meteorological parameters such as **u** and **K** in eq. 1, as well as temperature and humidity fields come from the MM5 application, while emission rates from SMOKE. CMAQ contains state-of-the-science descriptions of atmospheric processes and has a "one-atmosphere" approach for following the dynamics of gas-phase and particulate matter pollutants. The latter is an important characteristic to assess simultaneously O_3 and aerosols.

CMAQ, as MM5, allows for grid nesting. The horizontal grid structure used was described earlier. The vertical structure of all domains has 13 layers with its top at about 15.9 km above ground. Seven layers are below 1 km and the first layer thickness is set at 18 meters.

Initial and boundary conditions used for the mother domain were the same as the ones suggested by other RPO applications (Russell, 2008). Results from the simulations were compared to data from ground-based monitors for NO_2, O_3, CO, SO_2, PM_{10}, and $PM_{2.5}$. Observational data was obtained from the California Air Resources Board (CARB) for

monitoring stations in the State of California, and from Mexican border municipalities of interest (i.e., Tijuana and Mexicali). Observational data were also obtained from US EPA's Air Quality Data system. In each episode, the first two days were considered ramp-up days and were not further used for additional analysis.
CMAQ has been used extensively to study air pollutant dynamics in the continental USA (e.g., Tagaris et al., 2007; Liao et al., 2007) and in particular regions of that country (e.g., Dennis et al., 2010; Ying & Krishnan, 2010), as well as in other countries around the world (e.g., Che et al., 2011; Im et al., 2011). Only one additional application using CMAQ as the CTM has looked at trans-boundary air dynamics in the USA-Mexico border using fine scale grid resolutions. Choi et al. (2006) looked at high PM events over the sister cities of Douglas, Arizona (USA) and Agua Prieta, Sonora (Mexico). In that application, model performance was acceptable, and it was concluded that secondary processes contributed marginally to the modeled PM events. Primary local sources dominated high PM events.

3.6.2 Sensitivity analysis

Sensitivity analysis is an important tool that can be used to understand the impacts of emissions from various sources on ambient air concentrations of specific pollutants. The ability to conduct sensitivity analyses in an efficient fashion is critical to obtain robust descriptions of the response fields of pollutant concentrations to changes in model inputs (particularly emissions), which then are used in source attribution analyses and control strategy design (e.g., Bergin et al., 2007). Among the different choices to estimate the sensitivity fields, the direct decoupled method for three-dimensional models (DDM-3D) has proven to be superior to other techniques (Yang et al., 1997; Hakami et al., 2003). DDM-3D is an implementation of the Decoupled, Direct Method (Dunker, 1984; Dunker et al., 2002) for sensitivity analysis. The version of CMAQ used in our applications was extended with DDM-3D (Cohan et al., 2005). The method directly calculates the response of model outputs (concentrations) to parameters and inputs, i.e., the semi-normalized sensitivities S_{ij}:

$$S_{ij} = \frac{\partial c_i}{\partial e_j} \tag{2}$$

where c_i is the concentration of species i and e_j is the relative perturbation on parameter j –p_j- (e.g., NOx emissions) from its nominal value $p_j°$ (i.e., $e_i = p_j/p_j°$). This is an efficient approach for directly assessing the sensitivity of model results to various inputs and parameters, and replaces the need to use the traditional brute force approach of re-running a model after modifying a parameter. More importantly, it does not suffer from numerical noise problems that can overwhelm brute force approaches. In addition, it is a linear method. In prior studies, the atmospheric chemistry has been found to respond relatively linearly for emissions changes on the order of 25% or more (Dunker et al., 2002; Hakami et al., 2004).
Of particular interest was to explore the sensitivity due to variations on the emissions inventory. The implementation of DDM to CMAQ allows defining spatial- and source-specific emissions categories as the input being perturbed, and in a single model run the sensitivities of all species tracked by the model to changes in a set of emissions sources can be calculated. To accomplish this, the region of interest was divided into three different areas: Mexicali-Calexico (abbreviated as MXC), Tijuana-Tecate-San Diego (abbreviated as

TSD), and Los Angeles-Riverside-Orange-Ventura (abbreviated as LAR). O_3 sensitivities to area-, mobile-, and point-source emissions of NOx and VOC were calculated for each of the regions defined for both summer episodes. Additionally, $PM_{2.5}$ sensitivities to changes in the same source categories were calculated.

4. Results

4.1 Air quality model performance

Domain-wide episode performance statistics were determined to ascertain the confidence of the simulation results. Table 3 presents the average model performance for the 12 km domain in terms of Mean Bias Error (MBE), Root Mean Squared Error (RMSE), Mean Normalized Bias (MNB), and Mean Normalized Error (MNE). Established performance guidelines indicate that the model should have a MNB of ±5–15%, and a MNE 30–35% for O_3 (Tesche et al., 1990). Based on these guidelines, CMAQ performed well in predicting the observed O_3 concentrations. No guidelines exist for the rest of the gas-phase species, though the results are comparable to results obtained by others using different CTMs (e.g., Mendoza-Dominguez & Russell, 2001). Overall, the gas-phase species results indicate a tendency of the model to underestimate the pollutant concentrations. This is in line with the results obtained from mobile laboratory measurements that indicate an underestimation of the official emissions inventory for Mexicali (Zavala et al., 2009). PM proved to be more difficult to simulate correctly, which is a known setback of current CTMs (Russell, 2008).

		MBE	RMSE	MNB (%)	MNE (%)
August-01	O_3	−1.64E-03	1.60E-02	−0.21	19.7
	CO	−3.52E-01	6.56E-01	−18.6	63.2
	NOx	−1.25E-02	2.52E-02	−34.9	74.2
	SO_2	−1.40E-03	5.40E-03	−19.4	87.7
	$PM_{2.5}$	−6.76E+00	9.14E+00	−36.9	39.2
	PM_{10}	−3.00E+01	3.57E+01	−76.2	76.2
July-01	O_3	−9.19E-02	1.76E-01	−18.3	61.2
	CO	−1.43E+00	1.97E+00	−22.0	56.8
	NOx	−6.18E+00	7.58E+00	−31.0	60.0
	SO_2	−5.31E+00	6.53E+00	−29.2	60.2
	$PM_{2.5}$	−7.10E+00	8.72E+00	−33.3	63.0
	PM_{10}	−8.12E+00	9.97E+00	−35.3	59.5
January-02	O_3	2.86E-03	7.76E-03	7.0	14.1
	CO	−7.33E-01	1.33E+00	−26.5	73.1
	NOx	−3.99E-02	7.98E-02	−34.6	76.0
	SO_2	−1.10E-03	5.40E-03	−29.6	74.4
	$PM_{2.5}$	−2.54E+00	1.01E+01	−4.6	37.2
	PM_{10}	−2.53E+01	3.21E+01	−56.1	61.7

Table 3. Average performance metrics for the 12 km domain during the diferent episodes modeled. MBE and RMSE are in ppmv for gas-phase species and $\mu g/m^3$ for PM species.

Peak O_3 concentrations in Calexico during the August episode were observed at Ethel Street (MBE -4.0 ppbv, MNB -3.0%), and East Calexico (MBE 5.0 ppbv, MNB 12.0%) sites. Calexico and Mexicali, being adjacent to each other in the border region, experience similar O_3 concentrations. The inability to capture the minimums in Ethel Street and Calexico East sites can be attributed to the fact that both these locations are located close to roadways, hence experience strong O_3 sinks in the night time due to its reaction with NO which the model is unable to capture using the 12 km grid structure (Figure 6).

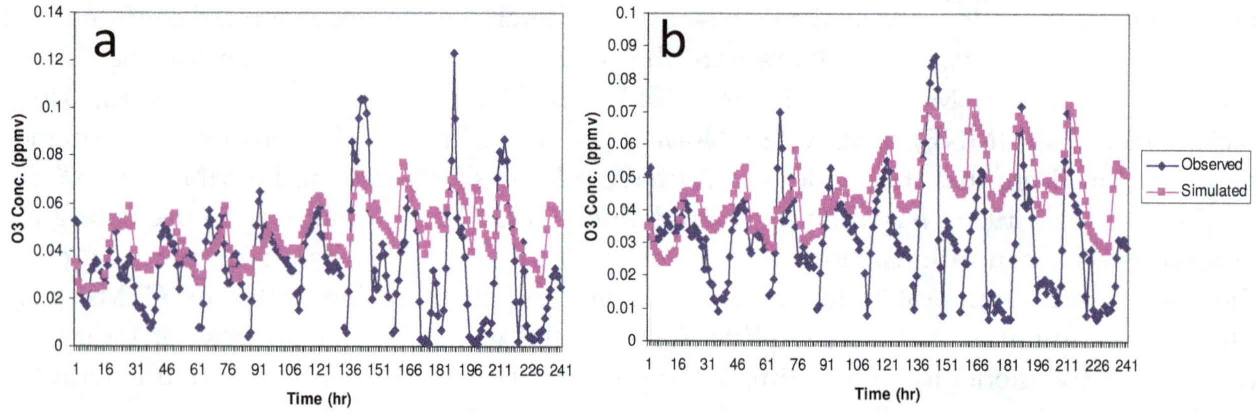

Fig. 6. Observed vs. simulated O_3 concentrations at representative sites in Calexico during August 2001 at (a) Ethel Street and (b) Calexico East site. Time scale represents hours of simulation.

Also, the LAR area was tested for model performance with respect to spatial as well as temporal variability (Figure 7). This region was chosen because of the high density of monitoring stations located in it, which can give a more realistic comparison with the modeled average hourly concentration values in the region. During the August episode, the peak O_3 concentrations occurred on August 26th, 2001: 189 ppbv at the Azusa site and 190 ppbv at the Glendora Laurel station, respectively. These sites are located in the San Gabriel Valley and come under the same 12 km grid cell. Simulated concentrations correlated well with the observed concentrations at Azusa (MBE 5.0 ppbv, MNB 12.2%) and Glendora Laurel (MBE 0.0 ppbv, MNB 3.5%) on most days. For the January episode, January 12 and 13, 2001 were the high concentration days with peaks of ~60 ppbv in the Los Angeles area, and ~75 ppbv in the Mexicali area.

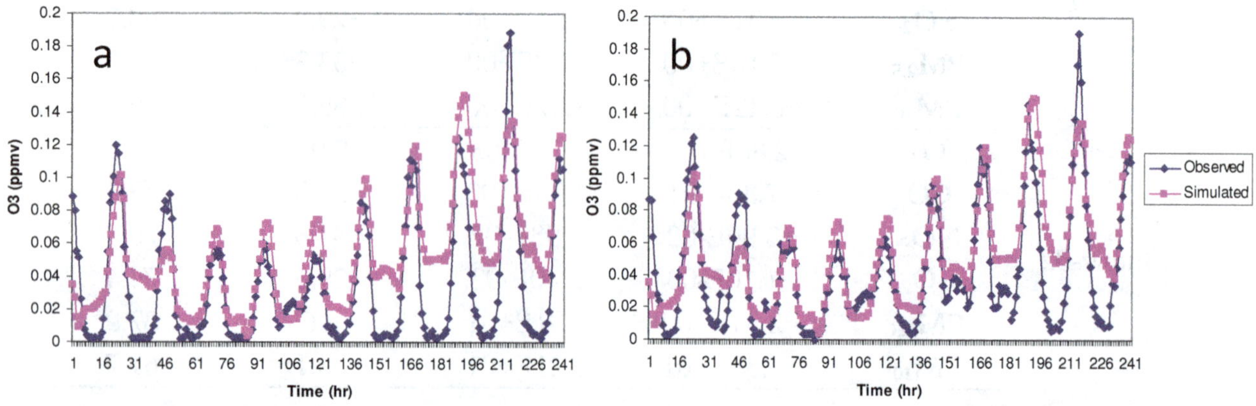

Fig. 7. Observed vs. simulated O_3 concentrations at representative sites in LA, during August 2001 at (a) Azusa, (b) Glendora Laurel. Time scale represents hours of simulation.

4.2 Modeled pollutant concentration fields

Resulting O_3 fields for the July and August episodes illustrate the influence of regional transport across the domain. During the July 2001 episode, a peak of 125 ppbv O_3 was simulated in the LA area on July 23, 23:00 hrs UTC (Figure 8d). The plume from LA can be seen transported towards the east (Figure 8 a-c). Plumes of up to 78 ppbv O_3 emerge from San Diego-Tijuana and travel eastwards and reach the Mexicali-Calexico region (Figure 8 a-c). Peak $PM_{2.5}$ concentrations of over 50 μg/m^3 were simulated in the LA area on July 15th, while $PM_{2.5}$ concentrations did not exceed 15 μg/m^3 in Tijuana-San Diego and Mexicali-Calexico during the July episode.

On August 24th (20:00 hrs UTC), strong O_3 plumes started to develop, and plumes from Mexicali-Calexico and Los Angeles almost converged (Figure 9a). At the same time plumes from Tijuana-San Diego build up as well and move eastwards towards Mexicali-Calexico (Figure 9b). Similar patterns start to emerge on August 25th (19:00 hrs UTC) (Figure 9c); peaks reach 162 ppbv in the LA area on August 26th (00:00 hrs UTC) (Figure 9d). A peak concentration of 162 ppbv is reached about 30 km northwestfrom the Glendora Laurel site where a simulated peak of 144 ppbv is reported. Similar to the July episode, O_3 plumes from San Diego-Tijuana border area are transported eastward towards Mexicali-Calexico during the August episode as well. Peak $PM_{2.5}$ concentration of 100 μg/m^3 are seen close to the LA area on August 25th. In the Mexicali-Calexico region, Mexicali showed a peak of 42 μg/m^3 on August 25th (14:00 hrs UTC).

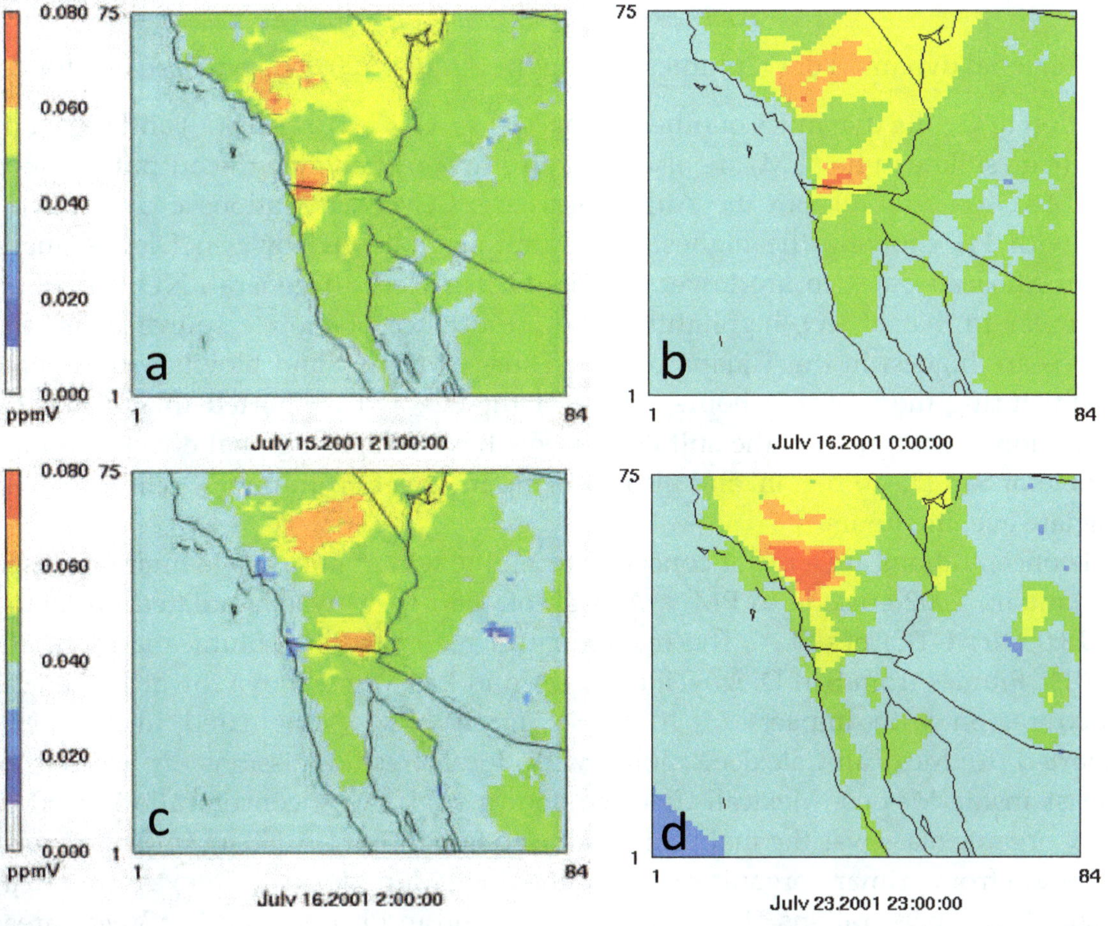

Fig. 8. Regional dynamics of O_3 plumes during the July 2001 episode (see text for details).

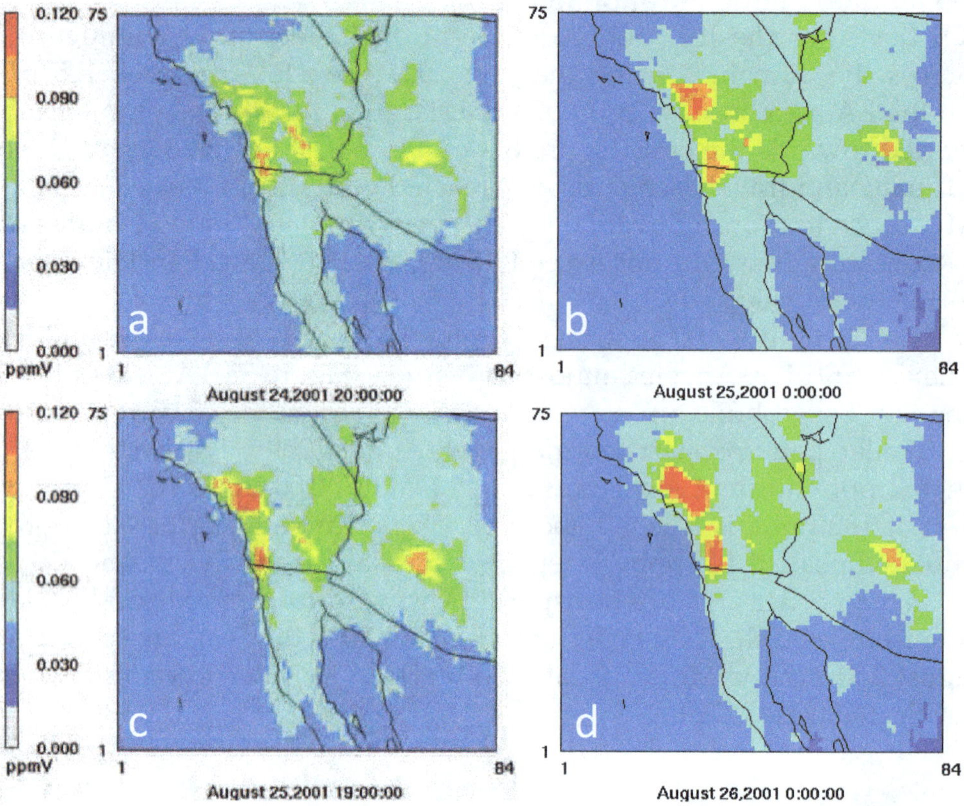

Fig. 9. Regional dynamics of O_3 plumes during the August 2001 episode (see text for details).

In addition to O_3, the dynamics of other primary (e.g., CO, NO_2, and SO_2) and secondary PM species (e.g., sulfate in fine PM) are also of interest when analyzing the output data obtained from CMAQ (Figure 10). For the August episode, CO concentrations peak around 7 PM (PDT), with LA showing the highest concentration, followed by San Diego-Tijuana. CO levels in Mexicali-Calexico are lower and more localized. In general, NO_2 distribution is very similar to that of CO, highlighting the importance of mobile source emissions. SO_2 emissions are highest in the Tijuana region. Thus, with the wind blowing in the northeast direction during the morning hours, much of the SO_2 is transported inland into the San Diego region. Consequently, the sulfate aerosols have a high regional effect encompassing the whole of San Diego region, and also showing its effect on Imperial Valley and Mexicali during late evening hours.

As PM concentrations are a major concern during the winter season, we limit our discussion of the January 2002 episode to $PM_{2.5}$. A peak of 188 $\mu g/m^3$ was simulated on January 12, 2002 (18:00 hrs UTC) near LA. The movement of regional $PM_{2.5}$ plumes is represented in Figure 11. Plumes from San Diego-Tijuana, LA and Las Vegas move towards the Mexicali-Calexico region with impacts of 10 to 35 $\mu g/m^3$. $PM_{2.5}$ originated in the USA and transported to Mexicali-Calexico, along with local fresh emissions, is carried further southeast inside Mexico. Mexicali-Calexico shows peak $PM_{2.5}$ concentration of 50 $\mu g/m^3$. Primary organic mass was the main contributor to fine PM in LA (98 $\mu g/m^3$). The maximum contribution from primary organic matter to the fine PM in Mexicali-Calexico was 10 $\mu g/m^3$. Peak soil dust concentrations of 40 $\mu g/m^3$ were found in Pheonix and Las Vegas areas. The soil dust contributions from LAR, TSD and MXC range between 5-25 $\mu g/m^3$ (Figure 11).

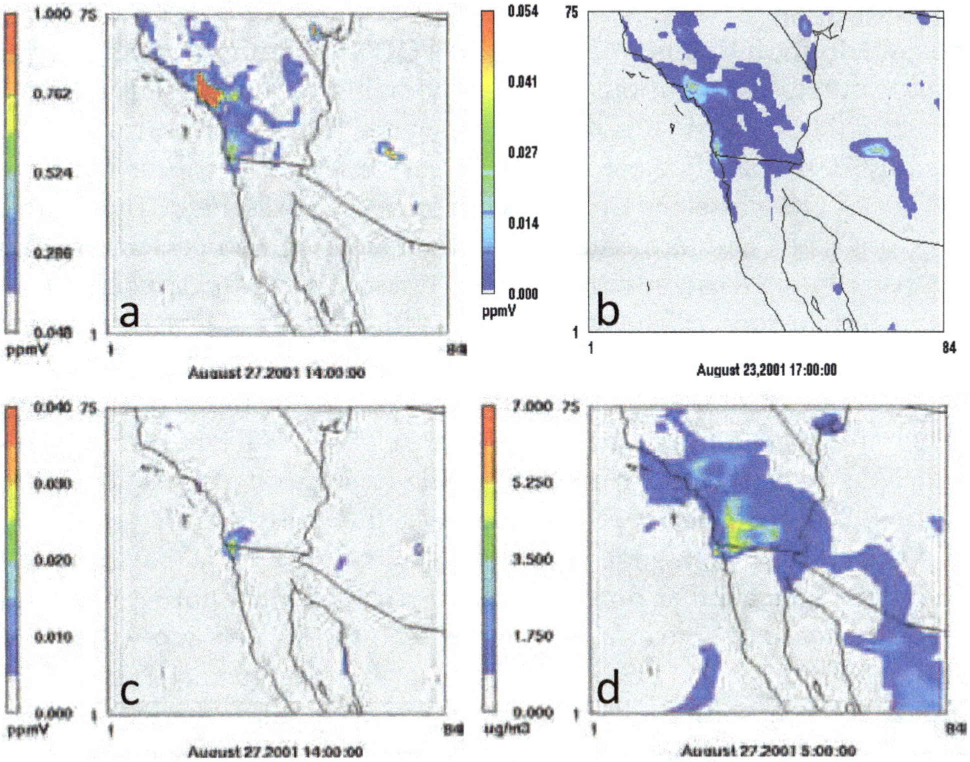

Fig. 10. Concentration fields for gas-phase and aerosol species (August 27, 2001): a) CO, b) NO_2, c) SO_2, d) sulfate $PM_{2.5}$.

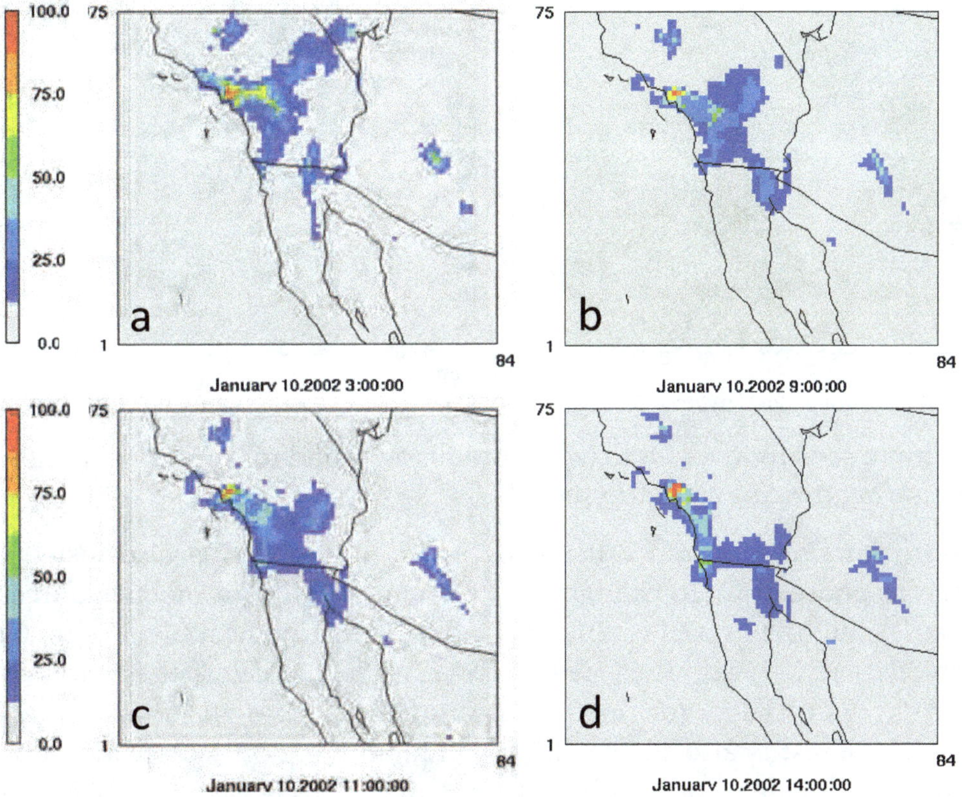

Fig. 11. Dynamics of $PM_{2.5}$ plumes during the January 2002 episode (see text for details).

4.3 Source contribution

4.3.1 Source contribution to specific NOx and VOC emission sources: August episode

To understand source contributions in the modeling domain, sensitivity fields were estimated using CMAQ/DDM. Results are presented only for the August 2001 episode; values for the July 2001 episode were similar. First, the sensitivity of the regional O_3 field to changes in NOx or VOC emissions from specific sources is presented. The figures presented are "response surfaces" and are interpreted as the amount of increment in pollutant concentration per 10% increase in emissions (or amount of reduction per 10% decrease in emissions) from certain source. The sensitivity coefficients are linear (first order) in nature and thus can be used in the manner described. In general, is reasonable to imply a linear response over a range of emissions perturbations (±30%) even for species that it is well known their non-linear response in the atmosphere (e.g., O_3; Hakami et al., 2003).

The impact of NOx emissions from MXC was the highest on August 26, with sensitivity response reaching 9 ppbv of O_3 per 10% change in the emissions (Figure 12). The area of influence of NOx emissions extends northwards into Imperial Valley, and partially into Arizona. The results indicate that down-wind the atmosphere is not NOx-inhibited; that is, an increase in NOx does not give as a response a decrease in O_3 down-wind as has been the case in other areas of the Mexico-US border (Mendoza-Dominguez & Russell, 2001). Change in VOC emissions from sources in MXC produce a smaller change -localized- in O_3 concentrations (maximum of 3 ppbv per 10% change), indicating the benefits of NOx control over VOC control in the region.

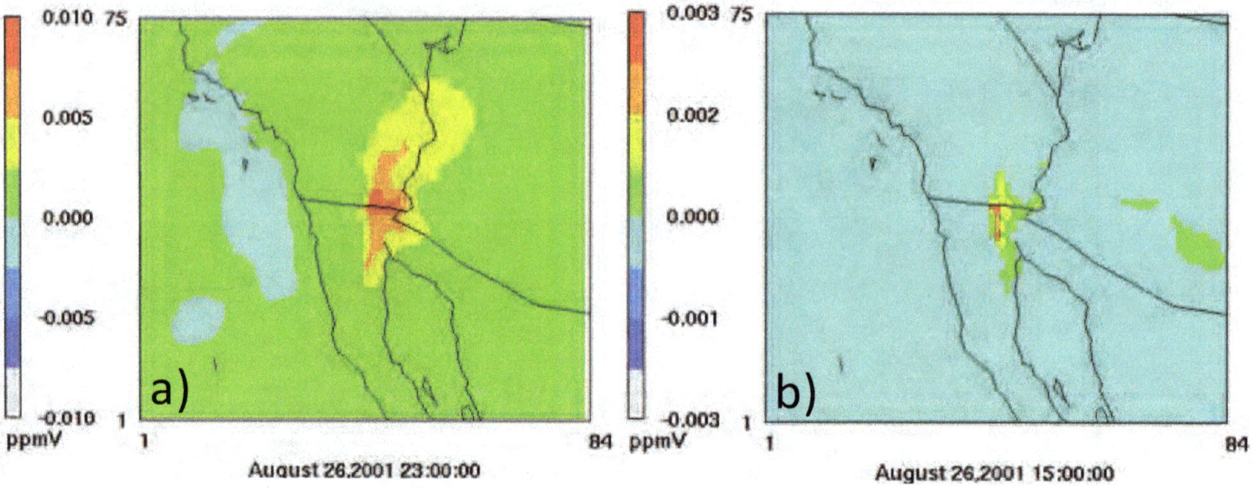

Fig. 12. Maximum sensitivity of O_3 to (a) NOx emissions and to (b) VOC emissions from the MXC region during the August 2001 episode.

The influence of emissions from other geographic locations was also tested. Figure 13a illustrates the response of O_3 to changes in NOx emissions from mobile sources located in the TSD area. The highest impact is almost 17 ppbv, occurring near San Diego, with a strong influence in the MXC region as well. This result indicates that emission controls implemented in San Diego (or increment in emissions) will impact the MXC area accordingly. Of interest is also the small NOx-inhibited region located down-wind of San Diego, toward the Tijuana border, that implies that a decrease in mobile emissions will result in an increase in O_3 concentrations. In contrast, and as expected, impact from VOCs emitted by mobile sources located in the TSD area is limited to less than 2 ppbv per 10% change in emissions (Figure 13b). The spatial extent of influence is also more limited than the sensitivity to NOx emissions, influencing the northern region of Imperial Valley County.

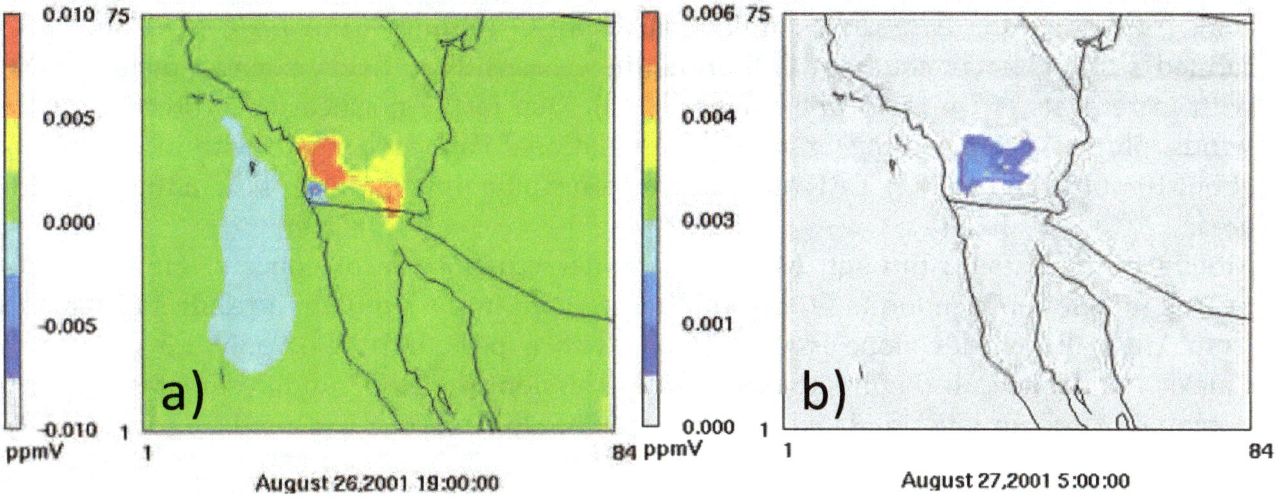

Fig. 13. Maximum sensitivity of O_3 to (a) mobile NOx emissions and to (b) mobile VOC emissions from the TSD region during the August 2001 episode.

Finally, example sensitivity values due to changes in NOx emissions from mobile and area sources from the LAR region are presented (Figure 14). For the case of mobile sources, the increment of NOx emissions results in a decrease in ozone (~7.5 ppbv per 10% increase in emissions) in downtown LA, with a corresponding increase (~30 ppbv) in neighboring counties of Ventura, Orange, and Riverside. From the extent of the sensitivity field, it is possible that under the right meteorological conditions, the influence can reach the Imperial Valley area. On the other hand, the sensitivity to area source NOx is smaller in value and extent because area emissions are smaller than the mobile emissions in Southern California.

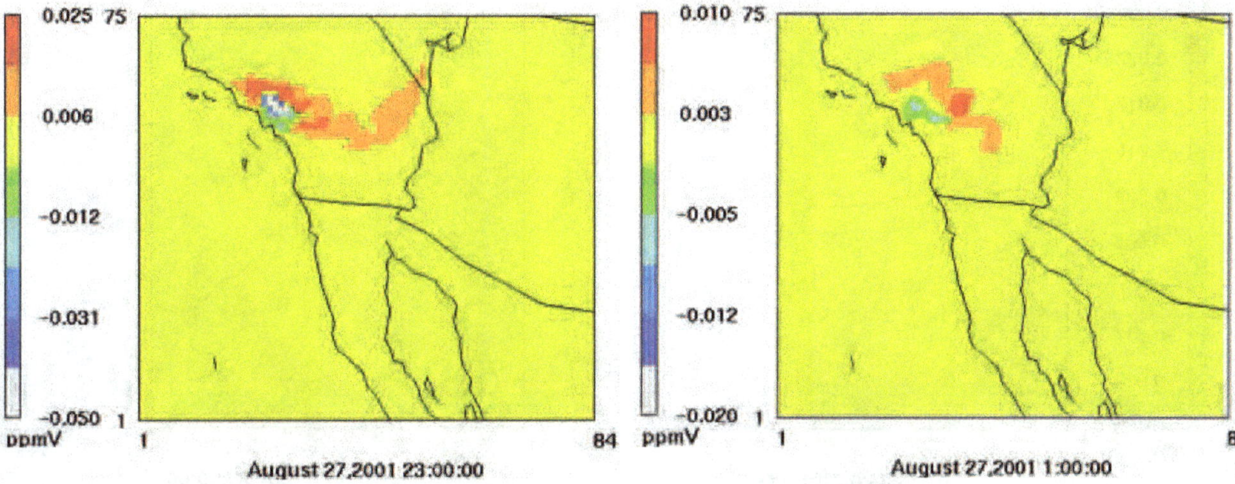

Fig. 14. Maximum sensitivity of O_3 to (a) mobile NOx emissions and (b) area NOx emissions located in the LAR area during the August 2001 episode.

4.3.2 Source contribution to overall emission sources: August episode

When considering the overall emissions from mobile sources, the LAR area made an overall contribution of 44 ppbv on the surrounding region i.e., east of the city of LA towards Glendora Laurel and Azuza on August 26 (00:00 hrs UTC) (Figure 15c). Presence of high concentrations of NOx results in negative sensitivities up to –46 ppbv in urban LA (Figure 15a). As seen in the base case simulations where O_3 plumes from LAR, MXC and TSD formed a triangle over southern California, the O_3 sensitivity fields extends towards MXC with increments of up to 10 ppbv (Figure 15a). Due to the northeasterly direction of the winds, plumes also reach the Grand Canyon National Park area, again with increments of about 10 ppbv (Figure 15d). LAR area sources contribute up to 8 ppbv of O_3 in the Riverside area.

Mobile traffic passing through Mexicali's border crossings is of concern. However, the overall mobile contribution to O_3 is found to be small in the simulation results. The impact from Mexicali vehicles alone is very small, with a peak impact of only 1.3 ppbv over Calexico and Mexicali (Figure 16a). Possible emission inventory underestimates can be a potential reason for low simulated impacts, and should be further explored.

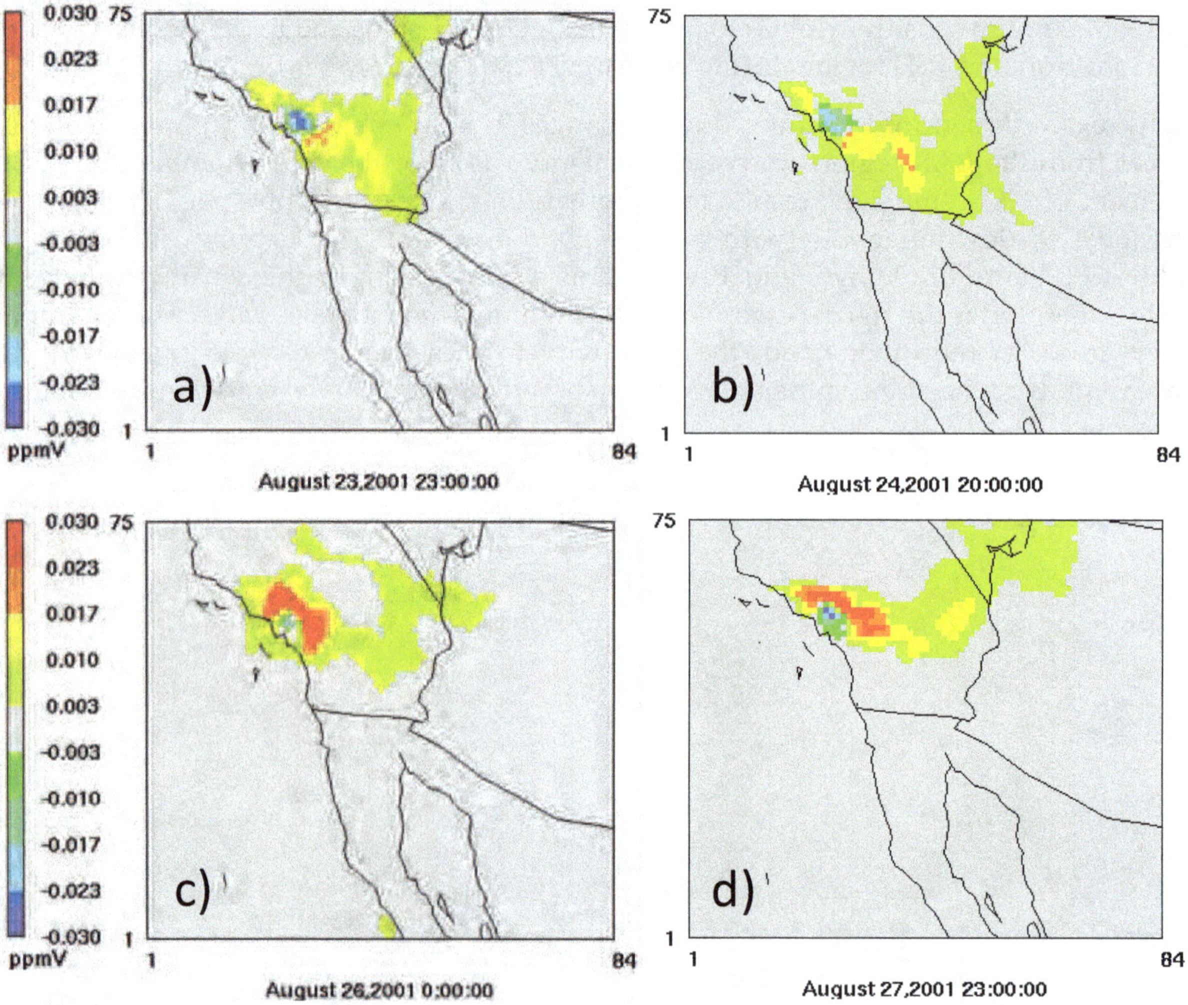

Fig. 15. O_3 sensitivity to LA mobile source emissions (see text for details).

The maximum impact from Calexico mobile sources is 2 ppbv of O_3 seen over the Calexico region itself, and the border between California and Arizona (Figure 16b). The primary areas of mobile emissions are the two border crossing areas (seen in blue as negative sensitivities). Area sources in MXC contribute a simulated maximum of 8 ppbv O_3 during the summer episode (Figure 17a). The area of influence can be seen encompassing California, and the border regions of California-Arizona. O_3 impacts up to 4 ppbv in the Grand Canyon area can be attributed to area sources in the Mexicali-Calexico region (Figure 17b).

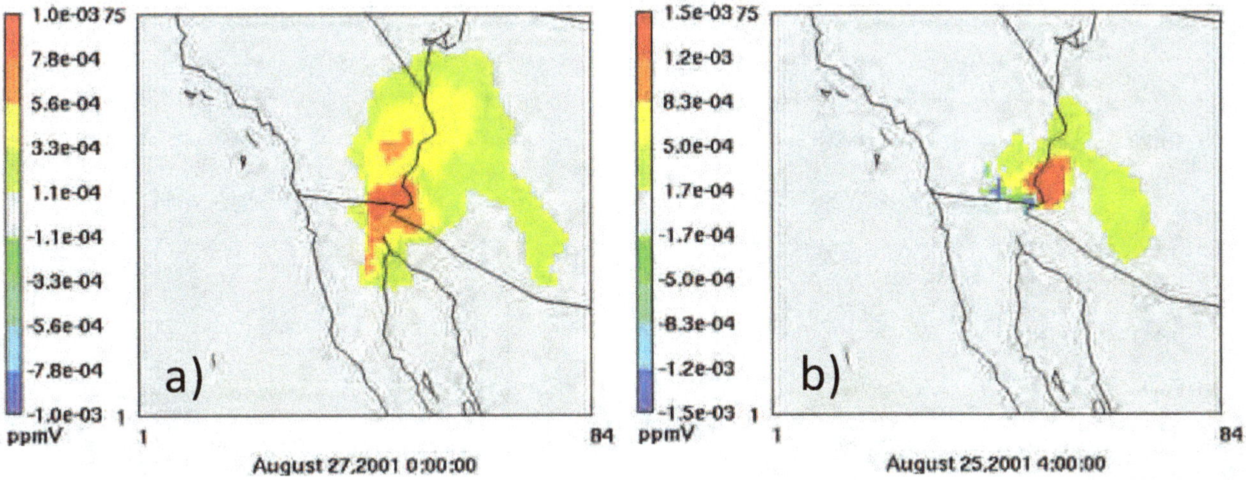

Fig. 16. Contribution to O_3 concentrations from (a) Mexicali and (b) Calexico mobile sources.

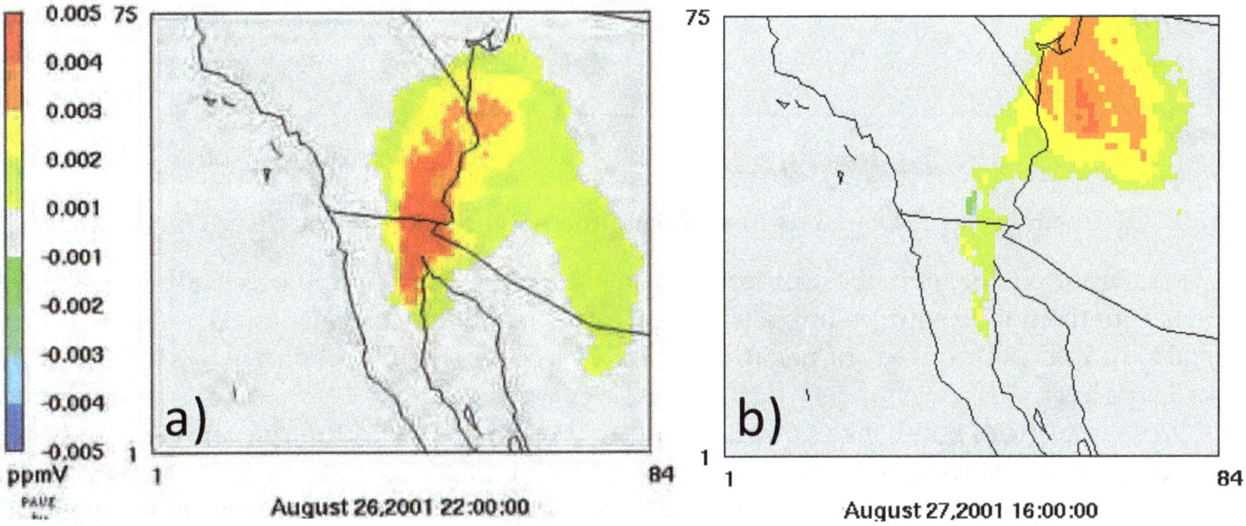

Fig. 17. Contribution to O_3 concentrations from MXC area sources (see text for details).

Area sources from TSD have a peak impact of 40 ppbv of O_3 over the San Diego area and this plume is carried eastwards into the USA close to the border region. The contribution of Tijuana emissions extends to the southeast into inner Baja California and impacting up to 20 ppbv of O_3 (Figure 18a,b). Also, on August 26th, the sensitivity field of O_3 from the TSD region extends eastwards towards Calexico, thus adding O_3 to the already polluted air in Calexico and Mexicali (Figure 18c,d).

Tijuana mobile source impacts reach up to 6 ppbv on both sides of the border depending on the wind direction (Figure 19b,c). Since the dominant wind pattern is towards the northeast,

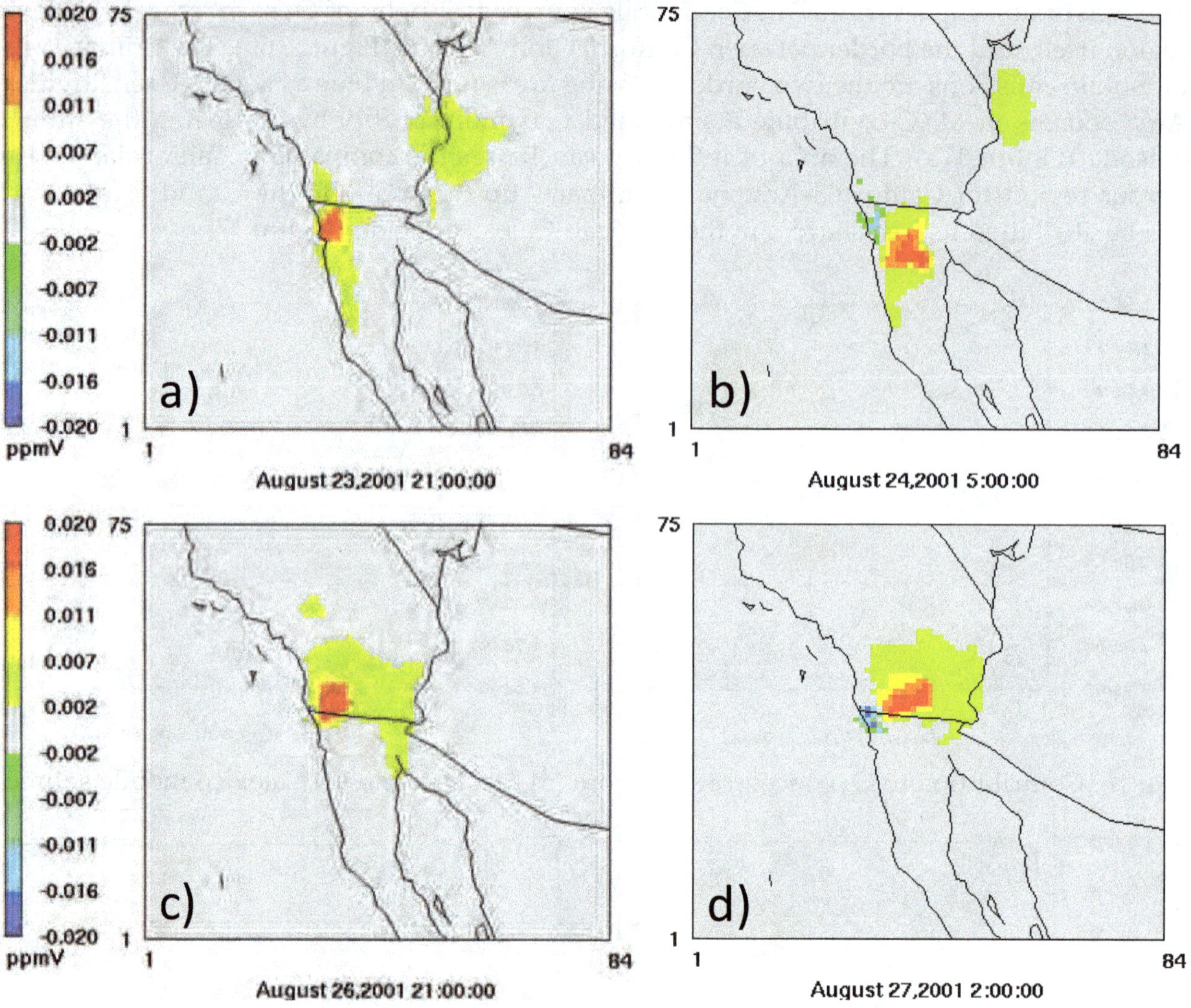

Fig. 18. O_3 sensitivity to TSD area sources during the August 2001 episode (see text for details).

O_3 is transported through the California-Baja California border towards Calexico (Figure 19a,c). Tijuana mobile sources impacts up to 3 ppbv of O_3 in Mexicali-Calexico (Figure 19a). It is also of interest the areas of negative sensitivity observed in downtown Tijuana of more than 3.0 ppbv.

Mobile sources from San Diego contribute up to 26 ppbv of O_3 in the region itself, and also over the park areas such as Anza Borrego Desert State Park located southeast of San Diego (Fig. 20a). The base case scenario showed O_3 plumes from TSD area transported to Mexicali-Calexico. A contribution of up to 11 ppbv of O_3 in Mexicali-Calexico can be attributed to the high density of vehicles in and around the San Diego region (Fig. 20b). This contribution is higher than the contribution from MXC mobile sources.

The peak $PM_{2.5}$ concentration simulated over the MXC region was 42 $\mu g/m^3$. Of this, MXC area sources contributed to 21 $\mu g/m^3$ of primary $PM_{2.5}$. Thus, 50% of the $PM_{2.5}$ levels in MXC can be attributed directly to MXC area sources during August 2001. $PM_{2.5}$ contribution from MXC mobile sources was very small, with peak contributions less than 1 $\mu g/m^3$. MXC point sources contributed the remaining share of up to 7 $\mu g/m^3$. Simulations found similar results for TSD with contributions of up to 33 $\mu g/m^3$ of $PM_{2.5}$ from TSD area sources, less than 2 $\mu g/m^3$ from mobile sources, while the point sources in the region contributed up to 13 $\mu g/m^3$.

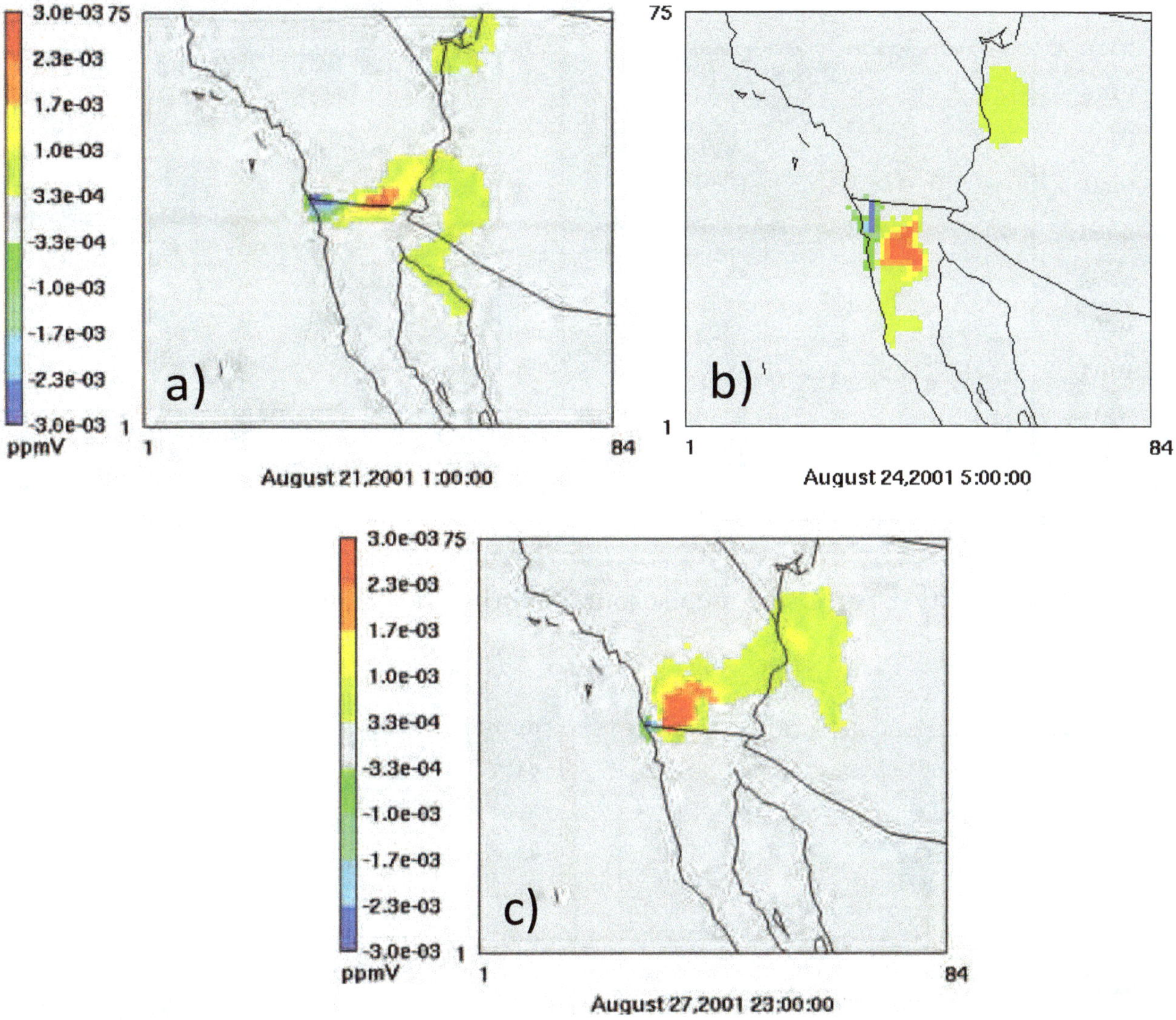

Fig. 19. O_3 sensitivity to Tijuana mobile sources during the August 2001 episode (see text for details).

4.3.3 Source contribution during the winter episode

Contributions to O_3 from various sources in the region were simulated for the winter episode. Impact of LAR mobile sources of up to 14 ppbv was seen over the Pacific Ocean. High contributions were also observed along the coast from Los Angeles to San Diego, which represents a major travel road route (Figure 21a). MXC area sources had simulated impacts of up to 6 ppbv over the southern regions of Baja California.

However, much of the time fresh NOx emissions led to decreases (negative sensitivities) over the local urban areas and positive impacts downwind. Peak impacts of 11 ppbv of O_3 were simulated over the Los Angeles area during the winter episode which originate from TSD area sources (Figure 21b). This same changes originated O_3 reductions of more than 7 ppbv over TSD.

Similar values as that in summer episode were simulated with peak impacts of up 2 ppbv O_3 on the Baja California region from Mexicali mobile emissions. Tijuana, San Diego and Calexico mobile sources contribute to less than 6 ppbv O_3 during the winter episode.

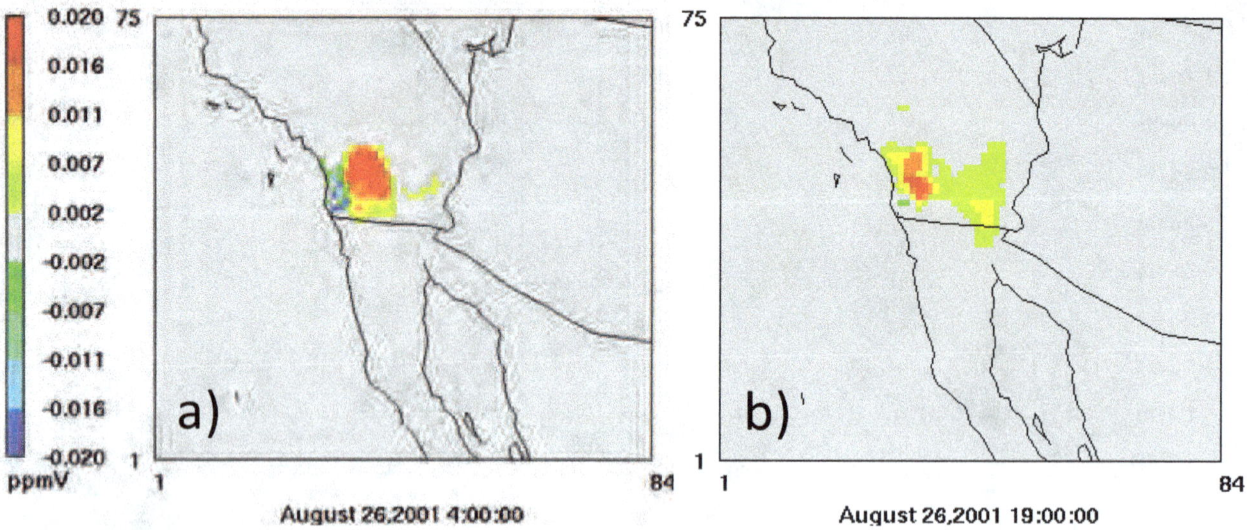

Fig. 20. O_3 sensitivity to San Diego mobile sources during the August 2001 episode (see text for details).

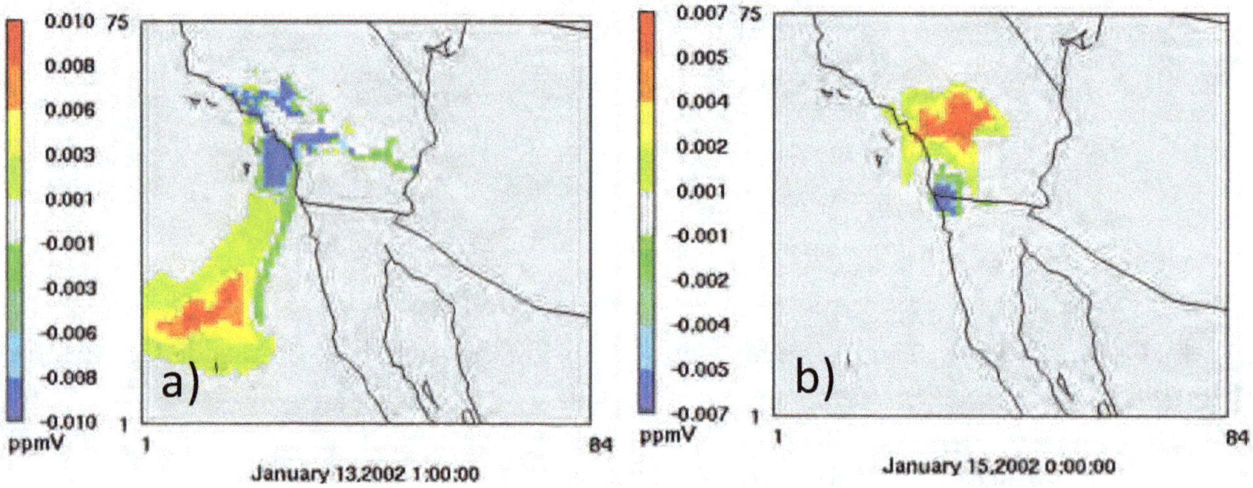

Fig. 21. Peak O_3 sensitivity to (a) LAR mobile sources and (b) to TSD area sources, during the January 2002 episode.

MXC area sources contribute to a simulated $PM_{2.5}$ maximum of 34 μg/m³ (Figure 22a). The pattern is very localized. Primary $PM_{2.5}$ emissions from MXC mobile sources contribute negligibly with peak contributions of 0.5 μg/m³. MXC point sources, primarily present in Mexicali contributed to a maximum of 12 μg/m³ over the border region. Area sources in TSD had very large contributions, ranging up to 52 μg/m³ (Figure 22b). However, once again the extent of the sensitivity field is constrained to the vecinitiy of the cities of Tijuana and San Diego. TSD mobile sources contributed to less than 3 μg/m³ of primary $PM_{2.5}$. Point sources in San Diego contributed to a maximum 13 μg/m³ of primary $PM_{2.5}$ in the region.

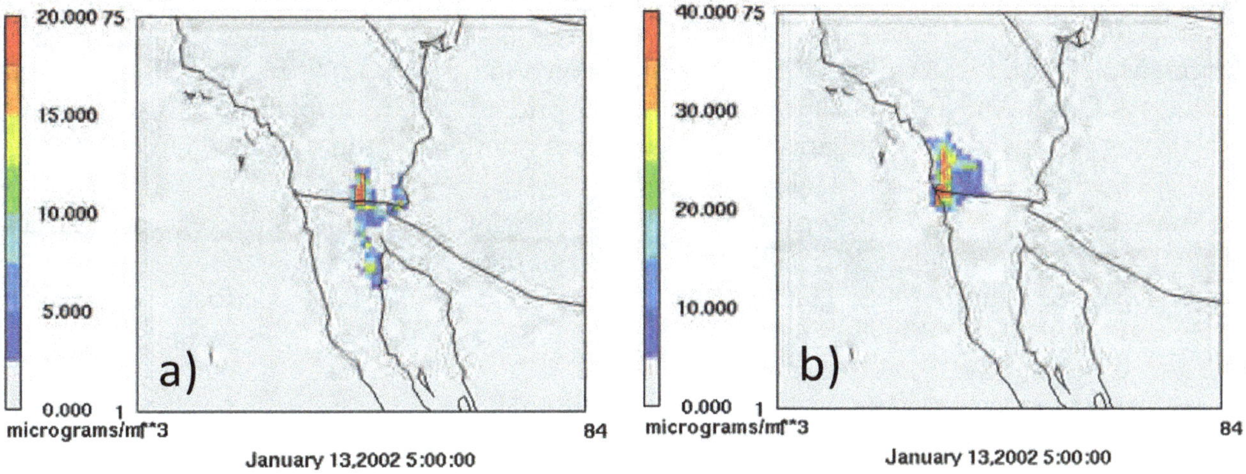

Fig. 22. Peak contribution to $PM_{2.5}$ from (a) MXC primary $PM_{2.5}$ area sources and (b) TSD primary $PM_{2.5}$ area sources, during the January 2002 episode.

5. Conclusion

Results suggest relevant information on trans-boundary impacts of air pollutants in the Mexicali-Imperial Valley border area. Simulated O_3 and $PM_{2.5}$ concentrations in the domain were the highest in the LA area, as expected. However, limited contribution of sources in the LAR area to O_3 and $PM_{2.5}$ levels in the border region was observed. Mobile sources, the most abundant sources in the LAR area contributed up to 10 ppbv of O_3 in MXC, but meteorological events that favored the transport of pollutants from LAR to MXC were few compared to prevailing conditions that favored transport to the east and northeast of LAR during the summer episodes or the southwest during the winter episode. Emissions from the TSD region play a much more important role in the air quality of the MXC area, particularly on the levels of O_3 during the summer episodes. Again, mobile sources contributed the most to the observed impacts from TSD to MXC. Even more, MXC O_3 levels were more sensitive to NOx changes in TSD mobile emissions than VOC changes in that same source. Even though, mobile sources are of concern in the MXC area, O_3 impacts from precursors emitted within the region were small. Area sources in MXC contributed the most: up to a maximum of 8 ppbv of O_3 during the summer episodes. O_3 plumes reached the border regions of California-Arizona and O_3 concentrations up to 4 ppbv in the Grand Canyon area can be attributed to area sources in the MXC region. The MXC region is more sensitive to NOx controls than to VOCs controls. In regards to $PM_{2.5}$, about 50% of the $PM_{2.5}$ in MXC during the summer episode can be attributed directly to area sources. During the winter episode, plumes from TSD, LAR and Las Vegas unite and move towards the MXC region with impacts of 10-35 $\mu g/m^3$. Soil dust contribution from LAR, TSD and MXC ranges between 5-25 $\mu g/m^3$. MXC area sources contribute a maximum of 34 $\mu g/m^3$ $PM_{2.5}$.

6. Acknowledgments

This study was supported by LASPAU: Academic and Professional Programs for the Americas, under its Border Ozone Reduction and Air Quality Improvement Program. Additional support was obtained from Tecnológico de Monterrey through grant CAT-186.

7. References

Bergin, M.S.; Shih, J.S.; Krupnick, A.J.; Boylan, J.W.; Wilkinson, J.G.; Odman, M.T. & Russell, A.G. (2007). Regional air quality: Local and interstate impacts of NOx and SO_2 emissions on ozone and fine particulate matter in the Eastern United States. *Environ. Sci. Technol.*, 41(13), 4677-4689.

Breiman, L., Friedman, J.; Olshen, R. & Stone, C. (1998) Classification and regression trees. Chapman & Hall/CRC, ISBN 0-412-04841-8, Boca Raton, FL.

Byun, D.W. & Ching, J.K.S. (1999). *Science Algorithms of the EPA Models 3 Community Multiscale Air Quality (CMAQ) Modeling System*. EPA/600/R-99/030, US EPA, Washington DC.

Carter, W.P.L. (2000). *Documentation of the SAPRC99 chemical mechanism for VOC reactivity assessment*. Center for Environmental Research and Technology, University of California, Riverside, CA.

Che, W.; Zheng, J.; Wang, S.; Zhong, L. & Lau, A. (2011). Assessment of motor vehicle emission control policies using Model-3/CMAQ model for the Pearl River Delta region, China. *Atmos. Environ.*, 45, 1740-1751.

Choi, Y.-J.; Hyde, P. & Fernando, H.S.J. (2006). Modeling of episodic particulate matter events using a 3-D air quality model with fine grid: Applications to a pair of cities in the US/Mexico border. *Atmos. Environ.*, 40, 5181-5201.

Chow, J. C.; Watson, J. G.; Green, M. C.; Lowenthal, D. H.; Bates, B.; Oslund, W. & Torres, G. (2000). Cross-border transport and spatial variability of suspended particles in Mexicali and California's Imperial Valley. *Atmos. Environ.*, 34, 1833-1843.

Chow, J.C. & Watson, J.G. (2001). Zones of representation for PM_{10} measurements along the US-Mexico border. *The Sci. of the Total Environ.*, 276, 49-68.

Cohan, D. S.; Hakami, A.; Hu, Y. T. & Russell, A. G. (2005). Nonlinear response of ozone to emissions: Source apportionment and sensitivity analysis. *Environ. Sci. Technol.*, 39, 6739-6748.

Currey, R.C.; Kelly, K.E.; Meuzelaar, H.L.C. & Sarofim, A.F. Eds. (2005). *The U.S.-Mexican Border Environment: Integrated Approach to Defining Particulate Matter in the Paso del Norte Region*. SCERP Monograph Series, No. 12, San Diego State University Press, ISBN 0-925613-47-9, San Diego, CA.

Deane, M. (1984). Epidemiological monitoring plan for geothermal developments. *Sci. Total Environ.*, 32(3), 303-320.

Dennis, R.L.; Mathur, R.; Pleim, J.E. & Walker, J.T. (2010). Fate of ammonia emissions at the local to regional scale as simulated by the Community Multiscale Air Quality model. *Atmos. Pollution Res.*, 1, 207-214.

Dunker, A.M. (1984). The Decoupled Direct Method for Calculating Sensitivity Coefficients in Chemical-Kinetics. *Journal of Chemical Physics*, 81, 2385-2393.

Dunker, A.M.; Yarwood, G.; Ortmann, J.P. & Wilson, G.M. (2002). Comparison of source apportionment and source sensitivity of ozone in a three-dimensional air quality model. *Environ. Sci. Technol.*, 36, 2953-2964.

Eatough, D. J.; Green, M.; Moran, W. & Farber, R. (2001). Potential particulate impacts at the Grand Canyon from northwestern Mexico. *Sci. Total Environ.*, 276, 69-82.

ERG (Eastern Research Group), Acosta y Asociados & TransEngineering. (2004). Mexico National Emissions Inventory, 1999: Six Northern States, Final Report prepared for SEMARNAT, INE, US EPA, WRAP and CEC; Report #3393-00-011-002.

Grell, G.; Dudhia, J. & Stauffer, D.R. (1994). *A description of the Fifth-Generation Penn State/NCAR Mesoscale Model (MM5)*. NCAR Technical Note, NCAR/TN-398+STR.

Gudiksen, P.H.; Ermak, D.L.; Lamson, K.C.; Axelrod,M. C. & Nyholm, R.A. (1980). Potential air quality impact of geothermal power production in the Imperial Valley. *Atmos. Environ.*, 14(11), 1321-1330.

Hakami, A.; Odman, M. T. & Russell, A. G. (2003). High-order, direct sensitivity analysis of multidimensional air quality models. *Environ. Sci. Technol.*, 37, 2442-2452.

Hakami, A.; Odman, M.T. & Russell A.G. (2004). Nonlinearity in atmospheric response: A direct sensitivity analysis approach. *J. Geophys. Res.*, 109 (D15), D15303.

Houyoux, M.R. & Vukovich, J.M. (1999) Updates to the Sparse Matrix Operator Kernel Emissions (SMOKE) Modeling System and Integration with Models-3. *Conference The Emission Inventory: Regional Strategies for the Future*, Air & Waste Manage. Assoc., Raleigh, NC.

Im, U.; Poupkou, A.; Incecik, S.; Markakis, K.; Kindap, T.; Unal, A.; Melas, D.; Yenigun, O.; Topcu, S.; Odman, M.T.; Tayanc, M. & Guler, M. (2011). The impact of anthropogenic and biogenic emissions on surface ozone concentrations in Istanbul. *Sci. Total Environ.*, 409, 1255-1265.

Kelly, K.E.; Jaramillo, I.C.; Quintero-Núñez, M.; Wagner, D.A.; Collins, K.; Meuzelaar, H.L.C. & Lighty, J.S. (2010). Low-Wind/High Particulate Matter Episodes in the Calexico/Mexicali Region. *J. Air & Waste Manage. Assoc.*, 60, 1476-1486.

Kenski, D.M. (2004). CART Analysis of Historic Ozone Episodes. In *Proceedings of the AWMA 97th Annual Conference and Exhibition*, Indianapolis, USA.

Liao, K.-J.; Tagaris, E.; Manomaiphiboon, K.; Napelenok, S.L.; Woo, J.-H.; He, S.; Amar, P. & Russell, A.G. (2007). Sensitivities of Ozone and Fine Particulate Matter Formation to Emissions under the Impact of Potential Future Climate Change. *Environ. Sci. Technol.*, 41, 8355-8361.

McMurry, P.; Sheperd, M. & Vickery, J. (Eds.) (2004) *Particulate Matter Science for Policy Makers: A NARSTO Assessment*, Cambridge University Press, ISBN 0-521-84287-5, New York, USA.

Mendoza-Dominguez, A.; Wilkinson, J. G.; Yang, Y.-J. & Russell, A. G. (2000). Modeling and Direct Sensitivity Analysis of Biogenic Emissions Impacts on Regional Ozone Formation in the Mexico-U.S. Border Area. *J. Air & Waste Manage. Assoc.*, 50, 21-31.

Mendoza, A.; Gutiérrez, A.A. & Pardo, E.I. (2009). Volatile organic compounds in the downtown area of Mexicali, Mexico during the spring of 2005: analysis of ambient data and source-receptor modeling. *Atmósfera*, 22, 195-217.

Mendoza, A.; Pardo, E.I. & Gutierrez, A.A. (2010). Chemical Characterization and Preliminary Source Contribution of Fine Particulate Matter in the Mexicali/Imperial Valley Border Area. *J. Air & Waste Manage. Assoc.*, 60, 258-270.

Morris, R.E.; Koo, B.; Lau, S.; Tesche, T.W.; McNally, D.; Loomis, C.; Stella, G.; Tonnesen, G. & Wang, Z. (2004). *VISTAS Emissions and Air Quality Modeling – Task 4: Model Performance Evaluation and Model Sensitivity Tests for Three Phase I Episodes*. Report prepared by ENVIRON International Corporation, Novato, California.

Mukerjee, S. (2001). Selected air quality trends and recent air pollution investigations in the US-Mexico border region. *Sci. Total Environ.*, 276, 1-18.

Olerud, D. & Sims, A. (2003). *MM5 sensitivity modeling in support of VISTAS (Visibility Improvement – State and Tribal Association*. Report prepared for the VISTAS Technical Analysis Workgroup by Baron Advanced Meteorological Systems, LLC.

Osornio-Vargas, A.R.; Hernandez-Rodriguez, N.A.; Yanez-Buruel, A.G.; Ussler, W.; Overby, L.H.; Brody, A.R. (1991) Lung cell toxicity experimentally induced by a mixed dust from Mexicali, Baja California, Mexico. *Environ. Res.*, 56(1), 31-47.

Pitchford, M.L.; Tombach, I., Barna, M.; Gebhart, K.A.; Green, M.C.; Knipping, E.; Kumar, N.; Malm, W.C.; Pun, B.; Schichtel, B.A.; Seigneur, C. (2004). *Big Bend Regional Aesorol and Visibility Observational Study (BRAVO) Final Report*, September 2004.

Quintero-Núñez, M.; Reyna, M.A.; Collins, K.; Guzmán, S.; Powers, B. & Mendoza, A. (2006). Issues Related to Air Quality and Health in the California-Baja California Border Region, In *The U.S. Mexican border environment: Binational Air Quality Management*, SCERP Monograph Series No. 14, R. Pumfrey (Ed.), 1-46, San Diego State University Press, ISBN: 0-925613-50-9, San Diego, California, USA.

Russell, A. & Dennis, R. (2000). NARSTO critical review of photochemical models and modeling. *Atmos. Environ.*, 34, 2283-2324.

Russell, A.G. (2008). EPA Supersites Program-Related Emissions-Based Particulate Matter Modeling: Initial Applications and Advances. *J. Air & Waste Manage. Assoc.*, 58, 289-302.

Sheya, S.A.N.; Meuzelaar, H.L.C.; Jeon, S.J.; Dworzanski, J.P.; Jarman, W.; Kasteler, C.; Lighty J.; Sarofim, A.; Li, W.W.; Valenzuela, V.; Anderson, J.; Banerji, S.; Perry, D.; Mejia, G.; Zavala, M. & Simoneit, B. (2000) Novel Analytical Dimensions in Exploratory Field Studies of Air Particulate Matter, *Proceedings of the 93rd Air & Waste Management Association's Annual Conference & Exhibition*, pp. 1104-1130, Salt Lake City, UT, USA, June 18-22, 2000.

Sweedler, A.; Fertig, M.; Collins, K. & Quintero-Núñez, M. (2003). Air Quality in the California-Baja California Border Region, In *The U.S. Mexican border environment: Air Quality Issues along the US-Mexico Border*, SCERP Monograph Series No. 6, A. Sweedler (Ed.), 15-58, San Diego State University Press, ISBN: 0-925613-38-X, San Diego, California, USA.

Tagaris, E.; Manomaiphiboon, K.; Liao, K.-J.; Leung, L.R.; Woo, J.-H.; He, S.; Amar, P. & Russell, A.G. (2007). Impacts of global climate change and emissions on regional ozone and fine particulate matter concentrations over the United States. *J. Geophys. Res.*, 112(D14), D14312/1-D14312/11.

Tesche, T.W.; Georgopoulos, P.; Seinfeld, J.H.; Cass, G.; Lurmann, F.L. & Roth, P.M. (1990). *Improvement of procedures for evaluating photochemical models*, Report prepared by Radian Corporation for the State of California Air Resources Board, Sacramento, CA.

US EPA. (2003) *User's Guide to Mobile6.1 and Mobile6.2 Mobile Source Emission Factor Model.* Office of Transportation and Air Quality, EPA420-R-03-010, Ann Arbor, MI.

Vanoye, A.Y. & Mendoza, A. (2009). Mesoscale Meteorological Simulations of Summer Ozone Episodes in Mexicali and Monterrey, Mexico: Analysis of Model Sensitivity to Grid Resolution and Parameterization Schemes. *Water, Air, & Soil Pollution: Focus*, 9, 185-202.

Vukovich, J. & Pierce, T. (2002). The Implementation of BEIS3 within the SMOKE Modeling Framework. *United States Environmental Protection Agency Emissions Inventory Conference*, Atlanta, GA.

Yang, Y. J.; Wilkinson, J.G. & Russell, A.G. (1997). Fast, direct sensitivity analysis of multidimensional photochemical models. *Environ. Sci. Technol.*, 31, 2859-2868.

Ying, Q. & Krishnan, A. (2010). Source contributions of volatile organic compounds to ozone formation in southeast Texas. *J. Geophys. Res.*, 115(D17), D17306/1-D17306/14.

Watson, J.G. & Chow, J.C. (2001). Source characterization of major emission sources in the Imperial and Mexicali Valleys along the US/Mexico border. *Sci. Total Environ.*, 276, 33-47.

Zavala, M.; Herndon, S.C.; Wood, E.C.; Jayne, J.T.; Nelson, D.D.; Trimborn, A.M.; Dunlea, E.; Knighton, W.B.; Mendoza, A.; Allen, D.T.; Kolb, C.E.; Molina, M.J. & Molina, L.T. (2009). Comparison of emissions from on-road sources using a mobile laboratory under various driving and operational sampling modes. *Atmos. Chem. Phys.*, 9, 1-14.

Zielinska, B.; Sagebiel, J.; Harshfield, G. & Pasek, R. (2001). Volatile organic compound measurements in the California-Mexico border region during SCOS97. *The Sci. of the Total Environ.*, 276, 19-31.

Zuk M.; Rojas Bracho, L. & Tzintzun Cervantes, M.G. (2007). *Tercer almanaque de datos y tendencias de la calidad del aire en nueve ciudades mexicanas*. Instituto Nacional de Ecología, ISBN 978-968-817-840-9, México, D.F.

10

Air Pollution, Modeling and GIS based Decision Support Systems for Air Quality Risk Assessment

Anjaneyulu Yerramilli[1], Venkata Bhaskar Rao Dodla[1] and Sudha Yerramilli[2]

[1]*Trent Lott Geospatial & Visualization Research Center @ e-Center, College of Science Engineering & Technology, Jackson State University, Jackson, MS,*

[2]*National Center For Bio-Defense @e-Center, College of Science Engineering & Technology, Jackson State University, Jackson MS, USA*

1. Introduction

Air Pollution is a state of the atmosphere with predominant presence of hazardous substances that are harmful to humans and animals. The air-borne pollutants degrade the air quality and constant exposure to polluted air may lead to several health problems such as cardiopulmonary disease, bronchitis, asthma, wheezing and coughing etc. Average composition of the atmosphere below 25 km indicates that nitrogen, oxygen, water vapor, carbon dioxide, methane, nitrous oxide, and ozone are the major constituents and their balance is important to the maintenance of the Earth's biosphere. An imbalance of these constituents for a considerable long time may lead to serious implications of air quality and weather and climate. The pollutants are categorized as primary and secondary, primary pollutants are directly emitted from a source and secondary pollutants result from reaction of primary pollutants in the atmosphere. The primary pollutants are the ash from volcanoes, carbon monoxide and sulfur dioxide emissions from vehicles and factories and combustion of fossil fuels etc whereas secondary pollutants are tropospheric ozone resulting from photolysis of nitrogen oxides and hydrocarbons in the presence of sunlight and smog from a mixture of smoke and sulfur dioxide. The sources of air pollution are classified as natural and anthropogenic. Volcanoes, forest fires and biological decay providing sulfur dioxide and nitrogen oxides, large barren lands providing dust, vegetation producing volatile organic compounds come under natural sources whereas anthropogenic sources are categorized as mobile and stationary sources. Different forms of transportation such as automobiles, trucks, and airplanes come under mobile sources whereas power plants and industrial facilities are the stationary sources. Stationary sources are further classified as point and area sources, wherein a point source refers to a fixed source such as a smokestack or storage tank that emits the pollutant and an area source refers to several small sources affecting the air quality in a region such as dry cleaners, gas stations, auto body paint shops and a community of homes using woodstoves for heating. Air pollutants are also categorized as 'criteria' and 'hazardous', where the criteria pollutants refer to the commonly and frequently observed six chemicals which are carbon monoxide, lead, nitrogen dioxide, ozone, particulate matter, and sulfur dioxide and hazardous pollutants are toxic pollutants which cause cancer and other serious health problems or lead to adverse environmental effects.

Anthropogenic primary pollutants such as carbon monoxide, particulate matter, nitrogen oxides and lead are detrimental to health as well as environment. Sulfur dioxide and nitrogen oxides get transformed as sulfuric acid and nitric acid in the atmosphere due to chemical reactions and may fall as acid rain. Some details of these pollutants are briefly described as follows:

Carbon monoxide is a colorless, odorless, poisonous gas produced from burning of fuels with carbon and so the major source is road transport vehicles. Due to oxidation process, CO will be transformed as carbon dioxide. The background levels of carbon monoxide are in the range of 10-200 parts per billion (ppb) and urban concentrations generally vary between 10 to 500 parts per million (ppm). Continuous exposure to higher levels (>500 ppm) for longer time periods (> 30 minutes) may lead to headache, dizziness and nausea and also death.

Nitric oxide (NO) is a colorless, odorless gas produced during burning of fuel at high temperatures in cars and other road vehicles, heaters and cookers. Mostly, nitrogen dioxide in the atmosphere is formed from the oxidation of nitric oxide (NO). Nitrogen dioxide reacts to form nitric acid and organic nitrates and plays an important role in the production of surface ozone. Mean concentrations in urban areas are in the range of 10-45 ppb reaching as high as 200 ppb. Continuous exposure to NO_2 leads to respiratory problems and lung damage.

Particulate matter comprises of both organic and inorganic substances, mainly from dust, fly ash, soot, smoke, aerosols, fumes, mists and condensing vapors and is regarded as coarse particulates with a diameter greater than 2.5 micrometers (μm) and fine particles less than 2.5 micrometers. The acid component of particulate matter (PM) generally occurs as fine particles. Primary sources of the particulate matter are from road transport (25%), non-combustion processes (24%), industrial combustion plants and processes (17%), commercial and residential combustion (16%) and public power generation (15%). In urban areas, secondary particulate matter occurs as sulfates and nitrates with mean values in the range 10-40 μg/m3 and may rise up to higher than 100 $\mu g/m^3$. Primary PM sources are derived from both human and natural activities which include agricultural operations, industrial processes, fossil fuel burning etc and secondary pollutants such as SO_2, NOx, and VOCs are considered as precursors as they help form PM. Measures to reduce these precursor emissions will have a controlling impact on PM concentrations. Fine PM will cause asthma, lung cancer, cardiovascular issues, and premature death and estimated to cause 20,000 - 50,000 deaths per year in US.

Sulfur dioxide (SO_2) is a colorless, nonflammable gas with an odor that irritates the eyes and air passages. The most common sources of sulfur dioxide are fossil fuel combustion, smelting, manufacture of sulfuric acid, conversion of wood pulp to paper, incineration of refuse and production of elemental sulfur. Coal burning is the single largest man-made source of sulfur dioxide accounting for about 50% of annual global emissions, with oil burning accounting for a further 25-30%. The most common natural source of sulfur dioxide is volcanoes. The mean concentrations are in the range of 5-15 ppb but hourly peak values may reach to 750 ppb. The health effects include asthma, respiratory illness and cardiovascular disease. Sulfur dioxide pollution can be more harmful when particulate and other pollution concentrations are high which is known as the "cocktail effect".

Secondary pollutants, resulting from the conversion of primary pollutants through complex chemical reactions, are potentially more harmful than their precursors. Since much of the pollutant chemistry is driven by the presence of sunlight, these are commonly referred to as photochemical pollutants. A well-known secondary photochemical pollutant is ozone (O_3) formed due to oxidation of benzene and other volatile organic compounds in the presence of nitrogen oxides. In urban areas with high NOx concentrations, ozone concentrations tend to be lower due to scavenging effect (NOx react with ozone to form NO_2 and O_2), but tend to be higher due to downwind transport of NOx and its reaction with VOCs under sunlight. Ozone is a colorless, pungent, highly reactive gas and is the principal component of smog, which is caused primarily by automobile emissions, predominantly in urban areas. Ozone concentrations in urban areas rise in the morning, peak in the afternoon, and decrease at night. Higher frequencies of higher concentrations are dependent on stable atmospheric conditions such as low level inversions which restrict the pollutants to the boundary layer. Two of the most important factors that affect human health are the concentration of ozone and duration of exposure.

1.1 Air Quality and Air Quality Index

Air Quality depends on the trace gas emissions from the biosphere, from human activities and the chemical reactions which govern the concentrations of trace species in the atmosphere. Air Quality is an assessment of extent of pollutants present in air in a given locality relative to permissible levels while Air Quality Index (AQI) is an index for reporting daily air quality. EPA (Environmental Protection Agency) of USA calculates the AQI for five major air pollutants regulated by the Clean Air Act: ground-level ozone, particle pollution (also known as particulate matter), carbon monoxide, sulfur dioxide, and nitrogen dioxide. For each of these pollutants, EPA has established national air quality standards to protect public health. Ground-level ozone and airborne particles are the two pollutants that pose the greatest threat to human health in USA. AQI as a yardstick ranges from 0 to 500 and higher the AQI value, greater the level of air pollution and greater the health concern. For example, an AQI value of 50 represents good air quality with little potential to affect public health, while an AQI value over 300 represents hazardous air quality. An AQI value of 100 generally corresponds to the national air quality standard for the pollutant, which is the level EPA has set to protect public health. AQI values below 100 are generally thought of as satisfactory. When AQI values are above 100, air quality is considered to be unhealthy at first for certain sensitive groups of people, then for everyone as AQI values get higher.

1.2 Spatial Temporal variability

Air pollution became a widespread problem in the United States (US) and, according to the Environmental Protection Agency (EPA), it is estimated that over 100 million individuals are routinely exposed to levels of air pollution that exceed one or more of their health-based standards [1]. The major air pollutants include oxides of sulfur and nitrogen, suspended particulate matter, oxides of carbon, hydrocarbons, lead. These gases are released into the atmosphere from either stationary sources such as industrial point sources or mobile sources like vehicular pollution. These pollutant gases will also produce secondary pollutant like surface Ozone, which has great oxidative capacity and is the primary component of urban smog. Criteria Pollutants Ozone and particulate matter (PM_{10} and $PM_{2.5}$) are main focus now as they manly drive the AQI (which is based on five criteria pollutants CO, NO_2, SO_2, O_3, PM). These pollutants can affect our health in many ways, with irritation to the eyes, nose and throat and more serious problems such as chronic respiratory disease and cardiovascular diseases. Understanding the long-term (yearly, decadal) trends of these criteria air pollutants is important for assessing chronic exposure to population and the efficacy of control strategies, emissions changes, and the year to- year influence of meteorology.

The air pollutant levels at any location can be explained by three factors: (a) the rate of emissions or production from all sources; (b) the rate of chemical or physical removal via reaction or deposition; and (c) the dispersion and transport of a chemical within, to, or from an area. These three factors are likely to be influenced by meteorological and anthropogenic factors such as temperature, precipitation, proximity to sources, or weekday - weekend emission activity differences. Meteorological conditions strongly influence air quality. These include transport by winds, recirculation of air by local wind patterns, and horizontal dispersion of pollution by wind; variations in sunlight due to clouds and season; vertical mixing and dilution of pollution within the atmospheric boundary layer; temperature; and moisture. The variability of these processes, which affects the variability in pollution, is primarily governed by the movement of large-scale high- and low-pressure systems, the diurnal heating and cooling cycle, and local and regional topography. Changes in climate affect air quality by perturbing ventilation rates (wind speed, mixing depth, convection, and frontal passages), precipitation scavenging, dry deposition, chemical production and loss rates, natural emissions, and background concentrations.

Further researches have shown that $PM_{2.5}$ chemical composition and concentrations vary by location and source [2]. Particulates from different sources also have varying amounts of correlation with adverse health effects [3]. Most epidemiological studies which examine relationships between air pollution levels and human health needs spatial and temporal variation. Effects of changes under different possible emission scenarios are also important in determining pollutant distributions and future air quality risks.

1.3 Air Quality Management

The damages caused by air pollution in many countries are large and it is generally accepted that there is an urgent need for reducing the emissions to the atmosphere. The damages are caused by high ambient air concentrations and depositions of many chemical components. Among the most important components are acidifying constituents (sulfur and both oxidized and reduced nitrogen compounds), photochemical components (including ozone), particulate matter and toxic compounds, such as metals, organic compounds and others. The concentrations and depositions are dependent on (i) the total mass of pollutants emitted to the atmosphere and its spatial and temporal distribution (ii) transport and transformation processes in the atmosphere and (iii) deposition processes.

Assessments of emission reduction strategies must consider all the three factors and the complexity of these problems call for the use of atmospheric models. Average exposure of the ecosystem to concentrations and deposition, emission scenario studies and linkage to economical aspects and cost effectiveness are all examples of areas where the models are needed. Air Pollution is a good example of how models can play an important role in the decision making process toward protocols on emission reductions. Regional scale models quantifying the trans-boundary fluxes of air pollution between the European countries and deposition to ecosystems [4] have successfully been applied together with knowledge on ecosystem critical loads of acidity and the costs involved in emission reduction in order to find optimal solution for the reductions.

2. Air Quality assessment

Pollutants in the atmosphere undergo transportation and dispersion, whose characteristics are dependent on the prevailing atmospheric conditions. Pollutants can be transported across the continents due to globally circulating winds on time scale of months to years, e.g. intercontinental transport of dust from Sahara desert in Northwest Africa and the Gobi desert in East Asia and dispersed over a few hundreds of miles in a few days under the influence of local scale wind circulations. While wind is primarily responsible for transportation and dispersion, topography and atmospheric stability also play an important role. Horizontal and vertical motions arise due to atmospheric pressure gradients and stability conditions. Atmospheric pressure, defined as the weight of the air column above a point, varies due to differential surface heating. Heating and cooling of the earth surface is dependent on the type of surface, for e.g. water bodies have larger specific heat capacity and so takes longer time to get heated or cooled as compared to land surface. Different surfaces have different heating/cooling rates, e.g. asphalt surface gets heated/ cooled faster than vegetation surface; heating/ cooling at the surface causes lower/ higher density vertical columns leading to lower/ higher pressures. As air tends to move from denser to lighter regions, atmospheric wind gets established as movement from high pressure regions to low pressure regions. Similarly vertical motions are dependent on atmospheric stability. Heating at surface causes air to rise mixing with cooler and denser air at higher levels leading to instability and most of the mixing takes place in the lowest part of the atmosphere referred to as atmospheric boundary layer. This is characterized by turbulent fluctuations of wind velocity, temperature and moisture due to energy exchange with earth surface capped by an infinitesimal transition layer below the free atmosphere with lower/ higher stability conditions associated with stronger/ weaker vertical mixing. Since the pollutants originate near the earth surface, characteristics of the atmospheric boundary layer play an important role in the dispersion of pollutants. This is the reason why winter nights which are cooler with higher stability have stagnation of pollutants causing health hazards. Atmospheric dispersion on the scale of a few days (synoptic time scale) is influenced by the transient surface low and high pressure systems. Moderate to intense low pressure systems characterized by winds > 5m/s, mass convergence, higher instability and upward motions contribute to dispersion over wider horizontal extent ranging up to few thousand kilometers thus reducing the pollution effect through mixing with larger environment volume. Conversely, high pressure systems with calm winds, subsiding downward motion and higher atmospheric stability contain dispersion to smaller volume enhancing the pollution effects. These emphasize the role of atmospheric stability criteria and prevailing atmospheric circulation on the atmospheric dispersion of pollutants. In the absence of synoptic forcing, mesoscale and local circulations developed due to topography and land use variations and with synoptic forcing admixture of synoptic scale and mesoscale circulations control the pollutant dispersion.

Assessment of atmospheric dispersion requires precise information of meteorological parameters such as wind, temperature, humidity and stability at spatial and temporal resolutions as high as possible. Since meteorological observations are made at discrete spatial locations and at synoptic times (twice daily at 0000 and 1200 UTC), these data are to be generated at the required spatial and temporal resolutions. Weather prediction models provide the required quantitative information of the meteorological fields, which are used as input to a pollutant dispersion model to derive the spatial distribution. The integration of the weather prediction and dispersion models is the basis of the air quality models.

3. Meteorological Models - Numerical Weather Prediction

In view of the importance of meteorological variables in estimating atmospheric dispersion, a brief description of weather prediction using numerical models is provided here. Weather is defined as the state of the atmosphere at a given time and place, with respect to variables such as temperature, moisture, wind velocity, and barometric pressure where atmosphere is a gaseous envelop covering the Earth. Weather prediction is the application of science and technology to estimate the future state of the atmosphere (in terms of pressure, temperature, wind, rainfall etc) with reference to a specified region or location. Atmosphere is influenced by transient air movement and predicting that flow provides the status of weather conditions. Although different methods are available, numerical models alone provide quantitative weather forecasting. These models use a closed system of mathematical equations of dynamics developed from fundamental equations based on conservation of mass, momentum and water vapor along with equation of state and thermodynamic energy equation for temperature in differential form and use of numerical methods to solve them to produce future state of atmosphere starting from an observed initial state. To solve these equations numerically, the study region is to be formulated as of rectangular form with equally spaced horizontal intersecting grid in the horizontal direction and enough number of vertical levels as suitable for numerical solution. Due to constraints of mathematical formulation, numerical set up and the computational resources, the horizontal grid resolution is often restricted to a few kilometers. Since atmospheric modeling includes both the dynamical and physical parts, the physical processes of atmospheric radiation, planetary boundary layer, convection, cloud microphysics and surface physics are to be parameterized as these processes occur on scales smaller than the resolvable scales of domain resolution. For the system to predict the weather for a certain area it needs initial conditions and boundary conditions as adjacent areas affect the area of interest. It is known that observations are available at non-uniform locations and are sparser than the model resolution; the atmospheric variables are to be interpolated to the model grid using objective methods. Errors in the observations, interpolation to model grid, limitations in the representation of physical processes and errors due to numerical methods of solution all lead to uncertainties in weather forecasting and increase of errors with the progress of prediction. Rapid advances in computational resources have lead to the design and development of atmospheric models as applicable for real time weather forecasting of various weather phenomena with scales of few kilometers to thousands of kilometers.
There are many meteorological models available for weather prediction and the mainly used are the WRF, MM5, RAMS. Brief description of these models as follows.

3.1 WRF modeling system

Weather Research and Forecasting (WRF) modeling system was developed and sourced from National Center for Atmospheric Research (NCAR), as the next generation model after MM5, incorporating the advances in atmospheric simulation suitable for a broad range of applications. It has two cores, NMM and ARW which are developed independently and have differences in dynamics, methods of solution and physical schemes. Here a description of ARW model is presented as it is widely used for air quality modeling studies. ARW (Advanced Research WRF) model has versatility to choose the domain region of interest; horizontal resolution; interactive nested domains and with various options to choose parameterization schemes for convection, planetary boundary layer (PBL), explicit moisture;

radiation and soil processes. ARW is designed to be a flexible, state-of-the-art atmospheric simulation system that is portable and efficient on available parallel computing platforms and a detailed description was provided by Skamarock et al. (2008) [5]. The model consists of fully compressible non-hydrostatic equations and the prognostic variables include the three-dimensional wind, perturbation quantities of pressure, potential temperature, geo-potential, surface pressure, turbulent kinetic energy and scalars (water vapor mixing ratio, cloud water etc). The model equations are formulated using mass-based terrain following coordinate system, and solved in Arakawa-C grid using Runge–Kutta third order time integration techniques. The model has several options for spatial discretization, diffusion, nesting and lateral boundary conditions. The ARW Solver is the key component of the modeling system, which is composed of several initialization programs for idealized, and real-data simulations, and the numerical integration program. ARW supports horizontal nesting that allows resolution to be focused over a region of interest by introducing an additional grid (or grids) into the simulation with the choice of one-way and two-way nesting procedures. ARW model system was used in this study for its accurate numerics, higher order mass conservation characteristics and advanced physics.

3.2 MM5 modelling system

MM5 is a regional mesoscale model used for creating weather forecasts and climate projections. MM5 modeling system software is mostly written in Fortran and has been developed at Penn State and National Center for Atmospheric Research (NCAR) as a community mesoscale model with contributions from users worldwide. MM5 is a limited-area, non-hydrostatic, terrain-following sigma-coordinate model designed to simulate or predict mesoscale atmospheric circulation [6]. The model is supported by several pre and post-processing programs, which are referred to collectively as the MM5 modeling system [http://www.mmm.ucar.edu/mm5/].

3.3 RAMS modeling system

Regional Atmospheric Modeling System (RAMS) is a highly versatile numerical code developed by scientists at Colorado State University for simulating and forecasting meteorological phenomena, and for depicting the results.
[http://rams.atmos.colostate.edu/rams-description.html].

4. Air Pollution Dispersion modeling

Air pollution models are the only method that quantifies the deterministic relationship between emissions and concentrations/depositions, including the consequences of past and future scenarios and the determination of the effectiveness of abatement strategies. Air pollution measurements give information about ambient concentrations and deposition at specific locations and times, without giving clear guidance on the identification of the causes of the air quality problem. This makes air pollution models indispensable in regulatory, research, and forensic applications. The concentrations of substances in the atmosphere are determined by 1) transport, 2) diffusion, 3) chemical transformation, and 4) ground deposition. Transport phenomena, characterized by the mean velocity of the fluid, have been measured and studied for centuries. For example, the average wind has been studies by man for sailing purposes. The study of diffusion (turbulent motion) is more recent.

Among the first articles that mention turbulence in the atmosphere, are those by Taylor (1915, 1921) [7,8].

Pollutants in the atmosphere get transported longer distances by large scale atmospheric wind flows and dispersed in the atmosphere by small scale turbulent flows and mix with environment. Dispersion is difficult to understand and estimate due to presence of different scales of eddies and their complex interaction. Atmospheric dispersion models shall include the physical and chemical processes of transportation, transformation and dispersion of pollutants in the atmosphere to provide estimates of pollutant concentrations with the information of emission sources and concentrations. Current models are designed to compute the pollutant concentrations using information of characteristics of emission sources, emission rates, terrain variations, meteorological variations and background concentrations. These dispersion models use mathematical based concepts of dispersion and diffusion with proper treatment of complex terrain and land use. The pollutants concentrations in the atmosphere constantly change influenced by weather conditions. Meteorological parameters play an important role in the atmospheric dispersion as transport is controlled by large scale wind flows and dispersion by atmospheric stability, vertical mixing and local mesoscale winds. A proper design and application of the dispersion models can be used for identification of contributors through establishment of source- receptor relationships, assessment of air quality compliance with government norms, planning of new facilities, management of emissions at sources, prediction of high concentration episodes, providing information for risk assessment and management. Above all these, modeling information saves cost of continuous monitoring over large areas and for longer times. All the model estimations are to be evaluated to understand the uncertainties due to inaccuracies in emission strength, meteorological input data, physics of dispersion concentration estimations and analysis before assessment of health effects.

There are two modeling approaches, following Lagrangian and Eulerian mechanics. In the Lagrangian approach, the path of an air parcel (or puff) is followed and the changes tracked along the trajectory path. Lagrangian modeling was mostly used for transport of SO_2 over large distances and longer time-periods [9, 10, 11, 12]. In Eulerian method, the 3-dimensional atmosphere is divided into grid cells, both in horizontal and vertical directions, and the time changes of the properties are identified at each grid cell. Eulerian modeling was adopted for urban area studies on ozone [13]; for SO_2 [14] and for regional scale sulfur [15, 16]. As such Eulerian modeling was used only for specific episodes of a few days. Hybrid approaches, in which particle-in-cell methods were employed, were used by Friedlander and Seinfeld (1969)[17], Eschenroeder and Martinez (1970)[18] and Liu and Seinfeld (1974)[19]. It may be stated that prior to 1980, Lagrangian models were generally used for transport studies of sulfur and other particulate matter whereas Eulerian models were adopted for episodic events of secondary pollutants such as ozone. After 1980, the basic concepts were fine tuned with development and use of 2-D and 3-D global troposphere models. AERMOD and CALPUFF computer packages were developed for simulation of non-reactive chemicals (e.g., SO2). AERMOD is a steady-state Gaussian plume model which uses wind field derived from surface, upper-air, and onsite meteorological observations combined with terrain elevations and land use. CALPUFF is a non-steady state Lagrangian puff dispersion model with advantage of realistically simulating the transport in calm and stagnant conditions and over complex terrain, and coastal regions with sea/land breezes. AEROMOD is suitable for short-range simulations whereas CALPUFF is appropriate for both long-range and short-range simulations.

There are different types of modelling approaches are available and they are photochemical modelling, Plume Rise models, particle models, Deposition Modules, Odor models and Statistical models and a brief description as follows.

4.1 Photochemical modeling

These photochemical models are large-scale air quality models that simulate the changes of pollutant concentrations in the atmosphere using a set of mathematical equations characterizing the chemical and physical processes in the atmosphere. These models are applied at multiple spatial scales from local, regional, national, and global. Some examples of photochemical models are to be CMAQ, CAMX, UAM, WRF/Chem etc.

The Community Multi-scale Air Quality (CMAQ) modeling system has been designed for air quality as a whole by including state-of-the-science capabilities for modeling multiple air quality issues, including tropospheric ozone, fine particles, toxics, acid deposition, and visibility degradation [http://www.epa.gov/asmdnerl/CMAQ/index.html]. CMAQ was also designed to have multi-scale capabilities so that separate models were not needed for urban and regional scale air quality modeling. CMAQ modeling system simulates various chemical and physical processes that are thought to be important for understanding atmospheric trace gas transformations and distributions.

The Comprehensive Air quality Model with extensions (CAMx) is a computer modeling system for the integrated assessment of gaseous and particulate air pollution [http://www.camx.com/]. This model is designed to simulate air quality over many geographic scales; treat a wide variety of inert and chemically active pollutants like Ozone, inorganic and organic $PM_{2.5}/PM_{10}$, Mercury and toxics and to provide source-receptor, sensitivity, and process analyses.

The Urban Airshed Model (UAM) modeling system, developed and maintained by Systems Applications International (SAI), is the most widely used photochemical air quality model [http://uamv.saintl.com/].

4.2 Plume rise models

The simplest way of estimating smoke concentrations is to assume that plumes diffuse in Gaussian pattern along the centerline of a steady wind trajectory. Plume models usually assume steady state conditions during the life of the plume, which means relatively constant emission rates, wind speed, and wind direction. For this reason, they can be used only to estimate concentrations relatively near the source or for a short duration with restrictions to the influence of topography or land use. Most air pollution models include a computational module for computing plume rise i.e., the initial behavior of a hot plume injected vertically into a horizontal wind flow, e.g., PRIME in AERMOD and HYSPLIT.

4.3 Particle models

Particle model simulates the source by the release of many particles over the duration of the burn. The trajectory of each particle is determined as well as random component that mimics the effect of atmospheric turbulence. This allows a cluster of particles to expand in space according to the patterns of atmospheric turbulence rather than following a parameterized spatial distribution pattern, such as common Gaussian approximations. These models tend to be the most accurate way of simulating concentrations at any point in time but restricted to individual point sources with simple chemistry or sources that have critical components such as toxins that must be tracked precisely. These models use Lagrangian coordinates for accurate depiction of place of each time of particle movement e.g. HYSPLIT.

4.4 Deposition modules

Many air pollution models include the computational module for computing the fraction of the plume deposited at the ground as a consequence of dry and wet deposition phenomena.

4.5 Odor modeling

Odor models include algorithms to simulate instantaneous or semi-instantaneous concentrations, since odors are instantaneous to human sensations. The mechanisms of dispersion of odorous chemicals in the atmosphere are the same as the dispersion of other pollutants. In multiple pollutants emission, the relationship between concentrations of individual chemicals and odor is not well defined and odor must be characterized in terms of an odor detection threshold value for the entire mixture of odorous chemicals in the air.

4.6 Statistical models

Statistical models are techniques based on statistical data analysis of measured ambient concentrations. These are non-deterministic, as they do not establish nor simulate a cause-effect, physical relationship between emissions and ambient concentrations. These models are used in air quality forecast and alarm systems and receptor modeling. Statistical techniques have been used to forecast air pollution trends a few hours in advance for the purpose of alerting the population. Receptor models are mathematical or statistical procedures for identifying and quantifying the sources of air pollutants at a receptor location. Receptor models use the chemical and physical characteristics of gases and particles measured at source and receptor to both identify the presence of and to quantify source contributions to receptor concentrations e.g., Chemical Mass Balance (CMB) Model.

Basically there are two approaches for the computation of pollutants dispersion in the atmosphere. One is the offline approach, in which meteorological fields are first simulated and dispersion models use these meteorological data for dispersion and deposition computations with no feedback between pollutant concentrations and meteorological fields. In the online approach, the meteorology and dispersion models are run simultaneously, with exchange of information both ways at each time step. Both these methods have their own advantages, offline method more frequently used for primary pollutants such as mercury, PM2.5 etc and online approach for secondary pollutants such as ozone. We briefly describe here outlines of HYSPLIT offline model and WRF/Chem online model as results from dispersion studies over the Mississippi Gulf coast region with these two models carried out at TLGVRC, Jackson State University under a NOAA supported research program are presented in later sections.

4.7 HYSPLIT model

HYSPLIT (Hybrid Single-Particle Lagrangian Integrated Trajectory) model version 4, developed jointly by National Oceanic and Atmospheric Administration (NOAA) Air Resources Laboratory (ARL) and Australian Bureau of Meteorology, is a computational tool designed to produce air parcel trajectories, and to carry out simulations on different spatial and time scales from local, regional and long-range transport, dispersion, and deposition of air pollutants [20]. The model can take inputs of meteorological data from any of the Mercator, Lambert and Polar map projections but the data is required on grids and at regular time intervals. The model handles the input data from different resolutions, interpolates to an internal terrain following sub-grid and performs the calculations using data from fine to coarse resolutions and also from different map projections, e.g. starting from a Mercator or Lambert conformal regional grid and switching to Polar stereographic global grid. The trajectories and the pollutant dispersion and concentrations are calculated using Lagrangian mechanics in which advection and diffusion are calculated independently and has computational advantages with single point source emissions without constraints on resolution as compared to the application of Eulerian method for complex emission scenarios requiring solutions at all grid points of a region. Hysplit model computes the trajectories based on Lagrangian approach which is the advection of each particle in time following three dimensional velocity fields. In this model, dispersion and deposition can be computed using either puff or particle model or a hybrid in which both puff and particle approaches are adopted. In puff model, pollutant puffs are released from the source at regular intervals with each puff containing an appropriate fraction of the pollutant mass. The puff expands in time according to the atmospheric dispersion and advected following the trajectory. The puffs will split into new smaller puffs when their size exceeds the meteorological grid. The expansion within grid cell represents dispersion due to sub-grid scale turbulence and puff splitting represents grid scale process. Concentrations are calculated on the grid following a defined spatial distribution within the puff. In the particle model, cluster of particles are released from the source which will expand in space and time following the atmospheric dispersion. Advection of each particle will have a component related to in situ atmospheric turbulence. Concentrations are calculated as a sum of all the particles in a grid cell. By default, Hysplit model uses a hybrid approach in which puff and particle dispersions are incorporated in the horizontal and vertical representations respectively, which has the advantages of using puff model for better representation of horizontal distribution limiting the number of particles and better particle representation in the vertical where large discontinuities are possible. Both the horizontal puff and vertical particle dispersions use turbulence velocity components obtained from meteorological data. Regardless of the approach, atmospheric stability, atmospheric mixing and rate of pollutant dispersion are necessary for simulating pollutant dispersion. Atmospheric stability is estimated from wind and temperature parameters, horizontal and vertical mixing coefficients are estimated from velocity deformation and the coefficients of heat respectively, and the dispersion is computed as dependent on the mean and turbulent velocity components. Concentrations are summed up at each time step at all the grid points. This methodology clearly indicates the importance of the input meteorological data towards best representation of the vertical atmospheric structure. This model is simple to use with menu driven operations with the computations requiring only a few minutes operable on both windows and Linux based computer systems. The current version 4.9, suitable for Windows, was released on February 24, 2009 and Beta version suitable for IBM AIX and Linux servers was released on August 25, 2009.

4.8 WRF/Chem model

The Weather Research and Forecasting - Chemistry model (WRF/Chem) is a new generation regional air quality modeling system developed at NOAA (National Oceanic and Atmospheric Administration) [21]. This model uses the Eulerian mechanics, in which meteorology and chemistry variables are predicted at all the grid points at each time step. Although the computations are time consuming, it is most suited for prediction of air quality which requires full and continuous interaction of chemical species with meteorological parameters. This modeling system has two modules for meteorology and chemistry. For meteorology part, ARW mesoscale weather prediction model described in section 3 is used. The chemistry module treats the processes of dry deposition, coupled with the soil/vegetation scheme; aqueous phase chemistry coupled to some of the microphysics and aerosol schemes; biogenic emissions; anthropogenic emissions; gas-phase chemical reaction calculations; photolysis schemes; and aerosol schemes with different choices. The air quality component of the model is fully consistent with the meteorological component; both components use the same transport scheme, the same horizontal and vertical grids, the same physics schemes and the same time step for transport and vertical mixing. The model is consistent with all conservative transport done by the meteorology model. The resolution of the model is flexible, ranging from a few kilometers to hundred kilometers. This modeling system is suitable as coupled weather prediction/ dispersion/ air quality model to simulate release and transport of primary pollutants such as particulate matter and the prediction of secondary pollutants such as ozone. The model can be run on parallel processing computer platforms to save computational time. The current version is 3.2.1 released on August 18, 2010.

5. Air Quality modelling over Mississippi Gulf Coast

Pollutant contamination in the environment is a serious environmental concern for US Gulf coast region as of many other coastal habitats elsewhere. The Mississippi coastal region is environmentally sensitive due to multiple air pollution problems originating as a consequence of several developmental activities such as oil and gas refineries, operation of thermal power plants, rapidly increasing traffic pollution etc. The Mississippi Delta encompasses the largest area of coastal wetlands in the United States and supports one of the most extensive developments of petroleum extraction of any coastal area in the world. This area has been experiencing ecological impacts from energy development related human activities for more than hundred years. Coastal wetlands are vital for protecting developed areas from storm surges, providing wildlife and fish habitat, and improving water quality. Mississippi Gulf Coast area has sensitive ecosystems like National Forests, State parks, wildlife management and refuge areas, conservation areas or wildlife sanctuaries. The growth of industrial and commercial operations has created a need for air pollution dispersion models that can handle complex meteorological conditions of the coastal environment. Differential heating, strong thermal gradients along the land-sea interface and topographic friction cause localized mesoscale phenomena such as land-sea breeze circulations, sea breeze induced convection and formation of thermal internal boundary layer. The horizontal and vertical extents of the land-sea breeze, the internal boundary layer and their spatial heterogeneity under varying synoptic meteorological settings typify the complex dispersion patterns in the coastal region. The Thermal Internal Boundary Layer (TIBL) limits the region of vertical mixing, heating/convection and the low-level circulation characteristics which influence the coastal area dispersion. These spatio-temporal effects are to be included in the dispersion assessment for realistic air quality estimations using appropriate meteorological and dispersion models.

During the recent past years, air quality modeling studies have been carried out under varied meteorological conditions using sophisticated mesoscale models and observational assimilation techniques which have shown significant success [22-31]. At Jackson State University, we have conducted several studies on multi-scale simulation of meteorological fields for wind, temperature, humidity, and planetary boundary layer (PBL) turbulence utilizing ARW mesoscale weather prediction model (described in section 3.1). The simulated high resolution meteorological fields were integrated with HYSPLIT dispersion model (described in section 4.7) for studies on the source- receptor relationships of mercury and $PM_{2.5}$ pollutants. Similarly WRF/Chem model (described in section 4.8) was used to study the evolution of surface ozone over MS Gulf coast region and Jackson city urban metropolitan region. We present here the results from our studies which indicate the importance and usefulness of offline models for study of primary pollutants (Hg and $PM_{2.5}$) and online models for surface ozone.

5.1 Case study of $PM_{2.5}$

The Mississippi Gulf Coast is a typical coastal urban terrain featuring several industries that have been identified as emission sources of $PM_{2.5}$ and its precursor gases. A case study was undertaken to assess the WRF-HYSPLIT modeling approach towards assessment of source-receptor relationships of $PM_{2.5}$ over MS Gulf Coast [32]. For this study, measurements of $PM_{2.5}$ sulfate and nitric acid collected as 6-h samples during 17-20 June 2009 at the two selected locations, Harrison County School (30.5N, 89.1W) and Wiggins Airport (30.8N, 89.13W) were used. During this period, the concentrations of $PM_{2.5}$ sulfate and HNO_3 at Harrison and Wiggins were in the range of 5–8 μg m^{-3} for sulfate and 1.5–3 μg m^{-3} for HNO_3. The experimental study constituted two parts, simulation of atmospheric fields using WRF model and deriving back trajectories and forward deposition concentrations using HYSPLIT model. WRF model was used to produce the necessary atmospheric fields for 24-h periods with the outputs were stored at 1-h interval as needed for input to dispersion model. Model-simulated atmospheric flow fields were validated by comparison with the North American Mesoscale (NAM)-12 km regional analysis as they are critically important for computation of back trajectories and forward dispersion. The model simulated wind flow showed anti-clockwise circulation agreeing with NAM analyses. The onset and extent of sea breeze along the coastal regions was well simulated. HYSPLIT model, driven by meteorological inputs from ARW model, was used to produce 24 numbers of back trajectories at 1-h intervals to identify the sources along the paths of trajectories. The back trajectories, drawn from each observation site (Figure 1) shown that the air parcels were mostly confined to heights below 1.0 km within the planetary boundary layer and had paths in the quadrant between south and west, all pointing the origination of parcels from land on west-side and from sea on south-side (i.e.) Gulf of Mexico. Specifically, Watson, Cajun, and Morrow power plants were identified to be possible land sources as located within the path of back trajectories. HYSPLIT model was run in the forward mode, driven by the ARW model-generated atmospheric fields at 1-h interval and the hourly emission rates, starting from each of the three identified coal-fired power plants to produce the 24-h atmospheric dispersion separately for SO_2 and NOx with the hourly emission rates derived from EPA annual emissions data. The forward 24-h atmospheric dispersions from the three sources for SO_2 (Figure 2) showed the dispersion to be towards east/northeast under the prevailing wind flow of southwesterly/westerly relative to the observation sites. The pattern of the pollutant dispersion was a slowly expanding plume due to the low magnitude of the wind speed (2–5 m/s). The contributions from each of the three power plants to the two observation sites have shown that Big Cajun and Morrow plants contributed to a smaller of extent <0.235 μg m^{-3} for Wiggins location while none of three sources contributed for the observations at Harrison although these two sites are only 65 km apart. These results lead to the conclusion that increasing activity in diesel-powered heavy duty vehicles (ship) over the Gulf of Mexico (as 60% of US energy imports are from this area) and ports around, diesel oil emissions from this area may also contribute to precursors SO_2 and NOx of $PM_{2.5}$. The source attribution from the Gulf of Mexico could also be due to the diurnal Gulf breeze; with the night time land breeze carry the land-borne precursor pollutants on to the sea and daytime Gulf breeze would bring back the pollutants from the sea to the land region, thus representing the Gulf of Mexico as a virtual pollutant source. Although the case study does not concern with an episodic event, it brings out the result that the integrated modeling approach does not produce spurious results.

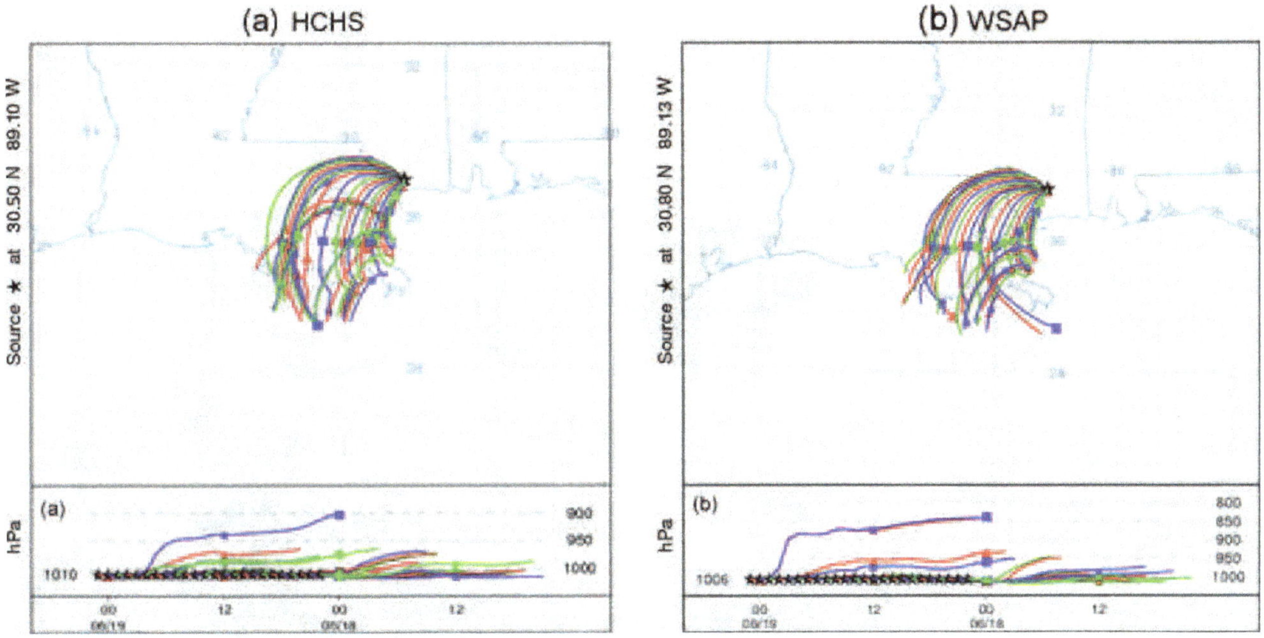

Fig. 1. Computed back trajectories at 1-h intervals for the 24-h period ending 0100 UTC on 19 June 2009 from the observation sites at (a) HCHS (left) and (b) WSAP (right). Top portion shows the horizontal path, and the bottom portion shows the vertical path of the trajectories.

5.2 Case study of Mercury

Mercury is known to be a potential air pollutant in the MS Gulf coast region in addition to SO_X, NO_X, CO and Ozone. An episode of high mercury concentration, with RGM value of 75 pg/m^3 at 1600 UTC of 5 May 2008 observed at the Grand Bay NERR location on Mississippi Gulf Coast (30.41N, 88.4W) was selected as a case study. WRF-ARW model was used to simulate the atmospheric fields at 4 km resolution and with the outputs stored at 1-h interval. The ARW model derived 10m wind flow over the study region was validated with EDAS- 40km analysis. WRF model produced sea breeze with upward vertical motions adjacent to the coast and downward vertical winds in the northeastern and central parts of the land region relatively stronger than EDAS fields. The HYSPLIT model was used to simulate Lagrangian back trajectories from NERR site for the 12-hour period ending at 1600 UTC 5 May 08 with each of inputs from EDAS 40 km, ARW 36 km and ARW 4 km data

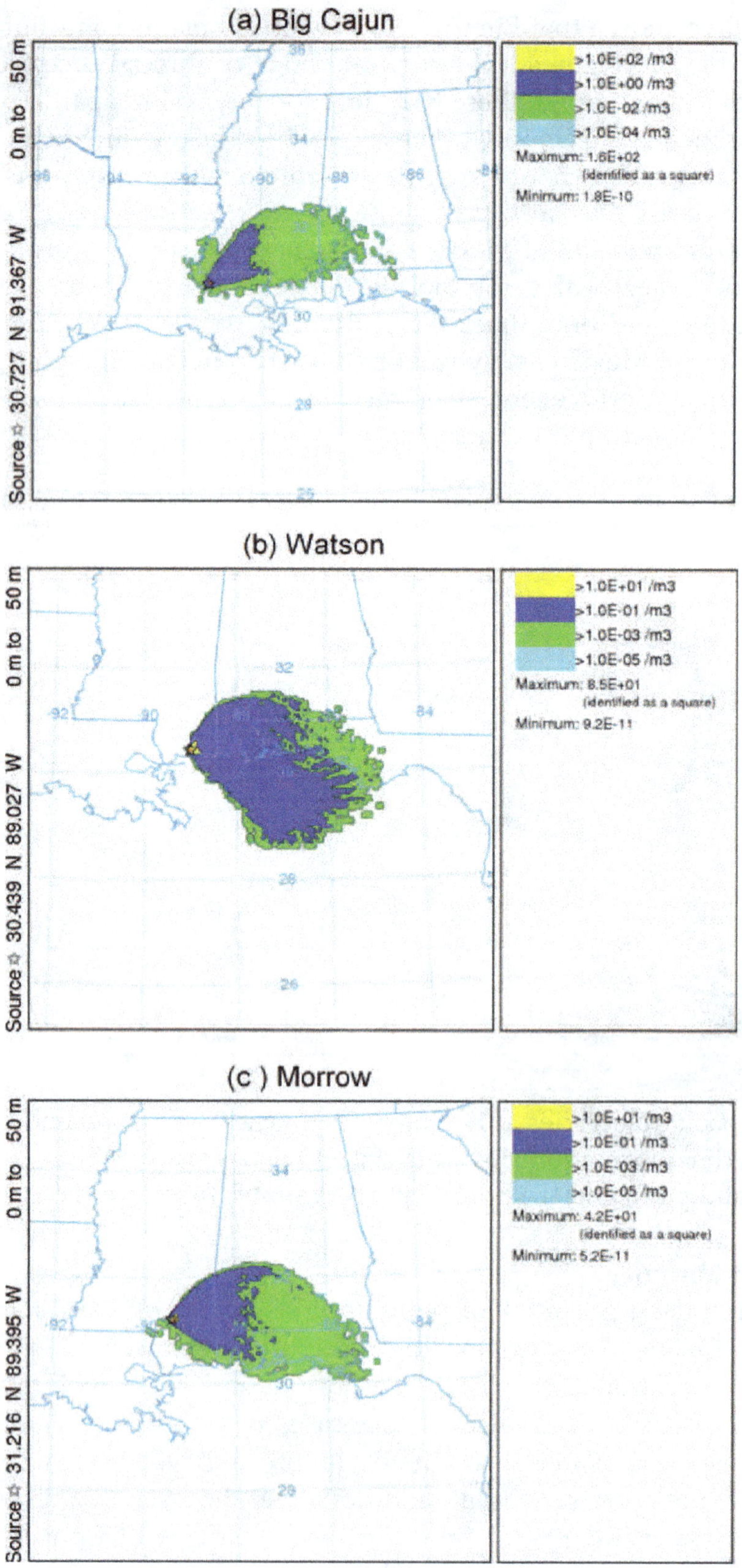

Fig. 2. HYSPLIT-generated SO2 concentration (μg m−3) averaged between 0 and 100 m levels and integrated for 24-h period between 0100 UTC of 18 June 2009 and 0100 UTC of 19 June 2009 sourced from the three identified coal-fired power plants.

(Figure 3). These back trajectories have shown that the air parcels originated towards north-northwest to north and the height of the parcels to be below 100 m level. The trajectories from ARW 4 km data were more towards east and realistic corresponding to mesoscale circulations and associated turbulence and vertical motion fields in the coastal region. The mean trajectory paths indicated four important sources i.e., Charles R Lowman (31.489N, 87.9W) and Barry (30.84N, 88.09W) power plants located in Alabama State and Jack Watson (30.44N, 89.026W) and Daniel (30.64N, 88.59W) power plants located in Mississippi State. The HYSPLIT model produced 24-hour ground level (0 -25 m) RGM concentration patterns with inputs from both the EDAS 40-km analysis (Figure 4) and the high resolution ARW 4km (Figure 5) data were compared. The deposition concentrations with ARW model have shown that Barry plant chiefly contributed with multiple peaks at different times during the day and with the highest concentration occurring 4 hours earlier than the actual measured peak and minute contributions from Daniel and Lowman plants. With EDAS data, Lowman and Daniel plants were identified to contribute chiefly. The concentration values with ARW fields were roughly one order higher than those obtained with EDAS fields. The representation of the wind field in the EDAS and the ARW data lead to major differences in the transport of the RGM from these two nearby sources (Barry and Daniel) in the two simulations. The multiple peaks in the simulated concentrations using ARW data was due to the diurnal circulation at the coast found in the ARW simulations indicating the impact of the sea- land breeze type mesoscale phenomena causing frequent plume transitions. This study helped to establish the sources and their relative contributions to the moderate mercury episode concentrations at NERR observation site using the integrated ARW-HYSPLIT modeling approach.

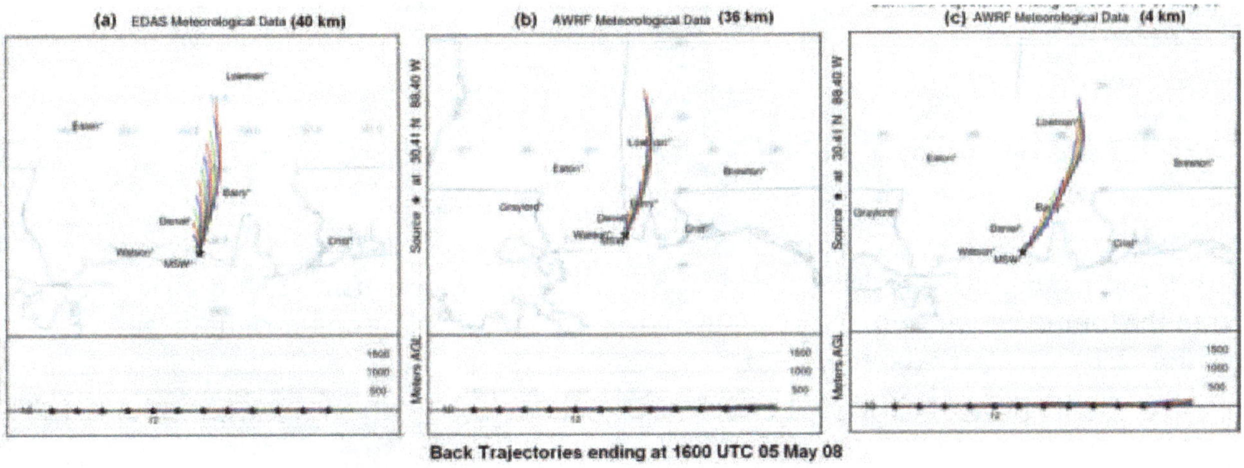

Fig. 3. Back trajectories produced by HYSPLIT model using EDAS-40 km (left); ARW-36km (middle) and ARW-4 km (right).

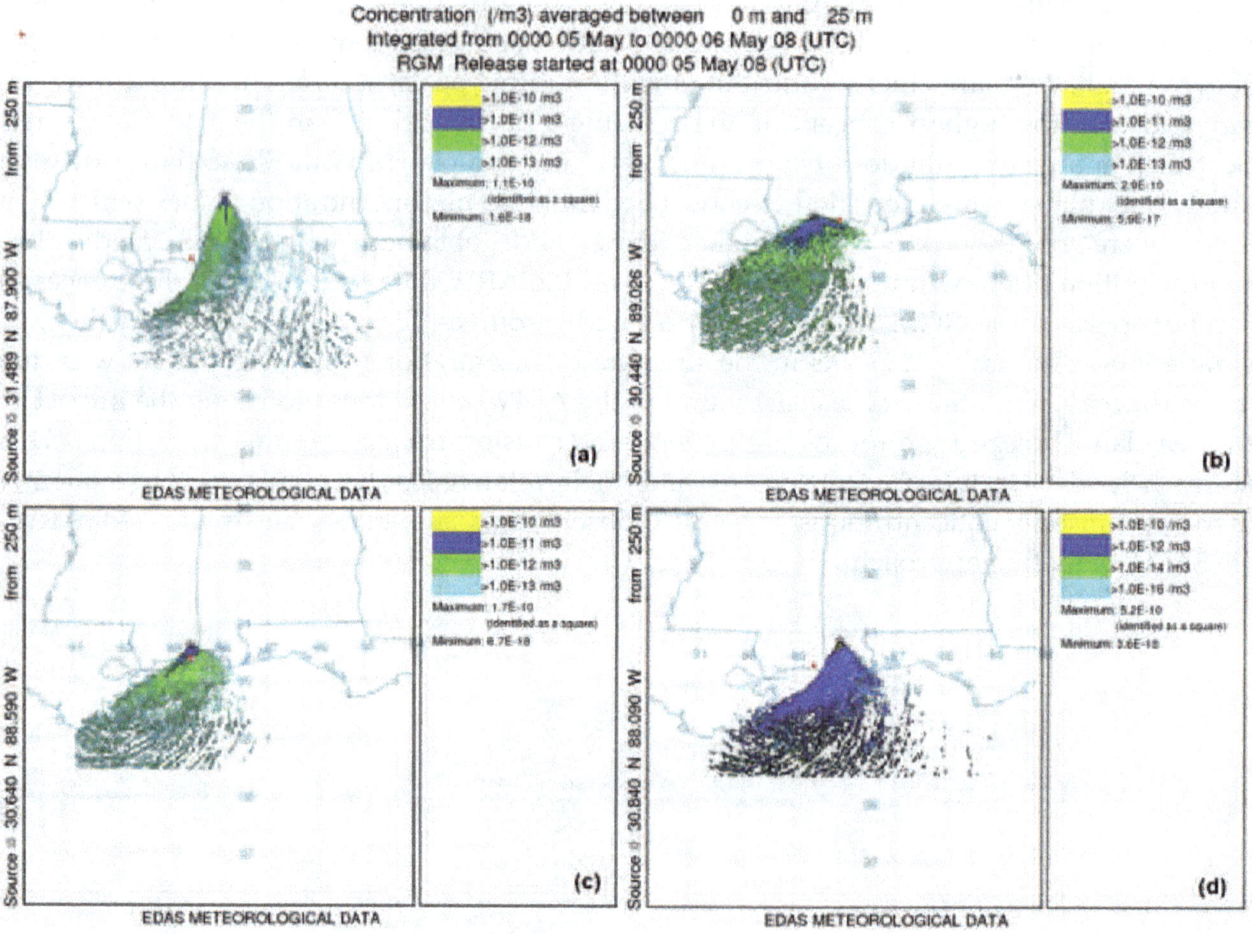

Fig. 4. Atmospheric dispersion as predicted by HYSPLIT model using EDAS-40 km data from the four different sources of Lowman (upper left), Watson (upper right), Daniel (bottom left) and Barry (bottom right).

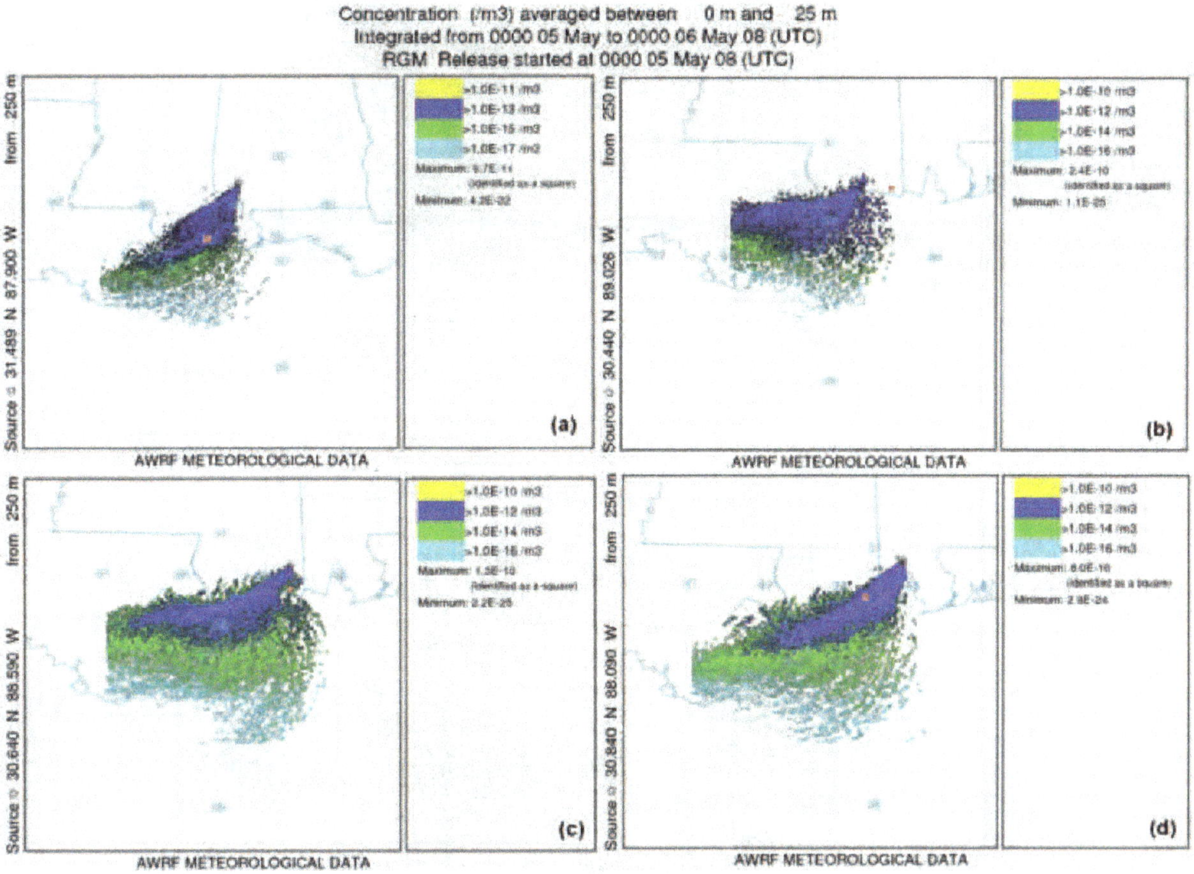

Fig. 5. Atmospheric dispersion as predicted by HYSPLIT model using ARW- 4 km data from the four different sources of Lowman (upper left), Watson (upper right), Daniel (bottom left) and Barry (bottom right)

5.3 Case study of surface ozone over Jackson metropolitan area

As per US EPA (United States Environmental Protection Agency), urban air quality in US with reference to ozone is a growing concern due to its oxidative capacity which will have great impact on the environment. Large number of people are likely to be exposed to the unhealthy ozone concentrations when ground-level ozone gets accumulated in urban metropolitan areas under certain weather conditions [33] and quantitative atmospheric dispersion models will be of great help to provide effective decision support systems for planners and administrators. We made an attempt to study the evolution of surface ozone and other precursor emissions like NOx and CO over southeast parts of US in general and Jackson, MS urban area in particular. We were motivated to take up this study to assess the performance of WRF/Chem in the simulation of a moderate ozone episode over an urban locality in which the atmospheric flow patterns are strongly influenced by local terrain and land cover patterns and the results from this study would provide useful information of urban air pollutants to air quality regulatory agencies and health administrators. WRF/Chem model was adapted to have nested four domains with the horizontal resolution of 36-12-4-1 km (Figure 6), with the innermost 1 km domain covering Hinds County. The chemistry was initialized with idealized profiles of anthropogenic emissions data which consists of area type emissions on a structured 4-km grid and point type emissions at latitude and longitude locations. WRF/Chem model was used to simulate the spatial and

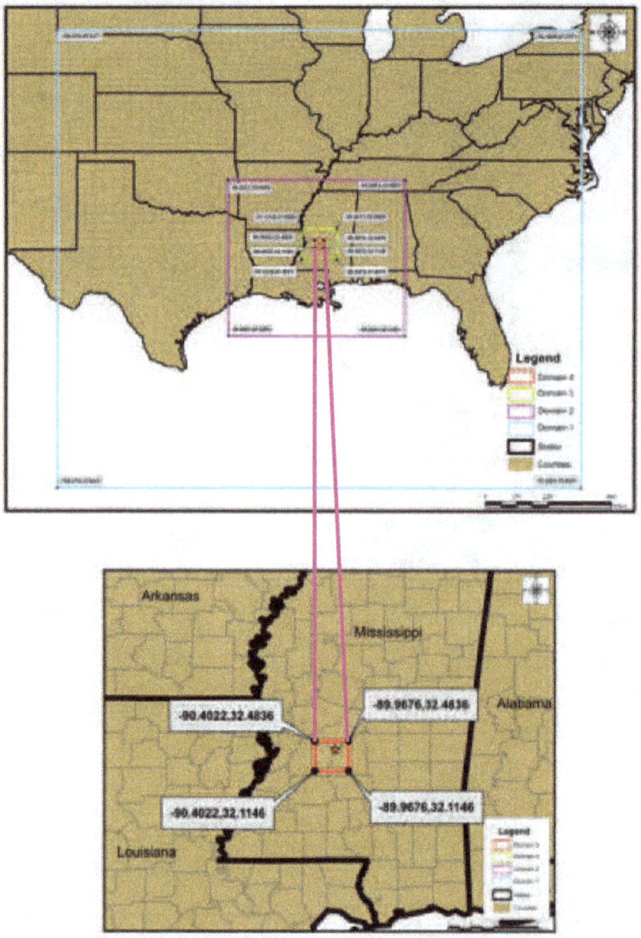

Fig. 6. Model domains. Innermost domain covering the study region of Jackson, MS is shown as red square. Observation site is shown as star.

temporal variations of surface ozone and other precursor pollutants CO (Carbon Monoxide), NO (Nitric oxide), NO_2 (Nitrogen dioxide) and HONO (Nitrous acid). The simulated time series at the specific location (32.38578 N, 90.141006 W) inside Jackson city during 7 PM 6 June to 7 PM 7 June 2006 showed that surface ozone gradually decreased from 7 PM onwards reaching minimum around midnight and stays nearly constant till dawn and gradually increased from dawn to noon reaching a maximum of 50 ppb around 1 PM. The model simulated the time variations and the time of attainment of maximum same as of observations with slight underestimation. The vertical variation of the model simulated ozone at peak time showed that the magnitude of ozone is nearly constant up to 900 hPa level, decreases rapidly up to 800 hPa level and then gradually increases upward. This means that there is good mixing of O_3 in the PBL region extended up to 900 hPa level as common for a summer afternoon. The time series of CO and NO_2 showed a maximum around midnight followed by gradually decrease and with increasing trends at certain times of daytime. The maximum during the first parts of the night could be attributed to these accumulations, decrease during later half to their dispersion and the increasing trends to the reaction between NO with VOC leading to NO_2 and other products. The occurrence of NO maximum at 11 AM, 2 hours prior to ozone maximum indicates the mechanism of NO_2 splitting as NO and O and the formation of O_3 through the amalgamated reaction of O and O_2. Similar explanation holds good for CO, as O_3 is produced in the troposphere by the

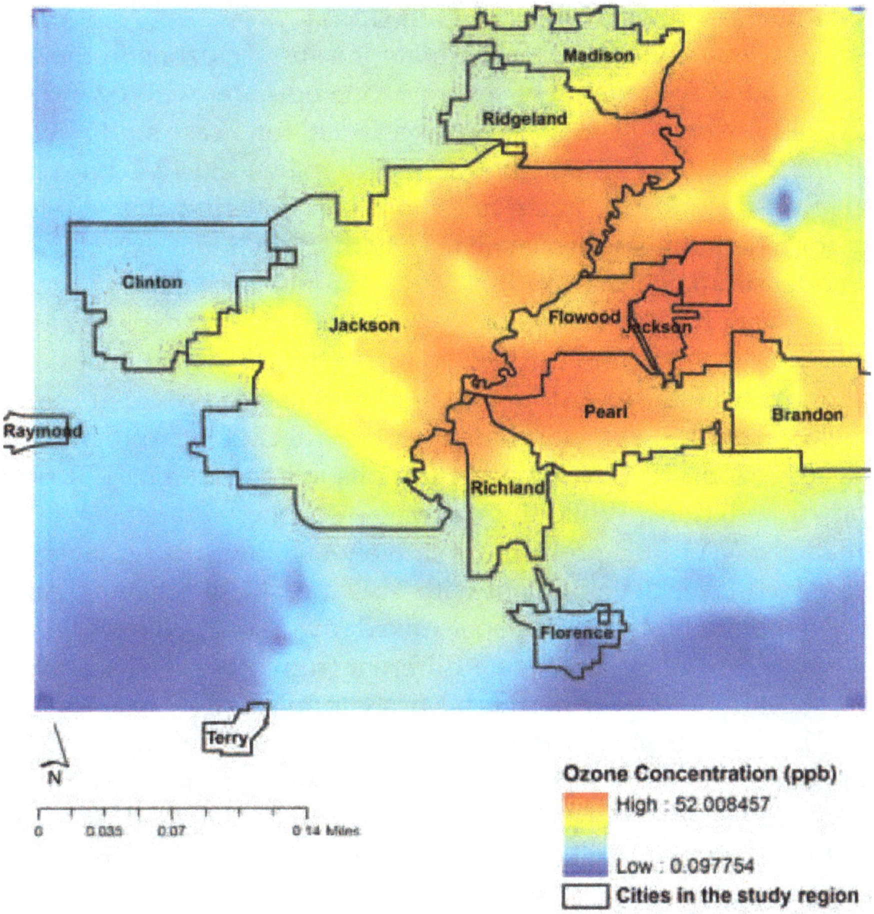

Fig. 7. Spatial distribution of model simulated surface ozone (ppbv) "over domain 4 covering Jackson city and neighbourhood at 1 PM CDT 7 June 2006.

photochemical oxidation of hydrocarbons in the presence of nitrogen oxides (NO and NO_2). The time series of HONO showed a maximum between 10-12 AM indicating contributions from nearby combustion sources and its decrease after 12 noon could be due to its breakdown through reactions with NO and OH in the presence of sunlight. All these indirectly supports formation of ozone as it leads production of NO and OH. The spatial distributions of O_3 corresponding to the peak time at 1 PM (Figure 7) showed that the central, east and northeast parts of model domain, covering the regions of Jackson, Flowood, Pearl and parts of Ridgeland and Madison have the maximum concentrations whereas the western and southern parts covering Clinton, Raymond and Florence were less affected. This spatial pattern indicates the maximum as associated with the identified urban locations due to mobile traffic and combustion sources. The spatial distributions of NO, NO_2, HONO and CO are similar indicating that mobile and combustion sources in Jackson, Ridgeland and Madison were contributing to the production of O_3 through chemical reactions. Back trajectories drawn from the observation site showed the air parcels to originate from west and south, where the road-ways are located which confirm that the pollutants of mobile origin are chiefly responsible for the production of ozone in sunlight hours. This study is important as ozone and other pollutant species were simulated at 1 km resolution over Jackson urban metropolitan region providing important information for assessment, management and mitigation of urban pollutants.

5.4 Case study of surface ozone over MS Gulf Coast

A regional scale study was undertaken to simulate the surface ozone in the central Gulf coast using WRF/Chem model. A moderately severe ozone episode, with ozone values exceeding 80 ppbv that occurred during 8-11 June 2006 was selected for a case study [34]. The model was configured with three two-way interactive nested domains (36-12-4 km resolution) and 31 vertical levels (Figure 8) with the inner finest domain covering the Mississippi coast. The default profiles for chemical species available with the model were used as the initial pollutants. The model simulated meteorological fields and ozone distributions were compared with available observations for validation. Several sensitivity experiments with different planetary boundary layer and land surface model schemes which revealed that YSU PBL scheme in combination with NOAH land surface model provided best simulation of meteorological fields in the lower atmosphere over the MS Gulf Coast region [35]. The patterns of simulated surface ozone and NO_2 concentrations with different PBL and land surface physics have shown variations. While the patterns of ozone were similar with the YSU and MYJ PBL formulations, MYJ PBL produced relatively shallow mixing layers as compared to moderately deep mixed layer development with YSU and ACM schemes. As of soil schemes, 5-layer soil model simulated relatively deeper mixed layers than NOAH LSM. The simulated ozone patterns with ACM PBL scheme were different from other PBL schemes as it produced localized higher concentrations over northern Mississippi, eastern Louisiana and west Florida coast (Figure 9). The peak ozone concentration were ≥ 65 ppbv with ACM PBL over the northern parts of Mississippi river and west Florida coast located to the south of Alabama. The convergence in the surface flow along the coast and along Mississippi river in runs with ACM PBL and RUC LSM caused high ozone formation in these areas. The combination of YSU PBL and NOAH land surface schemes gave best simulations for all the meteorological and air quality fields with least BIAS, RMSE and highest correlation values due to better simulation of PBL height, wind speed, temperature and humidity.

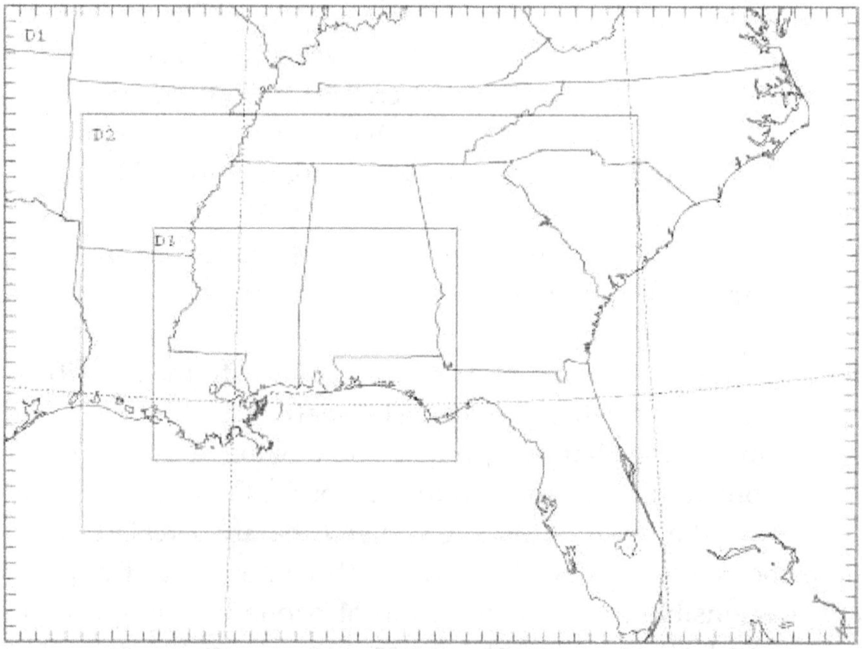

Fig. 8. Modelling domains used in WRF/Chem. Outer domains D01, D02 are coarse with resolutions 36 and 12 km and the inner finer domain (D3) is of resolution 4 km.

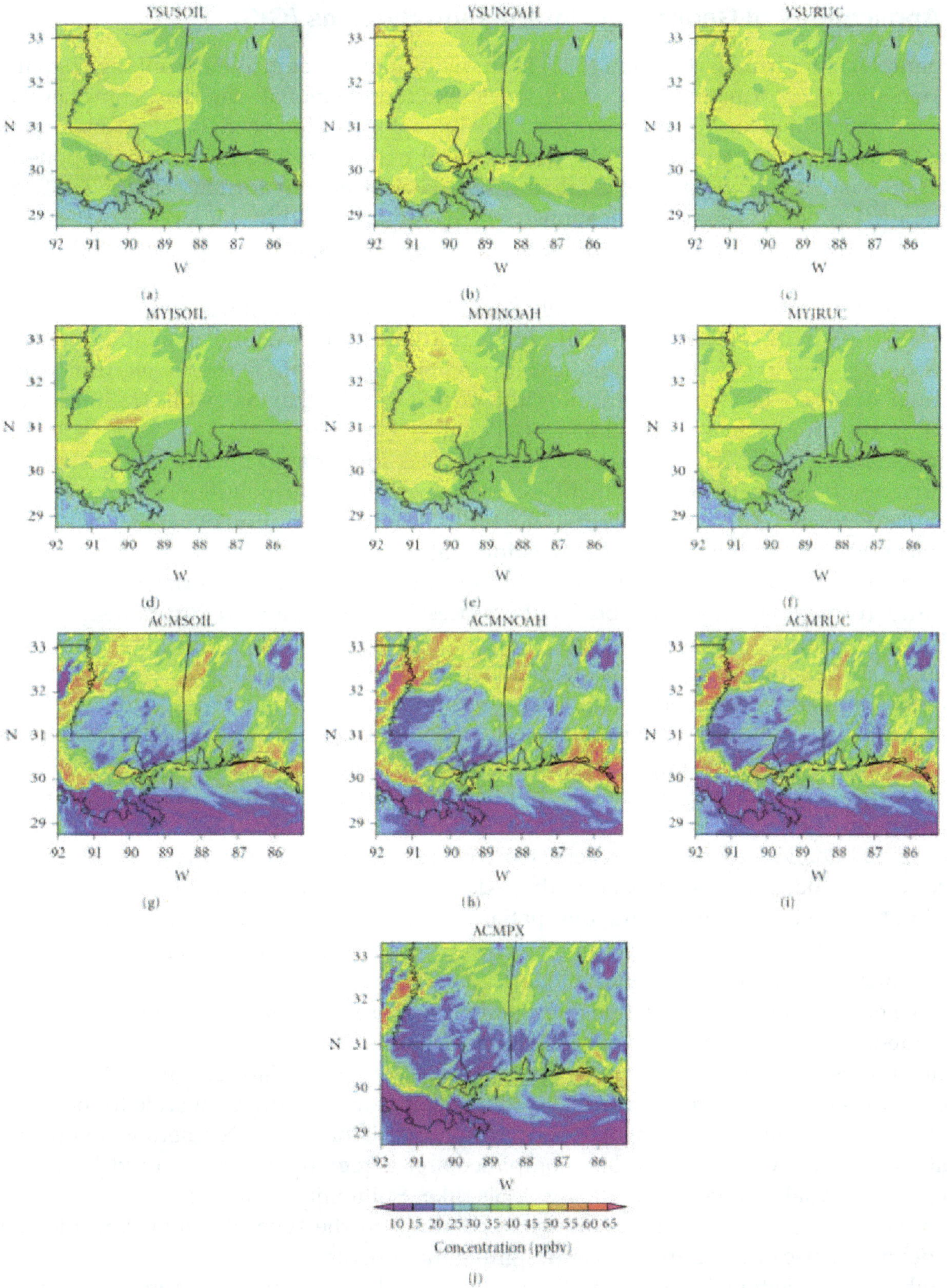

Fig. 9. Simulated ozone concentration (ppbv) in the model fine domain at the lowest level (30m) from experiments with different PBL and LSM options at 10 CST 9 June 2006 for the experiments YSUSOIL, YSUNOAH, YSURUC, MYJSOIL, MYJNOAH, MYJRUC, ACMSOIL, ACMNOAH, ACMRUC and ACMPX.

6. Applications of Geographic Information Systems (GIS)

Geographical Information System (GIS) is a powerful tool that facilitates linking spatial data to non-spatial information [36]. With its embedded relational database component, the system assists in storing, mapping and analyzing geo-referenced data in an organized structure [37]. The database and the geographical base form the two major components of the GIS system that helps in visualizing the data in a map format. Unlike the reports generated by stand alone applications, which summarize the tabular data, these maps illustrate the geographical connections among the spatial variables and visually communicate geo-specific information to a decision maker.

The conceptual approach of GIS not only provides the capability of querying the spatial data but also, with its inbuilt analytical tools, translates the existing spatial patterns into measurable objectives. These analytical capabilities of GIS offer a dynamic dimension to 'Spatial Analysis' factor in determining the principles of behavior or illustrating the inter-relationships among spatial and non-spatial data.

Understanding these relationships or dependencies in a spatial phenomenon is the major crux of any field; health, environment, geology, hydrology, air pollution studies, agronomy and many others. With its framework, GIS facilitates the integration of various field-specific applications into its interface and allows integrated ways to conduct research and develop new analytical approaches to relate their information to the terrestrial activities. These abilities (providing fully functional data processing environment) distinguish GIS from other information systems as it results in enhancing the applications with productive findings that are broadened and deepened in a geographic location.

To build a spatial data model, GIS Systems support three basic types of data (1) Vector Data: a) Events or Points: Pattern expressed as points in space, b) Lines: Patterns expressed as networks and c) Polygons: Patterns expressed as analytical units with defined closed boundaries, (2) Raster data: a) Grid-cell data and (3) Image: a)Satellite Imagery, photographs.

The spatial analysis modeling process involves interpreting and exploring the interactions, associations and relationships among these data types specific to a geographic location. The exploratory process in developing the spatial model is composed of a set of procedures.

- To identify/map the various data layers and describe the attributes associated with the spatial pattern (Representation model)
- To explore/model the relationships, association or interactions between the data layers identified in representation model (Process Model)

While representation model involves deriving necessary input datasets, in most of the cases, the datasets needs to be reclassified by setting a common measurement scale to the attribute variables by giving weight age depending on their influence. Numerous interactions/ relationships between various data components of a region can be captured by the wide range of spatial analytical modeling tools like Suitability analysis, Distance analysis, Hydrological analysis, Surface analysis. Depending on the type of interaction, the spatial model makes use of these analyses in designing the process.

Another dynamic tool GIS offers is the geo statistical analyst tools that include numerous techniques in exploring and modeling the relationships in the spatial data. The inclusion of interpolation methods provides a powerful environment to the users to assess the quality of their analysis. Many disaster mitigation spatial problems are addressed using the Network Analyst tool as required by different research applications. The capabilities of Network

Analyst tool allows to dynamically model realistic conditions and facilitates in examining the flows within and between the natural and man-made networks.
The broad range of many analyst tools, with their modeling and analysis features, not only makes the GIS applications penetrate through number of research fields (from environment, transportation, community *etc.*) but also exploring associations and relations ships between these fields.

6.1 GIS in air pollution studies

The quality of air is a very important factor in projecting or representing the status of environment and health of any region. Air pollution studies that analyze the quality of air provide strategic information to the decision making process and play a significant role in the implementation of the policies that influence the air quality of a region. Most of the air pollution models, in their pollutant distribution simulations, consider the physical characteristics of the pollution such as wind direction, speed, temperature etc., in determining the air pollution trajectory. Integrating these models with GIS presents a geographic dimension to the air quality information by relating the actual pollution concentrations to the plant and human life in that location. With its numerous analyst tools, GIS can demonstrate the relationship of poor air quality and occurrences of deficient human and environmental health [38]. GIS can portray the spatial correspondence between the air quality and the disease statistics in the area that is potentially impacted [39]. In this process, GIS can explore:

- The relative spatial phenomenon of the pollutant with respect to the geographical distribution in terms of location, extent and distance.
- The spatial extent of the pollutant dispersion and its intensity at any geographical location under the impact zone
- The socio-economic characteristics of the populations affected by these pollutants
- The data visualization from various perspectives by classifying and reclassifying the data by determining the class breaks.

Examining the relationships between high pollution concentrations across various demographic thematic layers helps in identifying *hotspots* that are in need of special investigation or monitoring. Data visualization, illustrating such information through a map provides an insight in a more dynamic way that helps the authorities to plan their future strategies.
The output generated from an air pollution dispersant model (Figure 10), illustrates the mercury pollution concentrations in the Gulf Coast region. The intensity and spatial dispersion of the pollutant are presented in a scientific terminology as units and latitude/longitude. The format of this scientific information generated by these models may not be suitable for directly integrating the data into the policy making process. The relevance of this information to a spatial location has to be derived and its implications need to be presented in a format that relates to the socio-economic or health characteristics of the people living in that region. The gap of presenting the scientific format from a geographical dimension can be filled by integrating these outputs with Arc GIS tools. Viewing the same data across various demographic themes projects the vulnerable populations under the impact zone and assists the authorities in making informed choices.
Figure 11 illustrates the HYSPLIT output shown in Figure 10 in a GIS environment. The output of HYSPLIT is obtained in the format of ASCII with relevant data on latitude, longitude and the mercury concentrations at various points in the dispersion trajectory.

Using data management and 3D Analyst tools, the ASCII file is converted into a point data and a TIN (Triangular Irregular Networks) file is created from this vector data. By interpolating the cell Z-values (Mercury concentrations) using natural neighbors method, the TIN is converted to a Raster file. The final mercury dispersion raster file is then mapped over the Gulf Coast states.

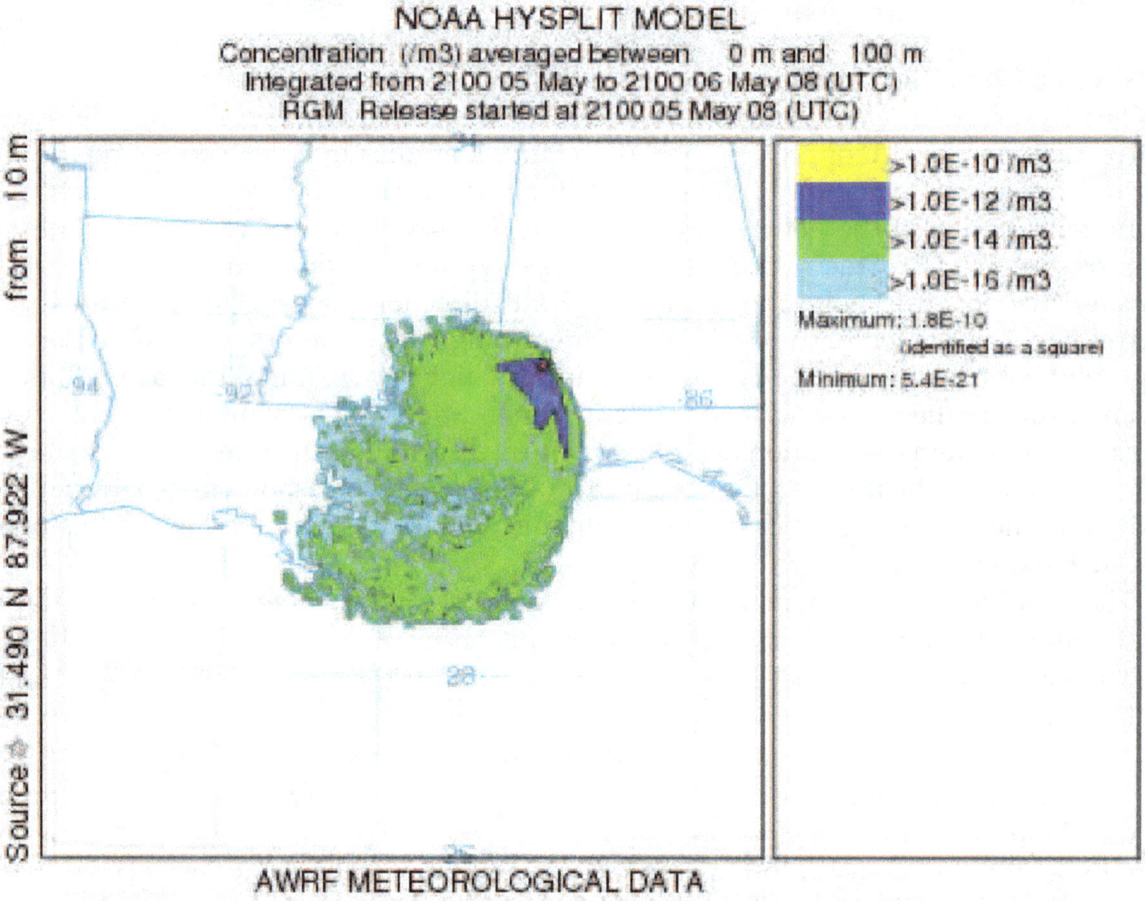

Fig. 10. Output generated from the HYSPLIT model developed by NOAA

Figure 11 describes that parts of Mississippi, Louisiana and Alabama were impacted by the emissions from the Barry Generating Plant in Alabama. The mercury dispersion raster file is overlaid on the classified demographic thematic layer (dark to light color indicates high to low dense population) shows most of the tracts that are under the impact zone are densely populated.

The impact of severe pollution zones: red (very high) and yellow (high) colors, on vulnerable populations (age above 65 years and less than 5 years) are presented in Figure 12 and Figure 13.

Integrating HYSPLIT model output with classified demographic GIS data gives a clear visual representation of the spatial location of highly dense vulnerable populations under impact zone. It assists in projecting the age of the vulnerable population (which in this case ages less than 5 years) under the risk due to the pollution impact.

A spatial phenomenon or a relation can be developed by examining the health statistics of these tracts against the pollution levels they are being exposed. Such spatial phenomenon may reveal hidden facts and can be a significant contribution in designing the policy strategies.

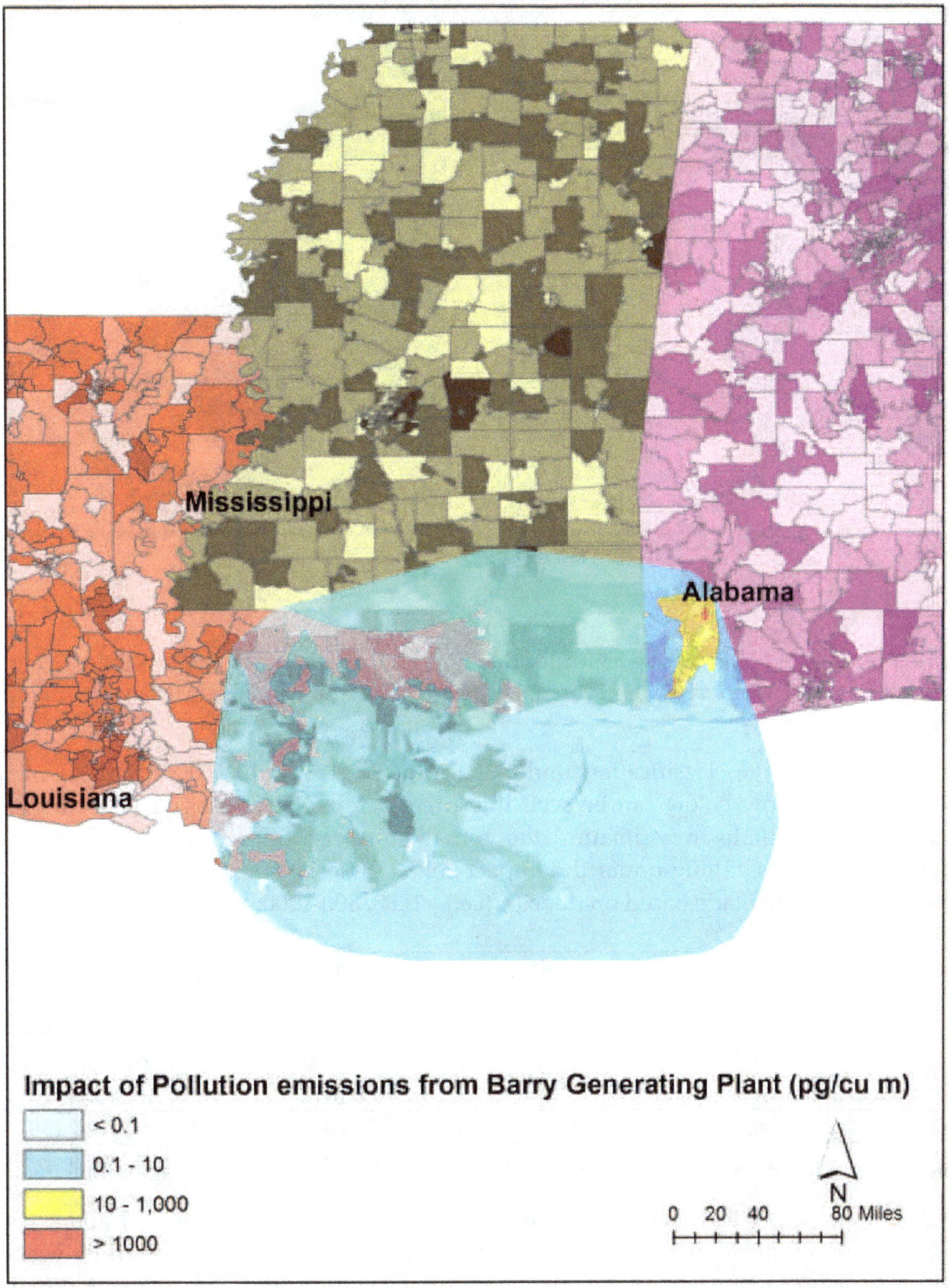

Fig. 11. Representation of HYSPLIT output in GIS environment

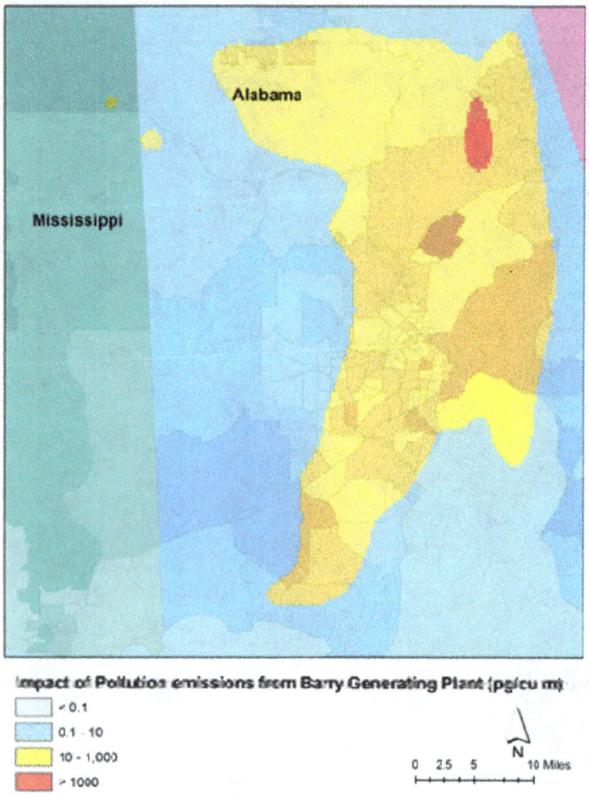

Fig. 12. Population of age above 65 under the impact of Barry Plant-mercury emissions.

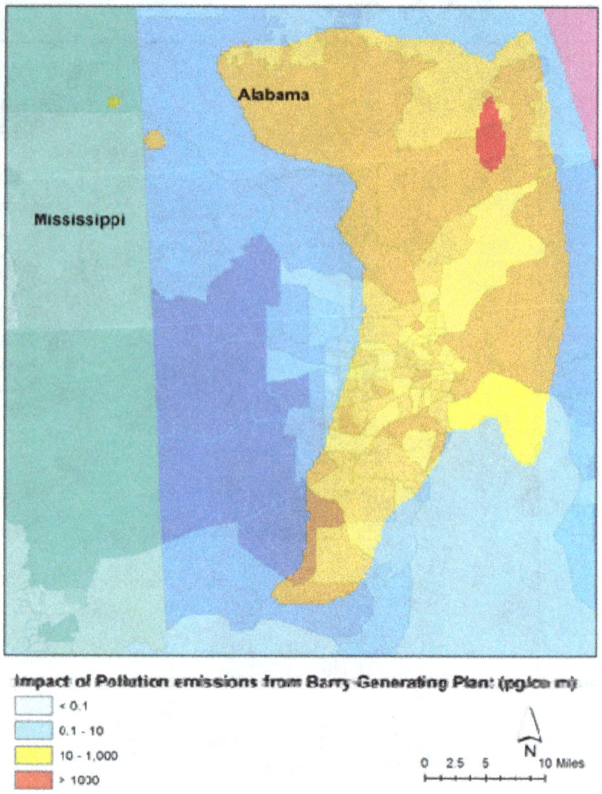

Fig. 13. Population of age less than 5 years under the impact of Barry Plant -mercury emissions.

The demographic age classification under the impact zone of 10-1000 pg/m^3 of mercury concentration (yellow color) can be well interpreted from the figures 12 and 13. The GIS functionalities facilitate in estimating the actual numbers: tracts, along with vulnerable population numbers, falling under the impact zone (Table 1). For convenience, four classes of populations were made based on density (i.e.) >1200, 800-1200, 400-800 and <400.

Population / Age	High density population (>1200) (no. of tracts)	Medium density population (800-1200) (no. of tracts)	Low density population (400-800) (no. of tracts)
5 years and below	1797(1)	15761(13)	11665 (28)
65 years and above	1640 (1)	5710 (6)	22652 (38)

Table1. Numbers of area tracts and population affected from pollutant dispersion

The pollutant dispersion region covered approximately 92 tracts in which about 338,439 habitants live. From the data in Table 1, it may be inferred that elders with 65 years and more are the most affected in the two categories of high and low density population concentrations whereas more children below 5 years were affected in the category of 800-1200. This emphasizes the need to detailed analysis using GIS to clearly understand the relative age groups affected, in this case due to pollution. Presenting such statistical information through GIS mapping leads to risk assessment and provides an idea of not only the areal extent of pollutant influence but also its impact on demographic data lines.

7. References

[1] US EPA, 2003. A Review of the Reference Dose and Reference Concentration Processes. US EPA/630/P-02/002F, December 1, 2002. Risk Assessment Forum, Washington, DC, 192pp. http://cfpub.epa.gov/ncea/raf/

[2] US EPA, 2004. Air quality criteria for particulate matter. Research Triangle Park, NC: Office of Research and Development, Report No. EPA/600/P-99/022aF-bF.

[3] Hannigan, M.P., Busby, W.F., Cass, G.R., 2005. Source contributions to the mutagenicity of urban particulate air pollution. *Journal of Air & Waste Management Association*, 55, 399-410.

[4] Barrett, K. and Seland, Ø. (Eds.), 1995. European Transboundary Acidifying Air Pollution: Ten years calculated fields and budgets to the end of the first Sulphur Protocol. EMEP Research Report No. 17, The Norwegian Meteorological Institute, Oslo.

[5] Skamarock, W.C., Klemp, J.B., Dudhia, J., Gill, D.O., Barker, D.M., Duda, M.G., Huang, X.-Y., Wang, W., and Powers, J.G., 2008. A Description of the Advanced Research WRF Version 3. *NCAR Technical Note, NCAR/TN-475+STR*. Mesoscale and Microscale Meteorology Division, National Center for Atmospheric Research, Boulder, Colorado, USA.

[6] Grell, G. A., Dudhia, J., Staufer, D.R., 1995. A Description of the Fifth-Generation Penn State/ NCAR Mesosscale Model (MM5), NCAR Technical Note, 6-1995.

[7] Taylor, G.I., 1915. Eddy motion in the atmosphere Phil. *Transactions of the Royal Soc. of London.*Series A, 215, 1.

[8] Taylor, G.I., 1921. Diffusion by continuous movements. *Proc. London Math. Soc.* 20, 196.

[9] Rohde, H., 1972. A study of the sulfur budget for the atmosphere over northern Europe. *Tellus*, 24, 128.

[10] Rohde, H., 1974. Some aspects of the use of air trajectories for the computation of large scale dispersion and fallout patterns. *Adv. in Geophysics,* 18B, 95, Academic press.

[11] Eliassen, A., and Saltbones, J., 1975. Decay and transformation rates of SO2 as estimated from emission data, trajectories and measured air concentrations. *Atm. Env,.* 9, 425.

[12] Fisher, B.E.A., 1975. The long-range transport of sulfur dioxide. *Atm.Env.,* 9, 1063.

[13] Reynolds, S., Roth, P., and Seinfeld, J., 1973. Mathematical modeling of photochemical air pollution. *Atm.Env.,* 7.

[14] Shir, C.C. and L.J. Shieh, 1974. A generalized urban air pollution model and its application to the study of SO2-distribution in the St. Louis Metropolitan area. *J. Appl. Met.,* 19, 185-204.

[15] Egan, B.A., Rao, K.S., and Bass, A., 1976. A three dimensional advective-diffusive model for long-range sulfate transport and transformation. 7th ITM, 697, Airlie House.

[16] Carmichael, G.R., and Peters, L.K., 1979. Numerical simulation of the regional transport of SO2 and sulfate in the eastern United States. *Proc. 4th Symp. on turbulence, diffusion and air pollution,* AMS 337.

[17] Friedlander, S.K., and Seinfeld, J.H., 1969. A Dynamic Model of Photochemical Smog. *Environ. Science Technol.,* 3, 1175.

[18] Eschenroeder, A.Q., and Martinez, J.R., 1970. Mathematical Modeling of Photochemical Smog. *Proc. American Institute Aeronautics and Astronautics, Eight Aerospace Sciences Meeting*, New York, Jan 19-21.

[19] Liu, M.K., and Seinfeld, J.H., 1974. On the Validity of Grid and Trajectory Models of Urban Air Pollution. *Atmos. Environ.*, Vol. 9, pp. 555-574.

[20] Draxler, R.R., Rolph, G.D., 2010. HYSPLIT (HYbrid Single-Particle Lagrangian Integrated Trajectory) Model. NOAA Air Resources Laboratory, Silver Spring, MD. Available at NOAA ARL READY Website http://ready.arl.noaa.gov/HYSPLIT.php

[21] Grell, G. A., Peckham, S.E., Schmitz, R., McKeen, S.A., Frost, G., Skamarock, W.C., and Eder, B., 2005. Fully coupled "online" chemistry within the WRF model. *Atmos. Environ.*, 39, 6957–6975.

[22] Indracanti, J., Srinivas, C.V., Hughes, R. L., Baham, J.M., Patrick, C., Rabarison, M., Young, J., Anjaneyulu, Y., and Swanier, S.J., 2007. GIS Assisted Emission Inventory Development for Variable Grid Emission Database for Mississippi Region. 16th Annual International Emission Inventory Conference on Emission Inventories: "Integration, Analysis, and Communications". EPA, NC.

[23] Srinivas, C.V., Jayakumar, I., Baham, J.M., Hughes, R.L., Patrick, C., Young, J., Rabarison, M., Swanier, S.J., Hardy, M.G., and Anjaneyulu, Y., 2008. Sensitivity of Atmospheric dispersion simulations by HYSPLIT to the meteorological predictions from a mesoscale model. *Environ. Fluid Mechanics*, 8, 367-387.

[24] Anjaneyulu, Y., Srinivas, C.V., Jayakumar, I., Hari, P. D., Baham, J.M., Patrick, C., Young, J., Hughes, R.L., White, L. D., Hardy, M.G., and Swanier, S.J., 2008. Some observational and modeling studies of the coastal atmospheric boundary layer at Mississippi Gulf Coast for Air Pollution Dispersion assessment. *Int. J. Environ. Res. Public Health*, 5, 484-497.

[25] Srinivas, C.V., Jayakumar, I., Baham, J.M., Hughes, R.L., Patrick, C., Young, J., Rabarison, M., Swanier, S.J., Hardy, M.G., and Anjaneyulu, Y., 2009. A Simulation Study of Meso-Scale Coastal Circulations in Mississippi Gulf Coast for Atmospheric Dispersion. *J. Atmos. Res.*, 91, 9-25.

[26] Anjaneyulu, Y., Srinivas, C.V., Hari Prasad, D., White, L. D. , Baham, J.M., Young, J.H., Hughes, R.L., Patrick, C., Hardy, M.G., and Swanier, S. J., 2009. Simulation of Atmospheric Dispersion of Air-Borne Effluent Releases from Point Sources in Mississippi Gulf Coast with Different Meteorological Data. *Int. J. Environ. Res. Public Health*, *6*, 1055-1074.

[27] Anjaneyulu, Y., Venkata, B.R. D., Hari, P. D., Srinivas, C.V., Francis, T., Julius M. B., John H. Y., Robert, L. H., Chuck, P., Mark G. H., Shelton J. S., Mark, D.C., Winston, L., Paul, K., and Richard, A., 2010. Source-receptor modeling using high resolution WRF meteorological fields and the HYSPLIT model to assess mercury pollution over the Mississippi Gulf Coast region. 90th American Meteorological Society Annual Meeting, Atlanta, USA.

[28] LaToya, M., William, R. P., Christopher, A. V., Anjaneyulu, Y., Venkata, B.R.D., Hari, P. D., Srinivas, C.V., Francis, T., Julius, M. B., Robert, L. H., Chuck, P., John, H. Y., and Shelton J. S., 2010. Evaluation of $PM_{2.5}$ source regions over the Mississippi Gulf Coast using WRF/HYSPLIT modeling approach. 90th American Meteorological Society Annual Meeting, Atlanta, USA.

[29] William, R. P., LaToya, M., Christopher, A. V., Venkata, B.R.D., Hari, P. D., Anjaneyulu, Y., Srinivas, C.V., Francis, T., Julius, M. B., Robert, L. H., Chuck, P., John, H. Y., and Shelton, J. S., 2010. Observation, analysis and modeling of the sea breeze circulation during the NOAA/ARL-JSU Meteorological Field Experiment, Summer-2009. 90th American Meteorological Society Annual Meeting, Atlanta, USA.

[30] William, R. P., LaToya, M., Christopher, A. V., Venkata, B.R.D., Hari, P. D., Anjaneyulu, Y., Srinivas, C.V., Francis, T., Julius, M. B., Robert, L. H., Chuck, P., John, H. Y., and Shelton, J. S., 2010. Analysis and perdition of the atmospheric boundary layer characteristics during the NOAA/ARL-JSU Meteorological Field Experiment, Summer- 2009. 90th American Meteorological Society Annual Meeting, Atlanta, USA.

[31] William, R. P., LaToya, M., Christopher, A. V., Venkata, B.R.D., Hari, P. D., Anjaneyulu, Y., Srinivas, C.V., Francis, T., Julius, M. B., Robert, L. H., Chuck, P., John, H. Y., and Shelton, J. S., 2010. Numerical prediction of atmospheric mixed layer variations over the Gulf coast region during NOAA/ARL JSU Meteorological Field Experiment, Summer-2009 - Sensitivity to vertical resolution and parameterization of surface and boundary layer processes. 90th American Meteorological Society Annual Meeting, Atlanta, USA.

[32] Anjaneyulu, Y., Venkata, B. R.D., Hari, P. D., Srinivas, C.V., Francis, T., Julius, M. B., John, H. Y., Robert, L. H., Chuck, P., Mark, G. H., Shelton, J. S., 2011. An Integrated WRF/HYSPLIT Modeling Approach for the Assessment of $PM_{2.5}$ Source Regions over Mississippi Gulf Coast Region. *Air Quality, Atmosphere & Health,* 1-12, doi 10.1007/s11869-010-0132-1.

[33] Paul, R.A., Biller, W.F., McCurdy, T., 1987. National estimates of population exposure to ozone. Presented at the Air Pollution Control Association 80th Annual Meeting and Exhibition, Pittsburgh, PA, 87-42.7.

[34] Anjaneyulu ,Y., Venkata, B.D., Srinivas, D., Srinivas, C.V., John, H. Y., Chuck, P., Julius, M. B., Robert, L. H., Sudha, Y., Francis, T., Mark, G. H., and Shelton J. S., 2010. Air Quality Modeling for Urban Jackson, MS Region using High Resolution WRF/Chem Model. *International Journal of Environmental Research and Public Health* (Accepted for publication).

[35] Anjaneyulu, Y., Srinivas, C.V., Venkata, B. R.D., Hari, P. D., John, Y., Chuck, P., Julius, M. B., Robert, L. H., Mark, G. H., and Shelton J. S., 2011. Simulation of Surface Ozone Pollution in the Central Gulf Coast Region Using WRF/Chem Model: Sensitivity to PBL and Land Surface Physics. *Advances in Meteorology,* Volume 2010, Article ID 319138, 24 pages, doi:10.1155/2010/319138.

[36] Matejicek, L., 2005. Spatial Modelling of Air Pollution in Urban Areas with GIS: A Case Study on Integrated Database Development. *Advances in Geosciences,* 63–68.

[37] Manjola, B., Elvin, C., Bledar, M., and Albana, Z., 2010. Mapping Air Pollution in Urban Tirana Area Using GIS. *International Conference SDI.*

[38] ESRI. (2007, December). GIS for Air Quality. Retrieved Febraury 2011. [http://www.esri.com/library/bestpractices/air-quality.pdf]

[39] Juliana, M., 2007. The Geography of Asthma and Air Pollution in the Bronx: Using GIS for Environmental Health Justice Research. *Health Care Services in New York: Research and Practice*. Greater New York Hospital Association-United Hospital Fund.

Permissions

All chapters in this book were first published in AAP, by InTech Open; hereby published with permission under the Creative Commons Attribution License or equivalent. Every chapter published in this book has been scrutinized by our experts. Their significance has been extensively debated. The topics covered herein carry significant findings which will fuel the growth of the discipline. They may even be implemented as practical applications or may be referred to as a beginning point for another development.

The contributors of this book come from diverse backgrounds, making this book a truly international effort. This book will bring forth new frontiers with its revolutionizing research information and detailed analysis of the nascent developments around the world.

We would like to thank all the contributing authors for lending their expertise to make the book truly unique. They have played a crucial role in the development of this book. Without their invaluable contributions this book wouldn't have been possible. They have made vital efforts to compile up to date information on the varied aspects of this subject to make this book a valuable addition to the collection of many professionals and students.

This book was conceptualized with the vision of imparting up-to-date information and advanced data in this field. To ensure the same, a matchless editorial board was set up. Every individual on the board went through rigorous rounds of assessment to prove their worth. After which they invested a large part of their time researching and compiling the most relevant data for our readers.

The editorial board has been involved in producing this book since its inception. They have spent rigorous hours researching and exploring the diverse topics which have resulted in the successful publishing of this book. They have passed on their knowledge of decades through this book. To expedite this challenging task, the publisher supported the team at every step. A small team of assistant editors was also appointed to further simplify the editing procedure and attain best results for the readers.

Apart from the editorial board, the designing team has also invested a significant amount of their time in understanding the subject and creating the most relevant covers. They scrutinized every image to scout for the most suitable representation of the subject and create an appropriate cover for the book.

The publishing team has been an ardent support to the editorial, designing and production team. Their endless efforts to recruit the best for this project, has resulted in the accomplishment of this book. They are a veteran in the field of academics and their pool of knowledge is as vast as their experience in printing. Their expertise and guidance has proved useful at every step. Their uncompromising quality standards have made this book an exceptional effort. Their encouragement from time to time has been an inspiration for everyone.

The publisher and the editorial board hope that this book will prove to be a valuable piece of knowledge for researchers, students, practitioners and scholars across the globe.

List of Contributors

Hwa-Lung Yu and Shang-Chen Ku
Dept of Bioenvironmental Systems Engineering, National Taiwan University, Taipei

Chiang-Hsing Yang
Dept of Health Care Management, National Taipei University of Nursing and Health Sciences

Tsun-Jen Cheng
Institute of Occupational Medicine and Industrial Hygiene, National Taiwan University

Likwang Chen
Center for Health Policy Research and Development, National Health Research Institutes, Miaoli, Taiwan

Panagiotis Nastos
Laboratory of Climatology and Atmospheric Environment, Faculty of Geology and Geoenvironment, University of Athens, Greece

Konstantinos Moustris
Department of Mechanical Engineering, Technological Educational Institute of Piraeus, Greece

Ioanna Larissi
Department of Electronic-Computer Systems Engineering, Technological Educational Institute of Piraeus, Greece

Athanasios Paliatsos
General Department of Mathematics, Technological Educational Institute of Piraeus, Greece

Galina Zhamsueva, Alexander Zayakhanov, Vadim Tsydypov, Alexander Ayurzhanaev and Ayuna Dementeva
Department of Physical Problems of Buryat Science Center, Siberian Branch of Russian Academy of Sciences, Ulan-Ude, Russia

Dugerjav Oyunchimeg and Dolgorsuren Azzaya
Institute of Meteorology and Hydrology of Mongolia, Ulaanbaatar, Mongolia

Galina Zhamsueva, Alexander Zayakhanov, Vadim Tsydypov, Alexander Ayurzhanaev and Ayuna Dementeva
Department of Physical Problems of Buryat Science Center, Siberian Branch of Russian Academy of Sciences, Ulan-Ude, Russia

Dugerjav Oyunchime and Dolgorsuren Azzaya
Institute of Meteorology and Hydrology of Mongolia, Ulaanbaatar, Mongolia

Nagaraja Kamsali
Department of Physics, Bangalore University, Bangalore, India

B.S.N. Prasad
University of Mysore, Mysore, India

Jayati Datta
Indian Space Research Organization, Bangalore, India

Danni Guo
Climate Change and Bioadapation Division, South African National Biodiversity Institute, Cape Town, South Africa

Renkuan Guo, Christien Thiart and Yanhong Cui
Department of Statistical Sciences, University of Cape Town, Cape Town, South Africa

Salvador Enrique Puliafito, David Allende, Rafael Fernández, Fernando Castro and Pablo Cremades
Grupo de Estudios Atmosféricos y Ambientales (GEAA) Universidad Tecnológica Nacional – Facultad Regional Mendoza Consejo Nacional de Investigaciones Científicas y Técnicas (CONICET) Argentina

Jasim M. Rajab, K. C. Tan, H. S. Lim and M. Z. MatJafri
School of Physics, Universiti Sains Malaysia, Penang, Malaysia

Boštjan Grašič, Primož Mlakar and Marija Zlata Božnar
MEIS environmental consulting d.o.o. Slovenia

Alberto Mendoza and Ana Y. Vanoye
Tecnológico de Monterrey, Campus Monterrey, Mexico

Santosh Chandru, Yongtao Hu and Armistead G. Russell
Georgia Institute of Technology, United States of America

Anjaneyulu Yerramilli and Venkata Bhaskar Rao Dodla
Trent Lott Geospatial & Visualization Research Center @ e-Center, College of Science Engineering & Technology, Jackson State University, Jackson, MS, USA

Sudha Yerramilli
National Center For Bio-Defense @e-Center, College of Science Engineering & Technology, Jackson State University, Jackson MS, USA

Index

P

R

S

T

U

W

CPSIA information can be obtained
at www.ICGtesting.com
Printed in the USA
BVHW062121290822
645775BV00003B/133

9 781639 890385